THE ENCYCLOPEDIA OF
HOW IT WORKS

FROM ABACUS TO ZOOM LENS

Edited by Donald Clarke

A & W Publishers, Inc.
New York

The Encyclopedia of How It Works from
Abacus to Zoom Lens

Edited by Donald Clarke
Designed by Jim Bamber

Copyright© 1977 Marshall Cavendish Limited
This volume first published 1977

Published in the United States of America by A&W Publishers,
Inc., 95 Madison Avenue, New York, New York 10016

Library of Congress Catalog Card Number 76-56962
ISBN 0 89479 002 1

Printed in Great Britain

INTRODUCTION

The history of technology has been a story of Man's effort to obtain mastery over his environment. For thousands of years, progress was made by trial-and-error; objects such as weapons and tools were made in a traditional way, often by the same individual who would use them, so that the design of everyday objects evolved into the most useful and pleasant shapes. Furniture, firearms and musical instruments were sometimes made with such care that they were works of art, simply because they were beautiful objects. Even today, a carpenter's tools would be easily recognized for what they are by a craftsman of thousands of years ago.

In the latter part of the eighteenth century, however, technology began to become synonymous with applied science. As electricity in particular began to be used, technicians had to apply scientific principles to solve their problems. As a result, technology has attained a bewildering pace of development. Diderot, the eighteenth century French encyclopedist, could aspire to produce a reference work which would contain everything worth knowing; such an aspiration would be foolish today. The microscope, a tool of fundamental importance to science, was three hundred years in a state of development; the electron microscope reached a similar state of development in only twenty years.

What this means to our everyday lives is that we are at the mercy of technicians and designers, to say nothing of salesmen. We are in danger of generating more technology than we need, and of using it for the wrong purposes. It has become fashionable nowadays in certain circles to forgo ownership of a television set, in reaction to the commercialism of broadcasting and to the more banal aspects of mass entertainment. Yet it is worth remembering that a television set itself is only an inanimate object, and not intrinsically a bad thing. Any mechanical or electronic device works according to the application of physical principles which have been available since the creation of the Universe itself.

We take our technology too much for granted. Our security and our pleasure in modern living are greatly increased if we know more about how things work; we can make better use of our technology and understand its limitations if we appreciate it at a mechanical level. A great deal of care has gone into the production of this book, in the hope that it will make a positive contribution to the understanding of technology, and therefore to the quality of life in an uncertain century.

TABLE OF CONTENTS

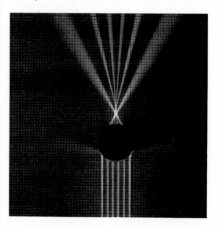

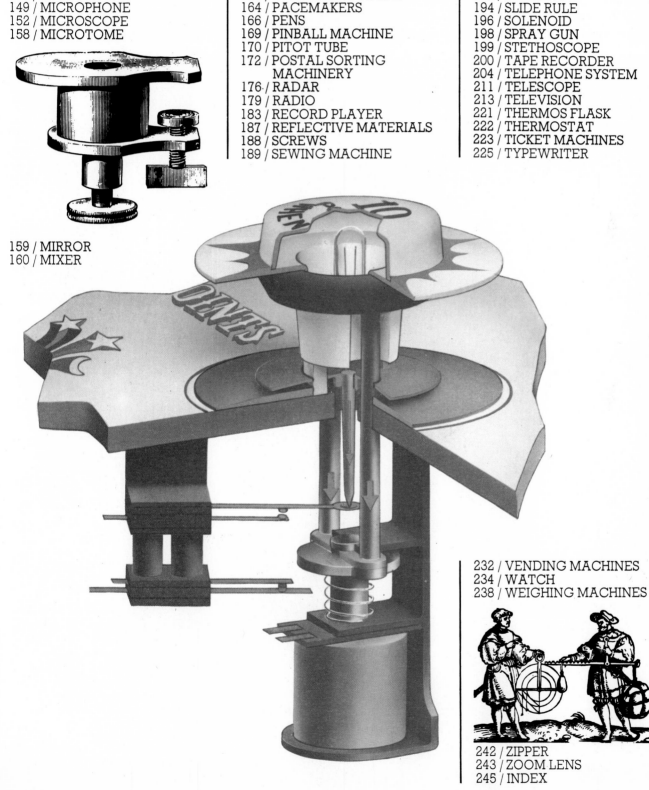

ABACUS

The abacus is a hand-operated calculating machine in which numbers are represented by beads strung on rods or wires set in a rectangular frame. It has been in use for thousands of years in various forms, and is still commonly used in China, Japan and parts of the Middle East. The original version was probably a tray of sand divided into strips by pieces of wood. Marks were made in the sand and wiped out again as required.

The most usual form of abacus is the Chinese *suan pan* ('reckoning board'). This has up to 13 columns of beads, each divided in two by a crossbar running right across the frame. In each column there are two beads above the crossbar and five below it. Some other types have one and four beads respectively, but these are less easy to use.

In the far right-hand column, each lower bead represents one, and each upper bead five. This is the 'units' column. The one to the left of it is the 'tens' column, and here each lower bead represents 10 and each upper bead 50. The next column is the 'hundreds' column, where each lower bead represents 100 and each upper bead 500. This continues leftwards with thousands, tens of thousands and so on. A 13-column abacus can register numbers up to 9,999,999,999,999.

Numbers are 'entered' on the abacus by moving beads to touch the crossbar. To enter the number 23, for example, three lower beads are slid up the 'units' column to the bar, and two are slid up the 'tens' column.

To add 6 to this, one *upper* bead is slid *down* the 'units' column to the bar (thus adding 5) and one lower bead is slid up to the bar (adding another 1, making 6 in all). The abacus now reads 29.

If another 1 is added by sliding the last lower 'unit' bead up to the bar, there will be no more lower beads against the outside of the frame in this column. The lower part of the column is considered as 'full' and must be 'cancelled' immediately. This is done by sliding all five beads back to the outside of the frame and moving down one upper bead (worth 5) to take their place.

In this case, there is already one upper bead against the bar, so moving the other one down 'fills' the upper part of the units column too. This is also cancelled and one more lower bead moved up the next column, which is the 'tens' column, to add ten to the total. There are now no beads entered in the 'units' column, and three in the lower part of the 'tens' column. The abacus therefore reads 30, the correct answer.

Of course, a skilled operator would foresee and skip most of the intermediate steps in the procedure. With practice, it is possible to calculate at very high speed on an abacus—at least as fast as an ordinary electric adding machine.

Subtraction is performed by exactly the opposite process: for example, three is subtracted by moving three beads away from the crossbar in the lower part of the column.

The abacus is basically an adding and subtracting machine, and cannot be used to multiply by any one-stage process. This was a source of difficulty to the ancient Chinese—and the Romans, who used a comparable type of abacus. Neither the Roman nor the traditional Chinese numeral system allowed them to write numbers in columns, so they could not write down multiplication sums in the way used today.

For multiplication by a number up to, say, five, the number to be multiplied can just be added to itself the right number of times. But with larger numbers, this would take an inconvenient length of time, so another method must be used.

The two numbers (say 478 and 35) are written side by side, on a separate piece of paper, and under whichever of them is more convenient, half of that number is written. If this is not a whole number ($35 \div 2 = 17\frac{1}{2}$) the fraction is discounted (17). The half is now halved and halved again by the same method, right down to one: 35; 17; 8; 4; 2; 1. The other number is now doubled and re-doubled in a similar way, the same number of times as the first number was halved—this is a very quick series of additions on the abacus.

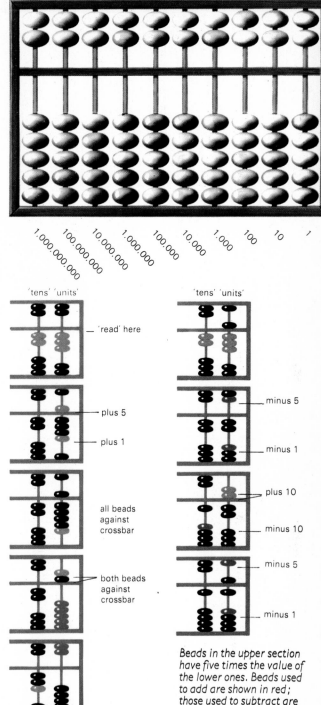

'tens' 'units'

— 'read' here

— plus 5

— plus 1

all beads against crossbar

— both beads against crossbar

'tens' 'units'

— minus 5

— minus 1

— plus 10

— minus 10

— minus 5

— minus 1

Beads in the upper section have five times the value of the lower ones. Beads used to add are shown in red; those used to subtract are shown in blue.

Addition: enter 23.
Add 6 to make 29.
Add 1: 'units' column lower section is now 'full' and abacus reads 20 + 10, not 30. 'Cancel' lower section by moving beads away from crossbar and move one upper section bead down instead. But upper section is now 'full'.
'Cancel' upper section, move one lower bead of 'tens' column up. Abacus now reads 30, the right result.

Subtraction: enter 28.
Subtract 6: abacus now reads 22. Before subtracting another 6, 'borrow' 10 from 'tens' column and give it to 'units' column.
Abacus reads 10 + 12 (=22). Now subtract 6 as before. Abacus now reads 16, the right result.

The sum now looks like this:

$$478 \times 35$$
$$956 \quad 17$$
$$1912 \quad 8$$
$$3824 \quad 4$$
$$7648 \quad 2$$
$$15296 \quad 1$$

The next stage is to strike out all the lines where the *halved* number is *even* (this would include the top line if it were even), leaving those where the halved number is odd:

$$478 \times 35$$
$$956 \quad 17$$
$$15296 \quad 1$$

The doubled column is now added up on the abacus:
$478+956+15296=16730$, which is the right answer.

AERIAL (antenna)

An aerial is a device for transmitting or receiving radio waves. A transmitting aerial converts the electrical signals from a transmitter into an electro-magnetic wave, which spreads out from it. A receiving aerial intercepts this wave and converts it back into electrical signals that can be amplified and decoded by a receiver, such as a radio, television or radar set.

A radio transmitter produces its signal in the form of an alternating electric current, that is, one which oscillates rapidly back and forth along its wire. The rate of this oscillation can be anything from tens of thousands of times a second to thousands of millions of times a second. The rate is known as the frequency and is measured in *kilohertz* or *kilocycles* (thousands of times a second) or, for higher frequencies, in *megahertz* or *megacycles* (millions of times a second).

The oscillating current in the transmitting aerial produces an electromagnetic wave around it, which spreads out from it like the ripples in a pond. This wave, which is shown in the diagram, sets up electric and magnetic fields. The lines of the electric field run along the aerial and those of the magnetic field around it. Both the electric and magnetic fields oscillate in time with the electric current.

Wherever this wave comes into contact with a receiving aerial, it induces a small electric current in it, which alternates back and forth along the aerial in time with the oscillations of the wave. Although this current is much weaker than the one in the transmitting aerial, it can be picked up by the

Above right: an aerial radiates both an electric field (shown in red and blue) and a magnetic field (green and brown). The polarity of these fields changes with the direction of the electric current in the aerial. The shape that the two fields take together is shown in picture 4.
Right: in this diagram, a simple indoor TV aerial with one reflector shows how the 'crest' of a wave strikes the aerial (1) and the reflector a moment later (2), and is reflected back with opposite polarity to reach the main aerial at the same time as the 'trough' (3).

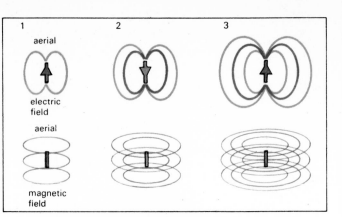

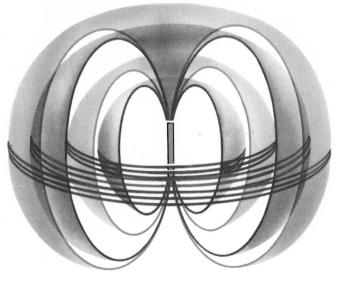

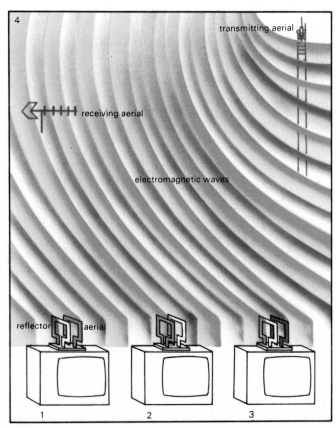

amplifier of the radio tuned to receive it.

The air is full of radio waves at all frequencies, which the aerial picks up indiscriminately. Each radio or television set has a means of selecting a narrow band of frequencies at any one time—this is what happens when a particular signal is tuned in. Each set can be tuned within a certain frequency range, and will only respond to signals in that range.

Each frequency is associated with a wavelength. This is because the waves, as they radiate out from the aerial at a certain frequency travelling at the speed of light, space themselves a certain constant distance apart. The higher the frequency, the shorter the wavelength (the product of the two being always equal to the speed of light). A transmission with a frequency of 100 kHz has a wavelength of 984 ft (300 m).

Electricity travels along a wire at a similar speed. It will therefore greatly increase the efficiency of an aerial if its length is correctly related to the wavelength of the signal it receives or transmits. Ideally, aerials are exactly one half or one quarter of the wavelength they receive or transmit.

Receiving aerials inside domestic radios cannot be even one quarter as long as the wavelength, and in any case have to work over a wide range of wavelengths. But fortunately, the signal from the transmitter is so powerful that it can be received on a comparatively inefficient aerial.

Types of aerial The same principles apply to transmitting and receiving aerials. The simplest form of aerial is a single elevated wire. This type of aerial was introduced in the early days of radio by Guglielmo Marconi, who found that by using a wire instead of a small metal cylinder as he had done previously, he increased the range of his transmitter from one hundred yards to one mile.

This type of single element aerial is called a *monopole*. It is connected to only one terminal of the transmitter; the other terminal is connected to earth. This arrangement does not stop current flowing in the aerial; it streams between the aerial and the ground as if across a capacitor, and sets up an electro-magnetic field between the two. The ground here is said to be used as a *counterpoise*. Car radio aerials use the car body.

Two-element aerials called *dipoles* are also used. These consist of two rods of equal length (again half-, quarter-, or eighth-wave) set end to end a few inches apart. One rod is connected to each terminal of the transmitter, but they are not connected to each other. The field forms about both rods, linking them. No earthing is needed, since the rods counterpoise each other; they are said to be *balanced-fed*.

A transmitting aerial may be set either vertically or horizontally, provided the receiving aerial is set the same way. Vertically set aerials transmit *vertically polarized* waves, which have little effect on a horizontal receiving aerial (and vice versa). For the best results, the receiving aerial should be set at exactly the same angle as the transmitting aerial.

This *directivity* (sensitivity to angle) of an aerial is very clearly shown in the comparatively inefficient aerials of portable radios. These may be of two types: the *loop aerial*, a long loop of wire wound many times around the interior of the cabinet, and the *ferrite rod* aerial, where the wire is wound around a magnetic material which increases its efficiency. For the best reception, the plane of the loop, or a plane at right angles to the ferrite rod, should pass through the transmitter. The performance of a portable radio or a television with an indoor aerial depends very much on the way it is pointing.

Television detector vans are used in Britain to detect households which have not paid the BBC license fee. They use the directivity of swivelling loop aerials connected to powerful receivers to locate the faint radio signals broadcast by the magnetic coils of a television set. The operator rotates the loop until the strongest signal is received. He can then tell in which direction the set lies by the way the loop is pointing.

Below: a row of various types of microwave aerial at the Post Office relay station, Ingleby Arncliffe, Yorkshire. The 'horn' aerials transmit narrow beams of radio waves aimed precisely at receivers miles away, making it hard to 'eavesdrop'.

Directivity has other uses, too. Reception of a broadcast is often impaired by interference from another transmitter with nearly the same frequency. Medium and long wave radio signals follow the curve of the earth and can travel hundreds, or even thousands of miles with comparatively little loss of strength, which can cause serious 'overcrowding' problems.

The shorter wavelengths or very high and ultra-high frequency transmissions (vhf and uhf), which are used for hi-fi radio and television broadcasts, will only travel in straight lines, and stop at the horizon. This means that there has to be a large number of vhf and uhf transmitters to cover a country, which can cause reception problems to someone half way between two transmitters sending different programmes.

Both problems can be solved by using a strongly directional receiving aerial lined up with the desired transmitter. The classic type of highly directional aerial is an ordinary domestic television aerial.

This consists of a half-wave horizontally polarized dipole aerial—the uhf band used in many countries for colour television has wavelengths ranging from 1 m (3 ft) down to 0.1 m (4 in). The dipole is lined up with the transmitter. In front of it (as seen from the transmitter) there is a row of *directors*, which are plain metal rods approximately the same length as the dipole, but not connected to it or the set. Behind it is a row of reflectors, which are similar in appearance.

The directors and reflectors pick-up the signal. This causes a slight current to flow in them, so that they re-radiate the signal, though very weakly and with a changed phase, i.e.

positive for negative and vice versa.

The nearest reflector behind the main element of the aerial is set one quarter of a wavelength away from it. This means that the 'peak' of each oscillating wave travels past the main element and strikes the reflector slightly later in time, making it one quarter of a cycle out of phase. It is re-radiated instantaneously by the reflector; the changed polarity makes it another one half cycle out of phase, or three quarters of a cycle in all. By the time it gets back to the main element, it has dropped back another quarter cycle, so when it reaches it it is one whole cycle behind, and is thus exactly in phase again. As a result, the reflected waves coincide exactly with the direct waves, and the signal is reinforced.

All the other reflectors and directors are spaced to act in the same way. But they will only have this effect when the aerial is lined up precisely on the transmitter. If it is not, the reflections will travel diagonally between the rods, and therefore through a greater distance. This will make them out of phase, so they will cancel each other out.

A typical rooftop UHF aerial for colour TV reception consists of a ladder-like row of directors—there may be between 6 and 18 depending on the strength of the signal in the neighbourhood—with the main dipole element behind them. Behind this is a *deflector*, an earthed grid which screens out unwanted reflections from other sources.

For the best directivity (needed for radar, radio astronomy, etc.) *parabolic* reflectors are used. These are shaped like the reflector of a car headlamp and focus the waves into a narrow beam in exactly the same way. Horn shaped reflectors are used for *microwave* (very short wavelength) transmitters and receivers, and have a similar effect. With longer waves, good transmitter and receiver directivity can be obtained with an *array* of aerials. This looks like an aerial with a row of directors and reflectors, but in fact all the aerials are 'live' and connected to the transmitter or receiver. The signals of the different members of the array reinforce each other in the same way.

New types of aerials have been developed which include their own transistor amplifier, separate from the rest of the radio. This arrangement enormously increases efficiency.

AEROSOL spray can

Aerosol spray cans have been used as convenient packages for an ever-increasing range of products since they first came on the market in the early 1950s. The enormous variety of products available in spray cans includes whipped cream, caulking compounds for sealing the seams of boats, and even the smell of leather.

A spray can is normally made of tinplate with soldered seams, though for products that are stored under high pressure, an aluminium can is used. At the top, there is a simple plastic valve to control the spray. From the bottom of this, a flexible 'dip tube' runs down to the bottom of the can.

The can is filled with the product to be sprayed and the propellant, a compressed gas such as butane or Freon. The gas is partly liquefied by the pressure in the can, but there is a layer of free gas above the liquid. As the can is emptied, more of the liquefied gas vaporizes to fill the space.

The valve is normally held shut by the pressure in the can, and by the coil spring directly below the valve stem. When the push button is pressed, it forces the valve stem down in its housing, uncovering a small hole which leads up through the stem to the nozzle in the button. This allows the product to be forced up the dip tube by the gas pressure in the can. The nozzle is shaped to give a spray or a continuous stream.

To produce a fine mist, a propellant is used which mixes with the product. The two leave the nozzle together and the propellant evaporates as soon as it reaches the air, breaking the product into tiny droplets. The same technique used with a more viscous liquid and a wider nozzle results in a foam. For a continuous stream of liquid, a non-mixing propellant is used, and the dip tube reaches into the product.

A different arrangement is used in cans containing very viscous substances. The product is enclosed in a plastic bag attached to the underside of the valve and the propellant fills the space between the bag and the can. This stops the product from sticking to the sides of the can and allowing the propellant to escape up the dip tube. Cans of this type can be used upside down; an ordinary can must be kept the right way up so that the end of the dip tube remains in the product.

Aerosol cans are filled on the production line by inserting the product, putting the lid and valve on the can and forcing the propellant in backwards through the valve. The bag type can, however, must be filled with propellant through a small extra valve in the base.

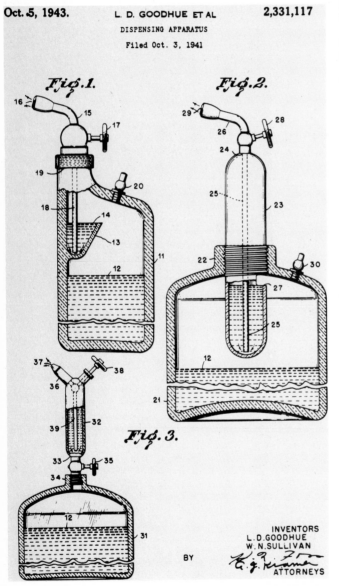

Oct. 5, 1943.　　L. D. GOODHUE ET AL　　2,331,117

DISPENSING APPARATUS

Filed Oct. 3, 1941

INVENTORS
L. D. GOODHUE
W. N. SULLIVAN
BY
ATTORNEYS

Above: diagrams from the original patent for the aerosol spray, which was filed in the United States in 1951. It differs from the modern type in being refillable and designed to dispense a metered dose, which was tipped into the upper section of the spray by tilting it.

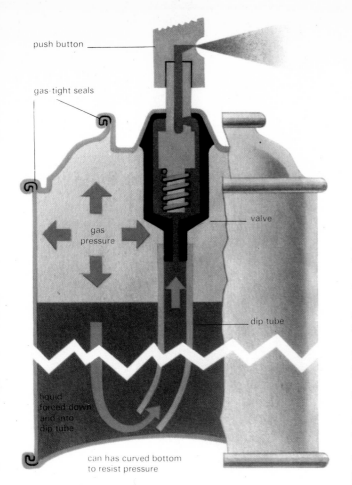

push button

gas-tight seals

gas
pressure

valve

dip tube

liquid
forced down
and into
dip tube

can has curved bottom
to resist pressure

Left: cross section of a typical modern aerosol spray. Gas pressure produced by the volatile propellant forces liquid down the can and up the dip tube to the nozzle when the valve is opened.

AIR CONDITIONING

Air conditioning is the creation of an artificial climate, making it possible to maintain constant, pleasant conditions inside buildings and provide a steady flow of purified air.

In cities, sealed double-glazed windows can be used so that noise and dust can be excluded, and a quiet, pleasant environment can be produced.

Air conditioning is essential in underground spaces, cinemas and theatres, crowded shops, hospitals, tall office buildings, and in many industrial processes which are sensitive to atmospheric conditions.

Methods used Air is purified, cooled or heated, humidified or dried, according to the need, by the air conditioning plant and circulated through the building by means of ducts, which may be of metal or may be formed out of the structure itself.

There are various stages in a large air conditioning plant: not all plants include every component, and in the smallest air conditioning unit the components are combined in one casing not much larger than a television set.

Air first enters a section where it mixes with re-cycled air from the building—only a certain proportion of fresh air is needed. Next, the mixed air passes through a filtering section, which may be in two stages. The first stage takes out coarse dust, and will be a fibrous medium, rather like cotton wool, either in the form of a screen of individual filter 'cells' which can be replaced when they become dirty, or an electrically driven roller screen. Following this is the second stage filter which is generally an electrostatic type and removes the finer particles such as cigarette smoke. In this, a high voltage is used to charge incoming dust particles which are then attracted to a grid of oppositely charged plates.

The air temperature is controlled by passing the air through two tube banks. One is supplied with hot water or steam, and the other with chilled water or a refrigerant fluid.

Inside the room to be ventilated is a temperature sensor—usually an electrical resistance thermometer—which is set to the desired value. The difference between the required and the actual temperature automatically determines whether the heating or cooling tubes are used.

The next stage is the odour filter, made of activated carbon, a substance which is capable of directly absorbing odour molecules from the air. This needs to be reactivated by heating from time to time to drive off the absorbed material.

Finally, moisture is added to produce the desired humidity, either by injecting steam into the air or by spraying a mist of very fine water droplets. This too is controlled from a sensor inside the room, the electrical resistance of which varies with the humidity. If moisture has to be removed from the air, the usual method is to arrange for it to be both cooled and then re-heated if necessary at the temperature control stage. The moisture will condense on the cooling tubes.

The air is normally moved through the system by a centrifugal fan, the rotor of which resembles a paddle wheel. Air enters at the centre and leaves around the edge of the wheel. This type of fan can move large volumes of air despite the appreciable drag of the plant and ducting.

Silencers are always placed after the fan to prevent the noise of the plant from reaching the room. These usually consist of a labyrinth of sound absorbing material.

Air is finally delivered through metal ducts to the room 'diffusers'. These take various forms, such as long slots or grilles in the walls close to the ceiling, vaned outlets flush with the ceiling, or perforated sections of the ceiling itself.

Air conditioning systems The same principles are used from the smallest to the largest system. Small room units contain a simple washable filter, refrigerating compressor, and electric air heater.

More powerful units are made to supply larger rooms, and frequently the refrigerating section (compressor and condenser) is placed outside the building.

For large buildings there are three main systems: *all air*, *air-water*, and *all water*. In the first, the plant supplies all the air that is needed at a fixed temperature. Local duct heaters are needed in different rooms or zones of a building to give final temperature control. An alternative is to have two ducts, one carrying cool air, the other warm air. The two air streams are blended, as in a mixing tap, to give the required temperature. In the 'variable volume' system, the temperature is regulated by controlling the amount of air supplied instead of its temperature.

In the air-water system, the central plant only delivers the minimum fresh air needed for ventilation. Each room then has a separate heating and cooling unit using heated or chilled water.

In the all water system, only the heating or cooling water is supplied from the central plant, and fresh air is brought in through individual ventilators in each room.

The air-water and all water systems can use small ducts, are more economical, and run quieter, but are less efficient than the first method, in which all the air is constantly being conditioned.

Air conditioning today Air conditioning is being used widely for offices, shops, supermarkets, restaurants, and entertainment centres, where the main problem is to keep the building cool in summer. Temperatures in modern buildings can become uncomfortably high through heat from lights, crowds of people, and sunshine through large windows. Sunshine through large windows can raise room temperature to

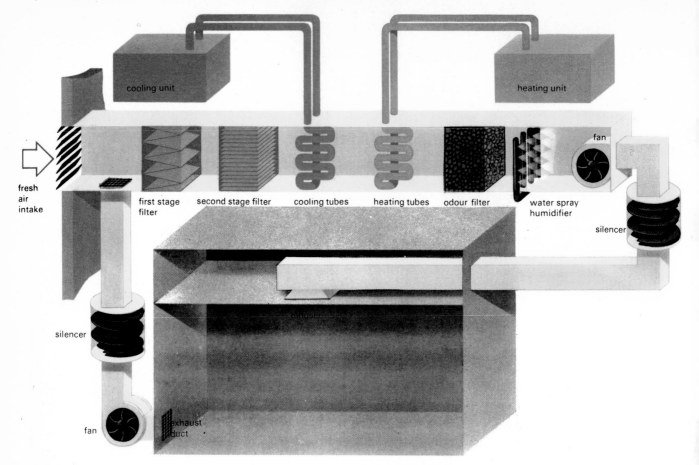

cooling unit

heating unit

fan

fresh
air
intake

first stage
filter

second stage filter

cooling tubes

heating tubes

odour filter

water spray
humidifier

silencer

silencer

exhaust
duct

fan

an equivalent of over 90°F (32°C) with only 60°F (16°C) outside.

The heat from artificial lighting is sometimes sufficient to keep a building warm in winter, and some buildings have been specifically designed to do just that: the hot air generated by the lights is picked up and returned to the plant.

In medicine, air conditioning is essential for the operating theatre. In this case the air must be sterile and must keep the area around the operating table free from contamination. In industry, air conditioning is needed to control the environment to a process or product. In so-called 'white' rooms, for example in the manufacture of transistors, the atmosphere is cleaner than ever occurs in nature.

Top of page: a schematic diagram of a typical 'all-air' system such as might be used for a large building. Air leaves a room through an exhaust duct and is mixed with fresh air from outside before being passed through two filters, cooling and heating tube banks, an odour filter and a humidifier. It is then returned to the room through a silencer. Only one room is shown here for the sake of simplicity.

AMPLIFIER

An amplifier is a device for increasing the strength of a weak signal fed into it. Electronic amplifiers, which are the best known and most important type, are used in a huge variety of devices such as radio and television receivers, record players, radar, analogue computers and electronic equipment generally. Other devices which amplify in a different way include hydraulic amplifiers such as the power brakes of a car, acoustic amplifiers such as the megaphone and magnetic amplifiers, used as theatre light dimmers and in computers.

All electronic amplifiers work in much the same way, though they differ widely in design and in the *gain* (degree of amplification) they produce. Gain can be measured as a proportional increase in voltage (the usual method for amplifiers), in current or in wattage—total electrical power.

The heart of an amplifier, and the device that actually does the amplifying, is either a thermionic valve [vacuum tube] or a transistor. Nearly all electronic amplifiers have several of these plus a set of resistors, capacitors, potentiometers and related devices to control the flow of electricity through the basic amplifying components.

Valves [tubes] and transistors use different principles to perform the same function. Basically, they act as variable switches where the flow of a small current through one part of the device controls the flow of a larger current through another part. When the small current flows, the large current, which is drawn from a separate power source, flows too. When the small current stops flowing, the large one is shut off, and when the small current flows at, say, half power, so does the large one.

The proportion of the smaller current to the larger one is constant (at least in a linear amplifier, the most usual type). So if the small current is modulated (varied) by adding a signal from a record, tape or other source, the signal will be reproduced more or less faithfully at a much higher power in the form of a modulation of the large current. This large

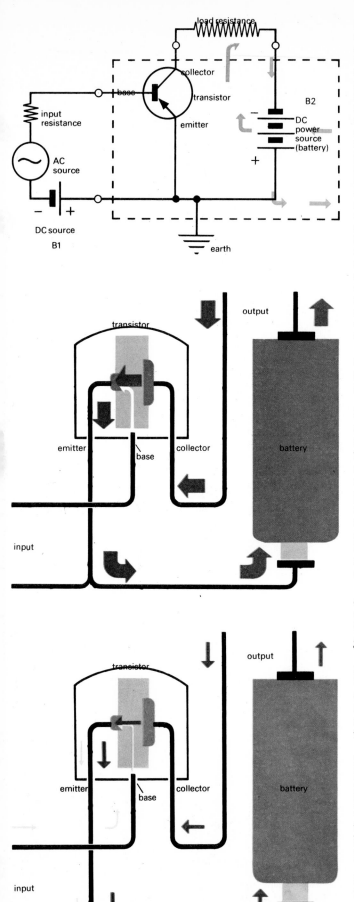

load resistance

collector

base

transistor

emitter

input resistance

AC source

DC source
B1

B2

−

DC power source (battery)

+

earth

transistor

output

emitter

base

collector

battery

input

transistor

output

emitter

base

collector

battery

input

Left top: a simple amplifier circuit. The signal voltage to be amplified is fed in from the left to the base-emitter circuit. The amplified signal voltage is found across the 'load' resistor at the top right of the diagram. The batteries B1 and B2 are required to provide the correct DC currents for the transistor to work. The battery B2 also provides the power source to produce the amplified signal across the load resistor.

Left centre and below: how the above circuit works. When the current to the transistor base-emitter circuit is large, the collector-emitter circuit current is correspondingly larger. When the emitter-base current is small, the collector-emitter current is correspondingly smaller. The input signal is always fed to the base-emitter circuit, and the output from the collector-emitter circuit.

current can then be fed to a loudspeaker to convert it into an intelligible sound (or elsewhere, depending on what the amplifier is used for).

In the case of a transistor, the small current is fed in between the terminals known as the emitter and the base, and the large current flows between the emitter and the collector. So a transistor has only three terminals, not four, to carry the ingoings and outgoings of two currents. The collector, emitter and base are represented in the thermionic valve by the cathode, anode and grid. The names and principles are different, but the function is the same.

Amplifiers in practice A simple amplifier as described would not normally produce enough gain for practical purposes. It might reach a 30 times increase in voltage. But an ordinary hi-fi amplifier operating in normal conditions would probably give a 100,000 times voltage increase. Some amplifiers used for other purposes have much higher gains than this.

Gains of this degree are produced by using an amplifier consisting of several stages. The output of the first stage is passed to the second stage and amplified further, and so on through as many stages as are needed to yield the necessary gain.

A hi-fi amplifier usually consists of two stages, the first a pre-amplifier with a fixed gain setting which boosts the incoming signal from the record, tape or radio to a manageable level at which it can be handled by the second stage, or main amplifier, which provides sufficient power to drive the speakers. This stage includes a volume control to adjust the final gain.

Amplifiers are normally designed with an inbuilt gain much higher than is actually needed or used. This is then moderated by the use of negative feedback, where a portion of the output signal is fed back to the input with a reversed polarity (current direction) to reduce the gain. In this way, the volume can be controlled by varying the amount of negative feedback. More importantly, distortion will be reduced and any changes in the supply voltage or the electronic components will have less effect on the gain.

If positive feedback were used, part of the output signal being fed back to the input with the *same* polarity to *boost* the gain, the result would probably be to produce unwanted oscillations, which are sometimes heard in public address systems as a loud howling noise. This is caused by the output boosting the input, the input increasing the output accordingly, the increased output further boosting the input, and so on up to uncontrollable levels, causing the amplifier to stop working. In public address systems this is caused by sound from the loudspeakers (the output end) reaching the microphone (the input end).

The quality of a linear amplifier is assessed by its ability to magnify the input signal 'faithfully', that is, without altering its essential shape. But amplifiers, like other physical systems, are not perfect. To give a faithful reproduction of, say, a musical instrument, an amplifier must respond to all the frequencies (pitches of sound) that the instrument produces, giving an equal response to all of them. In practice, this means that a hi-fi amplifier must respond equally to the whole range of audible frequencies, from about 30 Hz to 18 kHz (30–18,000 cycles per second). This range of frequency response is known as the bandwidth.

No actual amplifier can live up to this ideal, but high quality hi-fi amplifiers come closest. The frequency response can be partially altered by adjusting the treble and bass controls.

The video amplifiers used to form the pictures of TV and radar receivers have an enormous frequency bandwidth from 0 Hz (that is, direct current) to 6 MHz (6,000,000 cycles per second).

Amplifiers also suffer from harmonic distortion—output at frequencies twice, three or more times that of the signal—and from amplifier noise, a random jumble of different frequencies independent of the input signal. This is termed 'white noise' because it includes all frequencies just as white light includes light waves of all frequencies between red and violet.

Amplifier noise can never be totally eliminated. It can always be heard in a sound amplifier as a slight hiss. But a good hi-fi amplifier can have a signal to noise ratio better than 3,000,000 to 1. For a 10 watt amplifier this would mean a noise power of less than 3 microwatts.

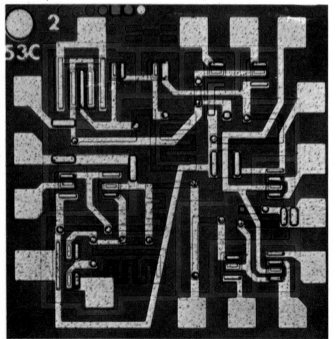

All the components of an amplifier stage can be built into a silicon chip which may be as little as 0.2 inches (0.5cm) square and 0.01 inches (0.025 cm) thick. The input and output connections of the circuit are made to the square terminals around the edges of the chip.

AQUALUNG

The aqualung [or SCUBA, short for self contained underwater breathing apparatus] is a system which allows a diver to carry his air supply with him when he dives. He is thus freed of any direct links with the surface and has much greater flexibility of movement than the 'helmet' diver, who needs air and safety lines.

The aqualung diver's stay under water is limited by the capacity of the air cylinders he carries. Nevertheless, the dive times afforded by the aqualung have made it an important tool in underwater science and technology, and in search and rescue operations, while it has opened the way for diving as a sport throughout the world.

The aqualung consists of five basic components: a demand valve or regulator, which reduces the high pressure air supply to ambient pressures (the same pressure as the water around the diver); the cylinders, which contain compressed air; the harness which keeps the apparatus in the correct position relative to the diver's body; the tubes for air delivery and exhaust; and the mouthpiece through which he sucks the air. All are vital, but it is the demand valve which is the most complex and ingenious.

The human body has evolved on dry land and therefore is designed to withstand the kind of pressures exerted on it by the air (one atmosphere or 14.7 psi at sea level). Water, though, is much more dense than air, and the diver's body is subjected to much greater pressure in water than on land. The deeper he goes, the stronger these pressures become. Seawater exerts a pressure of two atmospheres at 33 feet (10 m), and this increases by one atmosphere for every additional 33 feet (10 m) of depth.

The body consists largely of solids and liquids that are virtually incompressible, even under very great pressures. But it also contains cavities which are filled with air—the lungs, the sinuses, the inner ear, and the stomach—all of which connect with the respiratory system. Thus, if the air breathed in is not at the same pressure as the water around the body, these cavities will be forced to contract. As air is easily compressed, breathing will become extremely hard work even at relatively shallow depths, and if the diver goes deep enough, the pressure will crush the cavities flat and kill him.

Demand valve The job of the demand valve is to see that these problems do not occur. The simplest form is a circular box connected on one side to the outlet of the high pressure cylinder. The other side is open to allow seawater to enter, but the water does not flood the box—it is stopped by a rubber diaphragm inside the open face. The diaphragm is connected within the box to a valve controlling the air supply from the cylinder, and when the diaphragm is pressed inwards that valve is opened.

The external pressure, whether atmospheric or that of the sea, pushes the diaphragm in. This opens the air valve until just enough air has been let in on the cylinder side to balance the pressure. The pressures on both sides of the valve are then equal, so the air the diver breathes must be at ambient pressure.

When the diver breathes this air, he creates a partial vacuum inside the air chamber. The outside pressure opens the valve again, and the process continues.

This single stage design has the disadvantage that the high pressures to be controlled by the valve put a heavy strain on it. A modification to overcome this is the two stage type, in which the first stage brings the air down from around 3000 psi (200 bar) to about 100 psi (7 bar) above ambient pressure by means of a valve acting against pre-set spring pressure. The air can then be brought to ambient pressure by a device similar to the single stage type.

In the split-stage or single hose type, the first stage is mounted on the cylinder itself, which is connected by a single small-diameter pressure hose to the face mask, where the

The high pressure of the air in the cylinder is reduced in two stages to the level of the surrounding water by two sets of valves and chambers. In each set the valve is opened by the pressure 'downstream' of it falling below a certain level. The high pressure of the first stage is balanced by a spring.

Bottom of page: three stages in the working of the valve. Air at higher pressure than the water is shown in red, at roughly the same pressure in orange and at lower pressure in yellow. The diver opens the second stage valve by inhaling and so lowering the pressure in the second stage chamber.

second demand valve stage is located.

Some demand valves have inlets where air lines can be fitted to supply air at fairly low pressure from cylinders or a compressor at the surface. This allows the diver to stay submerged for a long time.

Cylinders and harness The cylinders which carry the air supply are relatively simple. They are made in various sizes,

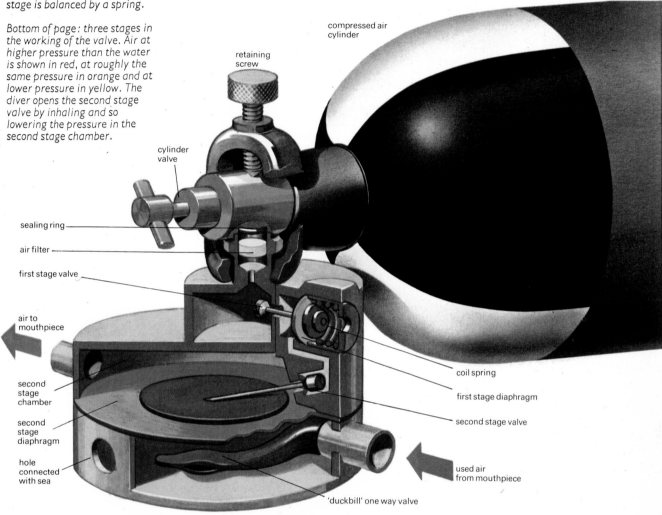

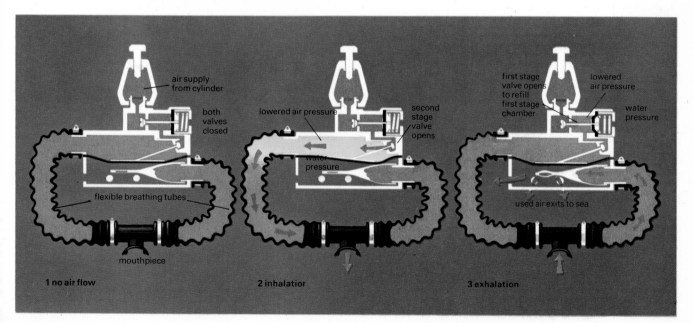

1 no air flow

2 inhalation

3 exhalation

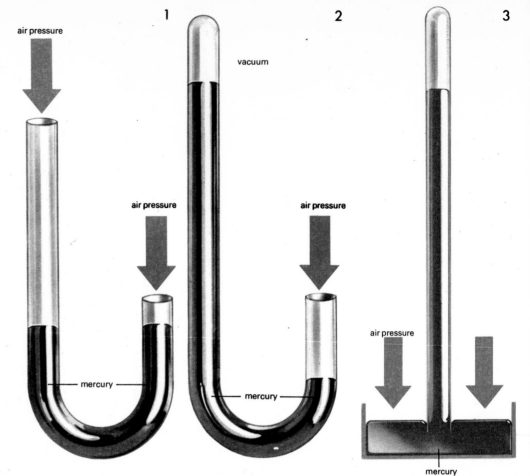

These drawings illustrate the principle of the mercury barometer.
I   both ends of the tube are open and the atmospheric pressure equal on both sides.
2   one tube end is sealed and evacuated. The pressure now affects one side only.
3   the open end of the tube is replaced by a shallow dish and the closed end by a single inverted glass tube.

and the divers like them to be buoyant when empty so that they will float to the surface after use. They are painted grey with black and white quarters at the top, a conventional code to show which gas they contain—in this case air. A typical cylinder might contain about 60 cu ft (1.7 m³) of air compressed at a pressure of around 3000 psi (200 bar).

The deeper the diver goes, the more air he uses for each intake of breath. Thus the depth of a dive has a bearing on its length in time. Cylinders which contain enough air for an hour's diving at 33 feet (10 m) will only support the diver for 30 minutes at 100 feet (300 m). Other variations in the time scale arise because the diver needs more air the harder he works, or when he is cold.

The harness is usually of nylon or cotton webbing with steel bands. Its task is not only to hold the cylinders in place on the diver's back, but also to make sure that the demand valve is as near the centre of ambient pressure in his lungs as possible.

Development   The first really dependable aqualung system was developed by Jacques Yves–Cousteau and Emil Gagnan, but their work was preceded and undoubtedly helped by many other attempts at providing the diver with a truly self-contained breathing system. Much of this was at first concerned with mine rescue work. As early as 1879, a British designer named H A Fleuss demonstrated a diving set in London in which air was carried in a flexible bag on his back. The diver breathed used air back into the bag, but on the way carbon dioxide was removed from it by caustic potash.

Since only a small percentage of the oxygen is used in each intake of air, a relatively small reservoir of air could last a long time. The oxygen actually used was replenished from a small cylinder which he also carried, and this allowed him to stay underwater even longer. The device was only suitable for shallow depths and even then must have made breathing hard work for it had no demand valve.

BAROMETER

Barometers are instruments for measuring atmospheric pressure. The atmosphere exerts a pressure because air has weight and is being pulled to the earth by the force of gravity. For this reason atmospheric pressure depends on the height of air above the point at which it is being measured and is lower on top of a mountain than it is at sea level.

Pressure is the force acting on a certain area and is commonly expressed in such units as pounds force per square inch (psi), newtons per square metre (N/m²), pascals (Pa) and bars (sometimes millibars or mbar). Another common unit of pressure is the atmosphere, which is defined as the average atmospheric pressure at 0°C (32°F), sea level and 45° latitude. One atmosphere is approximately 14.7 psi, 101,325 N/m² or 1.01325 bar.

The pressure at the bottom of a container of gas or liquid can be determined in the above units from the height (h) of the container column, the density (d) of the gas or liquid, and the acceleration (g) due to gravity, all expressed in compatible units. The pressure is h × d × g. The width of the container does not affect the pressure—in the case of atmospheric pressure the 'container' is worldwide.

In atmospheric pressure measurement, a column of a standard liquid (one with known density) in a vertical tube is used and as gravity is approximately constant at all points on the earth's surface, it is only necessary to record the column height. Mercury is the most common liquid used because it is extremely dense and so only needs a short column. One atmosphere is the pressure at the bottom of a column of mercury only 30 inches (760 mm) high. Consequently, one atmosphere pressure is often referred to as 30 inches (760 millimetres) of mercury (30 inch Hg or 760 mm Hg—Hg is the chemical symbol for mercury). Using water, whose density is about 1/13 that of mercury, one atmosphere is about 34 ft (10 m) of water.

Gases are not used because their densities are very much less than those of liquids and the column would have to be extremely tall. For example, normal atmospheric pressure would be the result of approximately five miles (8 km) of air, if its density were constant. There is, however, a further complication in that gases are compressible, while liquids are not, which means that the density increases with pressure.

Mercury barometers The fact that the atmosphere has weight and exerts pressure was first demonstrated by the Italian scientist Galileo in the early 17th century. It was his pupil Torricelli, however, who worked out the principle of the barometer.

In principle, the weight of a column of mercury is balanced against the weight of the air. To do this, a glass tube more than 30 inches (760 mm) long, sealed at one end, is completely filled with mercury. This is then placed upright with its sealed end uppermost and its open end dipped into an open bowl of mercury. The pressure on the open surface of the mercury is the atmospheric pressure, and this must balance the pressure created by the column of mercury. The mercury level in the tube therefore falls until the weight of the column exactly balances the atmospheric pressure; by falling, it creates a vacuum at the top of the glass tube. The fact that the bottom of the tube is submerged prevents air from entering the tube, which would let all the mercury run out.

As the atmospheric pressure changes, so will the height of the mercury column, and by measuring its height above the surface of the open bowl, the atmospheric pressure can be found.

All modern mercury barometers are based on Torricelli's basic design, and only differ by a few modifications which enable more accurate and consistent readings to be made. Two important types of mercury barometer are the Fortin and the Kew.

Fortin barometer In the simple barometer mentioned above, any change in the column height will mean a slight change in the mercury level in the bowl. Yet, as it is from this level that the column height is measured, this would mean using a moving scale to find the height of the top of the column. The Fortin barometer overcomes this problem by employing an adjustable container (called a cistern) so that the open mercury level can be raised or lowered to a fixed point. This point is where the mercury level just touches a fixed ivory pointer. The column height is then measured with a fixed vertical scale. For even more accurate readings an adjustable vernier scale is added which allows readings to be measured to within 0.002 inch (0.05 mm) of mercury.

Kew barometer The Kew barometer has a fixed cistern and allowance is made for changes in the open mercury level by altering the spacing of the column scale. If both the column and the cistern are perfectly cylindrical then the change in the column scale compared to the true inch or millimetre will always be in a fixed ratio. This ratio is made very close to equality by making the barometer column very narrow compared with the diameter of the cistern.

Kew barometers are used used in ships, where motion can affect the accuracy of the readings by causing oscillations in the mercury column. This is overcome by introducing a restriction in the column which dampens such oscillations without unduly affecting the sensitivity of the device.

A further modification is an air trap to prevent any air from entering the column from the cistern and reaching the vacuum at the top of the tube, which might happen if the barometer was tilted or shaken. Kew barometers are designed to be portable, usually being carried upside down with the mercury completely filling the glass tube.

Aneroid barometer It was another Italian scientist, Lucius Vidi, who in 1843 invented the aneroid barometer. The term 'aneroid' means without liquid, and although not offering quite the same degree of sensitivity or accuracy as the

The Fortin barometer was designed by Jean Fortin and first brought into use in the early 19th century. To obtain a pressure reading the adjusting screw is turned, compressing the flexible leather bag until the mercury level in the cistern reaches the tip of the ivory pointer. The difference between the two mercury levels can then be read.

pointer
glass cylinder
lower end of tube
boxwood frame
flexible leather bag
adjusting screw

mercury type, it has the advantage of being a robust instrument useful in such applications as altimeters, and generally where mobility is required (an aneroid altimeter is exactly the same machine as a barometer, and is used to measure height by air pressure).

The principle of the aneroid barometer is that a sealed metal chamber, sometimes called the bellows, expands and contracts with changes in atmospheric pressure. This expansion and contraction can be suitably amplified with a rack-and-pinion arrangement or levers to move a pointer on a scale.

The chamber is usually made of thin sheet nickel-silver alloy or hardened and tempered steel. High grade aneroid barometers have a series of steel diaphragms formed into a complete unit and corrugated to provide greater flexibility.

Temperature compensation is necessary with precision aneroid barometers, since the diaphragms expand and contract with temperature changes, and their elasticity alters.

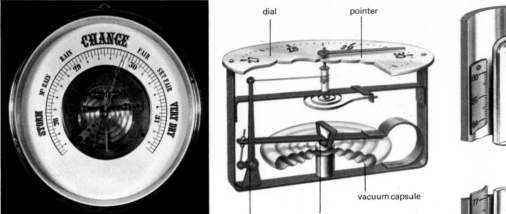

dial   pointer

vacuum capsule

crank   knife edge pivot

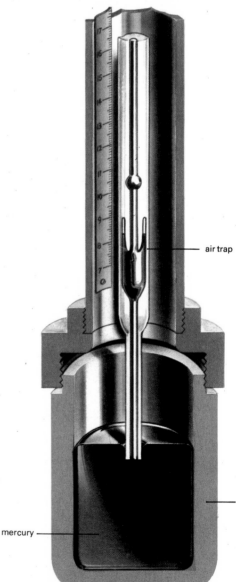

glass barometer tube

scale

air trap

barometer cistern

mercury

Any air within the chamber also leads to unwanted temperature effects from expansion and contraction, and so a high vacuum is generally created in the chamber. Collapse of the chamber under atmospheric pressure is resisted by the springy nature of the diaphragms.

The modern aneroid barometer is a precision instrument, which, when compensation is introduced for temperature and other errors, can easily give pressure readings to within 0.02 inch Hg (0.5 mm Hg) and can be estimated to within 0.001 inch Hg (0.025 mm Hg). When used as an altimeter this means height readings estimated to within one foot (30 cm) at sea level, and 1.5 ft (45 cm) at 11,000 ft (3350 m) where the air is thinner. Altimeters must be adjusted before use to take account of the prevailing air pressure in the area.

Aneroid barometers are calibrated against a standard mercury barometer, and require frequent readjustment.

Aneroid barograph A refinement of the aneroid barometer is the aneroid barograph, a self-recording barometer for meteorological and aeronautical applications where a continuous record of pressure variations is required. Instead of a pointer, a pen is attached which traces a graph on chart paper wrapped around a slowly revolving drum.

Because of the friction between the pen and the paper, its accuracy is not quite as good as the aneroid barometer. For better sensitivity, a refined version known as a microbarograph is used. In this device the graph is to a much larger scale, and it is possible to obtain readings to within 0.02 inch Hg (0.5 mm Hg).

Top left: an aneroid barometer dating from about 1876.
Top centre: the aneroid barometer consists of two corrugated metal diaphragms enclosing a vacuum. Any change in the air pressure on the outside of this chamber is transmitted mechanically to the pointer which registers the change directly on the scale.
Right: the English Kew pattern barometer has a fixed scale and a simple cistern. A change in atmospheric pressure will produce a change in the cistern mercury level and corresponding, but smaller, change in the level in the tube. In order to compensate for this difference the scale is very slightly contracted. A funnel-shaped air trap catches any small air bubbles which enter the mercury.

BATTERY, electric

The electric battery or cell produces power by means of a chemical reaction, although there are exceptions such as the nuclear battery. A battery can be either primary or secondary: the primary type (sometimes referred to as a 'dry' battery) is normally regarded as unchargeable whereas a secondary cell or storage battery can be recharged.

By far the best known electric battery is the unchargeable one which is used to power such things as flashlights, transistor radios, tape recorders, alarm bells and hearing aids. Both primary and secondary batteries provided the power for the equipment used to relay the Apollo TV pictures from the Moon. Several types of battery were used on this project including the modern fuel cell.

When one or more cells are connected together these are known as a battery. The single cell used in a flashlight, however, is also known as a battery although it is only one cell. When several cells are used, they are connected so that their individual voltages add up to give the voltage required to operate a particular appliance.

These days one cell can produce a voltage ranging from 0.9 V for a mercury-cadmium cell to 2.8 V in a lithium cell, but generally the voltage of the everyday type of cell is between 1.2 V and 1.6 V. Nowadays each type of battery is made to provide a particular source of power. This power requirement may have to withstand arduous environmental conditions such as the heat of the tropics and the cold of the Arctic Circle; it must also be capable of being dropped and jerked up and down when carried in a pocket. Another important feature of a modern battery is that it is leakproof.

Producing electricity A simple cell consists of two conductors, *plates* of metal or carbon, dipped into a water-based solution, the *electrolyte*. Pure water cannot be used because it is a very good electrical insulator and would block the flow of current, but when certain chemical substances are dissolved in it, water becomes a conductor although its conductivity is still not as good as that of a metal. Suitable chemical substances fall into three categories. There are the acids such as sulphuric acid; bases or alkalis such as caustic soda; and salts formed by the interaction of an acid and a base.

Electricity is generated in cells because when any of these chemical substances is dissolved in water its molecules break up and become electrically charged ions. A good example is sulphuric acid, the molecules of which consist of two atoms of hydrogen, one of sulphur and four of oxygen. When dissolved in water the molecules split into three parts; the two atoms of hydrogen separate and in the process each loses an electron, becoming a positively charged hydrogen ion. The sulphur atom and the four atoms of oxygen remain together as a sulphate group (SO_4), and acquire the two electrons lost by the hydrogen atoms, thus becoming negatively charged.

If one plate or *electrode* of zinc and one of either copper or carbon is dipped into a sulphuric acid electrolyte and each is externally connected to a load such as a light bulb, a current will flow through the bulb, lighting it. This is because one of the chemical elements chosen for the electrodes is electrically positive (that is, has a tendency to lose electrons and acquire a positive charge) with respect to the other, and when they are electrically connected the chemical equilibrium of the cell is upset and reactions start at both plates. Under these circumstances, the atoms of zinc each give up two electrons which flow through the external circuit and form the current. The positively charged zinc atoms left behind dissolve into the electrolyte and each one combines with one of the negatively charged sulphate ions. The result is a neutral zinc sulphate molecule. The two electrons originally given up by each zinc atom travel around the external circuit and reach the other plate. There they combine with and neutralize the positive charges on two hydrogen atoms from the electrolyte. These two neutral

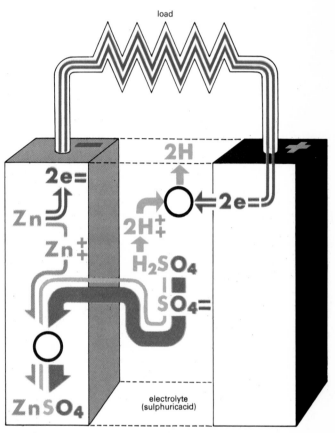

load

$2e=$

$2H$

Zn

$2e=$

Zn^{++}

$2H^{++}$

H_2SO_4

$SO_4=$

$ZnSO_4$

electrolyte (sulphuric acid)

negative plate (zinc)

positive plate (copper)

hydrogen atoms then combine to form a molecule of hydrogen gas, and gas bubbles are produced at that plate.

In theory, the chemical reaction would go on and electric current could continue to flow until all the zinc on the zinc plate (known as the negative electrode or *cathode*) has been used up. But in the simple cell a film of hydrogen bubbles begins to form on the copper or carbon plate (known as the positive electrode or *anode*), and as hydrogen has a much higher electrical resistance than the electrolyte proper, the internal resistance of the cell increases, reducing the current that can flow in the external circuit. At the same time, a voltage (a difference in electrical charge) is produced between the hydrogen and the zinc, and this voltage is in opposition to the main voltage produced between the zinc and the copper or carbon plate, further reducing the available voltage and current.

If the external circuit is disconnected, the hydrogen bubbles will gradually disappear and the cell can be used again, but the same thing will reoccur. In all modern batteries this effect, which is known as *polarization*, is greatly reduced by surrounding the positive electrode with a material known as the *depolarizer*. This works either by reacting with the hydrogen to form water or by taking over from the hydrogen the task of accepting the electrons as they arrive from the external circuit.

The Leclanché cell The 'dry' batteries used in flashlights and so on, which have a depolarizer, are of a type known as Leclanché with modifications to make the liquid electrolyte a semi-solid. In its original form the Leclanché cell was entirely 'wet' with an electrolyte consisting of a strong solution of ammonium chloride. A zinc plate was used for the negative electrode, and a carbon rod packed into a porous pot containing crushed carbon and manganese dioxide (to accept the electrons) formed the positive electrode and its depolarizer. Similar materials are used in a modern dry Leclanché cell. The electrolyte is not, in fact, dry but is made up in the form of a moist paste or jelly.

Additives include mercuric chloride introduced to inhibit what is known as *local action*. This is the name given to the chemical reactions that occur between zinc atoms and carbon and iron atoms which occur as impurities in the zinc plate. It can be overcome by a process known as amalgamation in which the mercury forms an amalgam or alloy with the zinc, preventing it from reacting with its impurities. Other additives include potassium dichromate which inhibits the corrosion of the zinc—an effect that would otherwise reduce the shelf life of the battery, that is, the length of time it can be stored without deterioration.

In the cylindrical Leclanché cell used in flashlights the zinc forms the outer casing and is also the negative electrode. The positive electrode consists of a mixture of graphite (carbon) and manganese dioxide depolarizer round a graphite rod.

Dry cells are supplied singly or in groups of two, three or more to give higher voltages. This is known as a series connection, and the positive electrode of one cell is connected to the negative electrode of the next. High voltage batteries, in which sixty or more individual cells are connected in series, are available but are very heavy and cumbersome. A lighter and more compact construction, where large numbers of cells are to be connected in series, is the flat or layer type. Batteries of this type consist of alternate thin, flat layers of zinc, electrolyte and the materials making up the positive electrode and its depolarizer.

Other batteries A disadvantage of Leclanché cells is that the current quickly falls, principally because hydrogen forms more quickly than it can be removed by the depolarizer. For this reason they are best suited to intermittent work. A more constant voltage and current is provided by the mercury cell widely used in hearing aids where almost continuous operation is necessary. In this type of cell, the full name of which is zinc-mercuric oxide, the electrolyte consists of potassium hydroxide. The negative electrode is zinc, as in the Leclanché cell, but the positive electrode and its depolarizer consist of graphite and mercuric oxide.

Development There are some indications that batteries may have been used by the Parthians, a tribe in what is now Iran, for electroplating jewellery in the 3rd century BC, but the work which led to modern batteries began with the discovery by an Italian, Alessandro Volta, that he could cause an electric current to pass through a wire by immersing two different metals in a salt solution.

In the year 1800 he described the first battery, known as a Volta pile or Voltaic cell. It consisted of a number of cells in series, each made up of a disc of silver, a disc of paper or cloth soaked in salt solution, and a disc of zinc. The first cell not subject to polarization was developed by John Daniell in 1836. His cell had a zinc negative electrode which was dipped into a dilute sulphuric acid electrolyte, and a positive electrode of copper in a saturated copper sulphate solution. The two liquids were separated by a porous membrane. A conventional depolarizer was first used in 1839 by Sir William Grove, who made a cell which consisted of a zinc negative electrode dipped in dilute sulphuric acid and separated by a porous pot from the depolarizer—nitric acid —which surrounded a positive electrode of platinum. Robert Bunsen later replaced the platinum with carbon. The Leclanché cell was patented in 1868 by Georges Leclanché. At first, these cells were principally used to recharge storage batteries or accumulators of the lead-acid type.

Despite their long history there is still a considerable amount of research being carried out on batteries. These include the fuel cell, the zinc-air cell, the lithium cell, sodium cells, the mercury-cadmium cell, the indium bismuth, sea water, magnesium and perchlorate electrolyte cells. Besides these there are other types of battery systems which are available commercially including manganese alkaline, mercury-zinc, mercury, silver-cadmium, and aluminium-air. But for most applications, the zinc-carbon cell is commonly used.

BEARINGS

A force is always required to slide one object over another in order to overcome the resistance or friction between the two surfaces. Friction is usually a hindrance in machines, absorbing power, producing heat, reducing efficiency and promoting wear which limits the life of the machine.

A bearing is a device which will reduce this friction while supporting a load. The moving member can be either rotating, such as the wheel of a bicycle, or it can be a flat surface such as a tool holder sliding along a lathe bed.

Most bearings are used for supporting rotating components, and can be classified into rolling element bearings, fluid film bearings, and rubbing bearings.

Rolling element bearings The ancient Egyptians used a type of rolling bearing when they moved the huge stones needed for the pyramids by rolling them along on logs. They realized that such a system meant less work. Modern rolling element bearings are precision mechanisms but work on the same principle using either balls or cylindrical rollers between two surfaces. As with the use of logs in earlier times, the force required to roll one part over the other is very small compared to the force which would otherwise be required to slide one part over the other.

Rolling element bearings are used today in automobile wheel hubs and gearboxes, electric motors, washing machines, ventilation fans, textile machinery and all types of industrial machines. Such bearings are comparatively friction-free (rolling friction between metals being about one hundredth of the sliding friction). Any energy losses that occur are because of the compression of the metals at the contact points and the motion of the lubricant.

The most common type of rolling bearing for rotating mechanisms consists of four basic parts: the inner ring or race; the rolling elements (either balls or rollers) of which there are several; a cage to retain and separate the balls or rollers; and an outer race. The inner and outer races and the rolling elements are made in a hard alloy steel to give a long life and to prevent permanent indentation of the races at the contact points when under load. The cage is made of soft steel, brass, or plastic resin.

Types of rolling bearings Ball bearings are normally used to carry radial loads, that is, loads perpendicular or at right angles to the axis of the shaft. Other types are available which will accommodate axial or end thrust loads, while some will carry combined radial and thrust loads. Depending upon the type of bearing, the rolling elements can be cylindrical (log shaped), convex (barrel shaped) or tapered. Tapered roller bearings are often used in automobile wheel hubs where the weight of the vehicle has to be supported, and the sideways cornering forces resisted.

Fatigue life When rolling bearings are subjected to load and rotation, the balls or rollers and the races undergo repeated application of stress at the contact areas, which may ultimately show signs of fatigue failure. Metal fatigue in rolling bearings gives rise to surface pits or craters which cause noisy operation and necessitate replacement. The length of time a bearing will operate until surface pitting begins is known as the *fatigue life*. The principal external

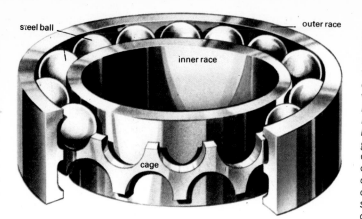

steel ball
outer race
inner race
cage

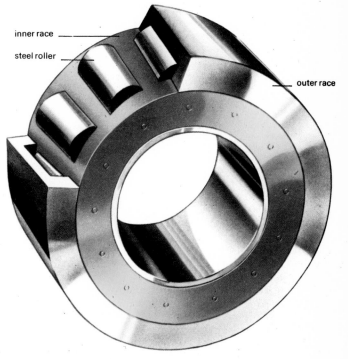

Left: a ball journal bearing. The inner race is surrounded by a reinforced cast metal cage which houses the ball bearings and the shaft can rotate freely within the outer race.
Below: a roller bearing. The rollers may be cylindrical, barrel-shaped, tapered or hour glass-shaped. This is a combined form of journal and double thrust bearing: it is able to take both a radial load and thrust from either end. The shaft must be mounted very accurately.

inner race
steel roller
outer race

factors which affect the fatigue life are therefore the *load* which, for a given size and type of bearing determines the contact stress, and the *rotational speed* which determines the frequency of stress application. The size and type of bearing are also important in controlling the internal stresses and therefore the fatigue pattern. For example, a big bearing will carry a heavier load than a small bearing or alternatively will have a longer fatigue life for the same load, and a roller bearing, with rollers that have 'line' contacts will support a greater load than a ball bearing of the same overall size, which has 'point' contacts.

Tests have shown that as a general approximation the fatigue life of a rolling bearing varies inversely with the cube of the applied load, and inversely as the rotational speed. Expressed simply, if the load is reduced by one half and the speed unchanged then the life is increased eight-fold. Alternatively if the load is unchanged and the speed is halved, then the life is doubled. Consequently the load carrying capacity of a rolling element bearing is a very important factor in determining its suitability for a given application, and is always given in terms of a specific load which the bearing will carry without fatigue failure for a given number of revolutions.

Because of variations in materials and conditions, the fatigue life prediction is usually based on a rotational life—usually one million revolutions—which will be reached or exceeded by 90% of identical bearings operating at the same load and speed.

Lubrication Rolling bearings must be lubricated for long life and quiet operation. The majority, including many automobile and household equipment bearings, are partly filled with grease which in some cases lasts the life of the machine. In most industrial applications where heat is developed in the bearings, circulating oil is used as the lubricant and as a means of removing the heat. In communication satellites, which are required to remain in space for 7 to 10 years, lubrication of the ball bearings in gyroscope and solar paddle mechanisms is critical because the grease tends to evaporate and dry up in the vacuum of space. In this and other difficult situations—for example in high temperature applications—greases made from synthetic materials are used.

Fluid film bearings In fluid film bearings the frictional forces are reduced by putting a film of fluid instead of balls or rollers between the two surfaces. The fluid is usually oil, such as in automobile engine bearings, but it can be water (pump bearings) or even air or gas, depending on the application.

The two main types of fluid film bearings are the *hydrodynamic*, in which the fluid film pressure is generated by rotation, and the *hydrostatic*, in which the fluid is supplied under pressure from an external source.

The majority of fluid film bearings are hydrodynamic. Here fluid film pressure required to separate the loaded surfaces is generated by the movement or rotation of one part relative to the other. Automobile engine crankshaft bearings operate on this principle and although there is an oil pump, its purpose is only to supply oil to the bearing, and the pressure is not sufficient to support the load. When the engine is stopped, the oil film is squeezed out and the metals touch. Water skiing uses a similar hydrodynamic principle (aquaplaning) where the movement of the skis relative to the water produces a pressure sufficient to support the weight of the skier. If the speed is too low then the skier will sink. Hydrodynamic bearings are used, therefore, in applications where the speed of rotation is high enough to produce a supporting fluid film, such as in automotive engines, steam turbines, alternators for producing electricity, steel rolling mills (though some modern mills use rolling element bearings), and paper making machinery. The rotating member is usually the shaft and the stationary part is the bearing, which in some cases is made in the form of a sleeve or bush shaped like a tube, but is more often split in two halves for ease of assembly.

Because there is no film separation when the machine is being started and stopped, bearings of this type have to be made in materials which will permit some rubbing contact with the shaft (which is usually of steel) without causing damage. Generally soft bearing materials are preferred, provided their strength is adequate. Soft materials allow dirt to be embedded which would otherwise score the surfaces.

Moreover soft metals generate very little heat when rubbing occurs and do not weld or seize when in rubbing contact with a steel shaft. The softest metal bearings for light loads are made from 'white metals' (also known as Babbitt metal). These are alloys, mainly of lead and tin. Next in hardness are the copper-lead alloys, and harder still are the bronzes which are basically alloys of copper, tin and lead. Phosphor bronze is the hardest alloy and is used for very highly loaded bearings like diesel engine piston pin bushes. Other metals such as antimony, nickel and aluminium are also used in some of these bearing alloys.

There are many applications of slow speed bearings such as chemical drum driers, plate shears, dock gates, large blanking presses and so on, where full film generation is not obtained. In such cases the bearing materials given above are normally used and the system is lubricated with grease instead of oil.

In some bearing applications where air is used as the separating medium rather than oil or water, the principle of operation is aerodynamic and not hydrodynamic. Self acting air bearings are being developed for textile spinning machinery where the spinning spindles rotate at very high speeds, say 50,000 revolutions per minute, and where oil lubrication is undesirable because of possible staining of the yarn. Other problems associated with spinning machinery, such as noise, friction and wear, are minimized by using air bearings. Air bearings of this type are only suitable for high speeds, light loads and in small sizes, say up to one inch (25 mm) diameter.

Hydrostatic bearings Hydrostatic bearings are not widely used as they require a pumped supply of fluid under pressure to separate the two surfaces. The pressure must be high enough to support the load. The bearing itself is not very different from a hydrodynamic bearing except for the addition of pockets or recesses in the bearing surface into which the fluid is pumped, and as there is never any metallic contact, steel or similar materials can be used throughout. The system, however, is expensive to make, principally because of the need for a high pressure pump and motor with a complicated control system.

Hydrostatic bearings are used where heavy loads have to be supported at very slow speeds or even when stationary, and where under these conditions friction must be kept to a minimum. Some precision machine tools use this type of bearing in the form of flat pads to support the heavy table on which the work piece is carried. A good example of the application of hydrostatic bearings is in large telescopes, such as the famous Hale telescope at Mount Palomar in California. The moving parts weigh 500 tons and are supported on three hydrostatic bearing pads. The whole apparatus is driven by a 1/12 horsepower motor, and in fact could be moved by hand.

Bearings of this type but using air as the pressurized support film are properly called *aerostatic* bearings. Air cushion vehicles work on this principle: air is blown into the under-side of the hull at sufficient pressure to support the weight and to lift the hull. One of the first aerostatic bearings was developed for dental drills which rotate at about 500,000 revolutions per minute and are driven by a small air turbine which also supplies air to the bearings. Another example where air (or rather gas) bearings have been used is in some gas cooled nuclear reactors, where, because of radioactivity and the need to prevent contamination of the circuit with lubricating oil, the gas circulators have been completely sealed, and their bearings continuously pressurized with circuit gas. In this case carbon dioxide is used.

Rubbing bearings In recent years a great variety of plastic materials have been developed which have proved to be very useful in many bearing applications. The materials include phenolic, epoxy, and cresylic resins which are usually reinforced with cloth, nylon, acetal, and polytetrafluoro-

ethylene (PTFE). PTFE, also called Teflon, is also used as a coating on non-stick pans.

These materials have the advantage over metal bearings of being able to operate dry—that is unlubricated. This is a valuable property in some applications, for example in the food and pharmaceutical chemicals industries where contamination of the product has to be avoided. Other advantages are their cheapness and ease of machining, or in some cases moulding into the required shape. They do not seize or cause damage to the rubbing shaft, and being softer than metals they accept more misalignment.

The life of plastic bearings in dry conditions is limited by wear. Because of this they are generally only suitable for slow speeds with intermittent operation, or light loads. Lubrication of these bearings greatly increases their wear life, and there are today many applications where the performance of lubricated plastic bearings—usually bush or sleeve bearings—is better than that of lubricated metal bearings, particularly where lubrication is neglected.

Examples are the steering linkage bearings in automobiles

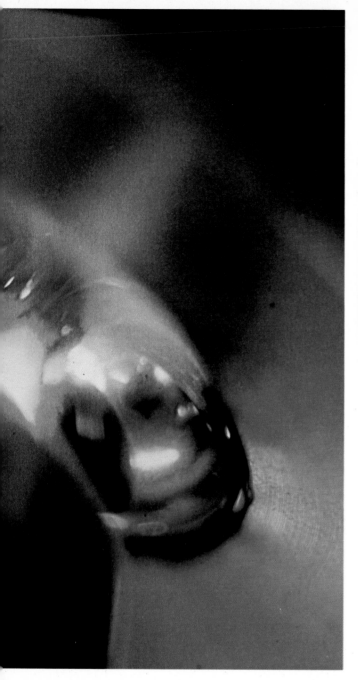

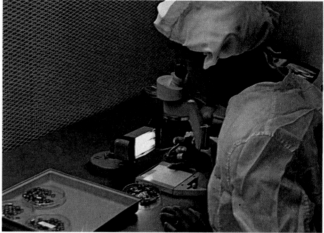

Left: a ball bearing undergoing a seizure test. The fatigue life-span of a rolling bearing varies inversely as the fourth power of the applied load and inversely as the rotational speed.
Below: the bearing components are assembled and inspected in a dust-free atmosphere.

BELL, electric

The electric bell found in many houses is made to ring by a very simple device that makes a clapper vibrate against a bell or gong.

The mechanism consists of the bell itself, the clapper that strikes it, which is mounted on a spring, an electromagnet (most bells have one consisting of two coils) and a simple adjustable electrical contact.

When the bell-push is pressed, electricity flows through the contact into the electromagnet. This attracts the iron arm of the clapper, which moves out on its spring and strikes the bell. As it does this, however, it swings away from and 'breaks' the electrical contact, stopping the flow of current through the electromagnet. With the magnetism gone, the arm is pulled back by the spring. When the arm falls back, it touches the contact again, restarting the flow of current to the electromagnet so that the arm moves out again and hits the bell—the cycle repeating itself. The speed of vibration, and to a certain extent the loudness of the bell, can be increased by reducing the distance that the clapper must travel. This is done by adjusting the electrical contact mechanism.

Electric buzzers work on the same principle, but the clapper hits the outer casing of the buzzer instead of a bell, producing a dry sound.

Chimes The two-tone chimes that some houses have instead of bells are worked in a different way. There is an electromagnet that pulls a clapper over, but no contacts. Instead, the end of the arm has a flexible joint. When the bell-push is pressed current flows through the electromagnet causing the flexible end to swing across and hit the first chime. When the bell-push is released and its arm falls back the flexible end swings back farther than the arm and hits the other chime. It comes to rest roughly halfway between the chimes ready for next time.

Some sets of chimes have a second bell-push for the back door connected to the electrical circuit through an electrical resistance. Current to the electromagnet is consequently weaker and the magnetic field produced pulls the arm with less force. If the arm is set off-centre between the chimes, it will only hit one of them, thus making it clear whether the caller is at the front door or the back. More complicated

which are grease lubricated; dock gate bearings, also grease lubricated; central heating pumps and many other water pump bearings which are water lubricated; and sliding door wheels and furniture castors, which normally operate dry.

Some rubbing bearings are self-lubricating, usually made from sintered metal powders based, mainly, on aluminium, copper, brass, gun-metal or bronze. They are a result of the technology of powder metallurgy, in which very fine powders of metal are precision compressed to the required shape and dimensions. Their porosity depends upon their degree of compactness.

In essence they can be thought of as metal sponges having holes ranging from 1 to 30 thousandths of a millimetre, which can retain up to 30% of their volume of oil. Typical applications are as clutch release bearings and main bearings for automobiles. The main advantage is that they do not depend upon an external oil supply system, are cheap, and can be easily installed. Mainly, however, they are restricted to light duty applications since their compressive strength is less than that of a solid metal.

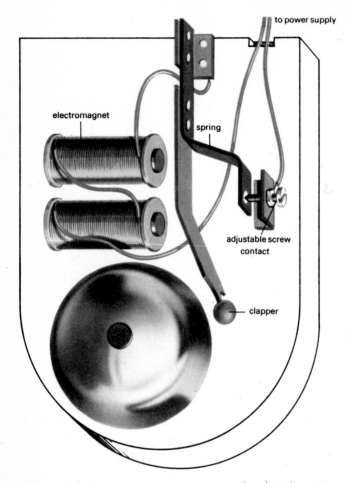

The working voltage of the simple electric bell depends on the fineness of the wire used in the coils; it operates equally well with AC or DC electricity.

to power supply

electromagnet

spring

adjustable screw
contact

clapper

chimes with three or more notes are rung by electric motors turning a striker that rings them in order.

Alarms Some fire and burglar alarm bells on the outside of shops and offices use electric motors rather than electromagnets. Attached to the motor shaft is an arm with a jointed hammer end. The spin causes the hammer to fly outwards, striking the bell, but the joint allows the hammer to bend back after striking so that it can pass the bell.

Amplified bells playing through loudspeakers are also found on ambulances and police cars in some countries.

BINOCULARS

A pair of binoculars is essentially two telescopes mounted side by side. Since both eyes are used to look at a scene, a stereoscopic view is obtained so that nearby objects are seen in depth, and there is less eye strain than with a single telescope.

The simplest binocular instrument uses two Galilean telescopes, each of which has just two lenses. Known generally as an 'opera glass' or 'field glass', this has a rather small field of view—like looking through a tube—and a low magnification, rarely more than four times.

Better telescopes usually have achromatic lenses and are much longer and heavier than the Galilean type. To make a powerful pair of binoculars using ordinary telescopes without any modification would mean that they would be very long and clumsy. But by using a pair of right-angled prisms

in each optical system, the light can be 'folded' so that the distance it travels between the front object glass and the eyepiece is lengthened quite considerably without lengthening the body of the instrument. This also allows the object glasses to be offset so that they can be further apart than the viewer's eyes, giving a greater stereo effect and allowing larger lenses to be used without touching at the centre. The basic telescope design gives an upside down image, but the prisms turn it the right way up so no extra lenses need be used.

Focusing is usually carried out by a central wheel on a bridge which links the two eyepiece tubes. Turning this wheel moves the two tubes in or out simultaneously. There are often differences in strength between an individual's eyes, so a separate twist focusing thread is provided on the right hand eyepiece to allow for this. These centre focus (CF) models can be quickly focused on objects at different distances, and so are popular. They are not as robust as independent focusing (IF) models, however, in which each eyepiece is fitted directly to the main body and has to be focused separately. Binoculars intended for military or marine use are of the independent focus type.

The two halves of a pair of binoculars are hinged, so that each user can alter tham to suit the distance between his eyes. A scale marked on the bridge shows this separation in millimetres. A value of 64 mm ($2\frac{1}{2}$ in.) is common.

If the two light paths of the separate halves of the binocular are not parallel, a double image will be seen. Even a slight misalignment is noticeable, if only by the discomfort produced when the eyes try to bring the image together. This is actually bad for the eyes, and binoculars should always be perfectly *collimated* so that the beams are parallel.

The prisms are held in place in recesses by metal straps, and are adjusted to collimate the beam by using wedges or screws. Fine adjustment is carried out by rotating the 'cells' which carry the object glasses. These cells are made so that the lens is slightly off centre; the screw thread they fit into is also slightly off centre so that by turning one inside the other the lens can be brought to any position, including dead centre.

Cheap binoculars, though they may at first appear to be every bit as good as more expensive ones, often suffer from poor prism mountings. The prisms can be knocked out of alignment quite easily.

Specifications The magnifications given by binoculars range from 6 to 20 times—written 6× to 20×. The diameter of each object glass is called the *aperture*; 30 mm to 80 mm apertures are widely available, though stand mounted military reconnaissance models with apertures as large as 150 mm have been produced.

The usual way to specify the performance of a pair of binoculars is by a pair of numbers, such as '8×30'. This means that the magnification is 8× and the aperture is 30 mm. There are certain popular combinations of magnification and aperture. The commonest specifications are 8×30, 7×50 and 10×50.

It might be thought that the higher the magnification, the more useful a pair of binoculars will be. This however, is often not true, and the ideal specifications vary with the circumstances.

With a high magnification, it is more difficult to hold the image steady—any slight trembling of the hands will be amplified. Another difficulty is that the more an image is magnified, the fainter it gets, because the same amount of light is spread over a larger area; the field of view also gets smaller unless a wide angle eyepiece is used. For these reasons, magnifications greater than 10 are rarely satisfactory. Zoom eyepieces are made, in which the magnification can be varied continuously from, say 7× to 15×. The optics and mechanics of these are relatively complex, and must be particularly robust if they are to work well.

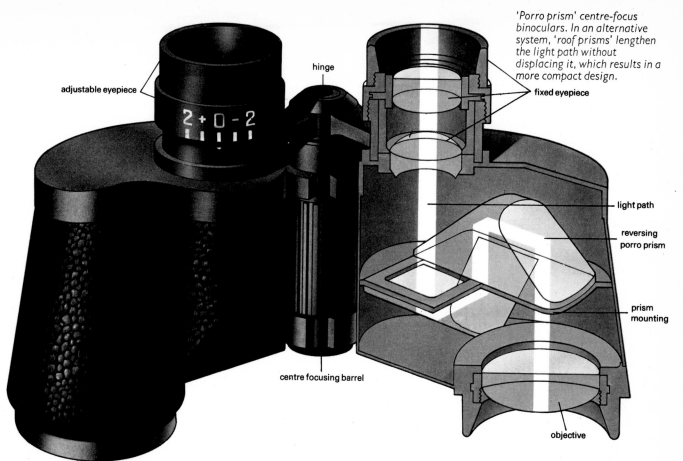

adjustable eyepiece

hinge

2 + 0 - 2

centre focusing barrel

fixed eyepiece

light path

reversing porro prism

prism mounting

objective

Another factor is the size of the *exit pupil* of the binoculars. This is the size of the light beam at the eyepiece end, and its diameter in millimetres is found by dividing the aperture by the magnification. For example, the exit pupil of an 8×30 instrument is $30/8 = 3.8$ mm, while that of a 7×50 is $50/7 = 7.1$ mm. In broad daylight, the pupil of the human eye is about 2 mm across, while in dim light it opens up to a maximum of about 7 mm. This means that although there is too much light for the eye to use during the day, so that some is wasted, a 7×50 instrument will give as bright an image as possible at night.

More light can be provided by increasing the aperture, but this means that the instrument will have to be made longer and heavier, with larger prisms. Convenience of use therefore puts practical limits on the power of binoculars. The 8×30 size is popular as a lightweight general purpose binocular, while the 7×50 type is almost always used for military purposes where the good light grasp is preferred to high magnification.

Two values of the field of view can be quoted for binoculars. One is the sector of the surroundings that the binoculars actually see—usually 6° to 8° of the 360° circle of the horizon. But magnification makes this seem wider than it is, so the apparent field of view can range from 40° to 70° or over, the higher values giving almost 'wrap-round' vision. This field of view, and the magnification itself, are produced by the eyepiece alone. Eyepieces may consist of five or more lenses combined to give a wide field of view, freedom from false colour and lack of distortion. On cheap binoculars, it is often found that straight vertical lines, such as the corners of buildings, tend to become curved at the edge of the field of view.

Other faults on binoculars are the use of prisms which are too small, in order to produce a more compact instrument, and badly made bridges between the two halves, resulting in jerky or unequal focusing between the two images.

BOOMERANG

Throwing sticks have been used since prehistoric times, and their use has persisted until recently in many parts of the world. Most throwsticks, such as the *knobkerrie* used by South African tribes, are straight wooden clubs, one to three feet (0.3 to 1 m) long, with a roughly circular cross-section and a knobbed head. They have a simple ballistic trajectory, like that of a thrown stone, and a maximum range of about 50 yards or metres. In some localities, however, throwsticks of more advanced design have been developed, which use aerodynamic lift to help keep them in the air. Typical of these throwsticks is the Australian *boomerang*.

Though it is commonly believed that all boomerangs are intended to return to the thrower, most boomerangs are designed to fly a nearly straight trajectory, and a good straight flying boomerang can be thrown about 200 yards or metres. The much greater range of the boomerang, when compared with a simple throwstick of similar size and weight, is due to the fact that it has a cross-section of aerofoil shape, which gives it lift as it flies through the air. This lift force does not act through the centre of the aerofoil, and if the boomerang were straight it would twist until it was broadside on to the airflow. This would give no lift and a great deal of drag. Boomerangs are therefore curved and thrown with spin so that the two wings rotate like the spokes of a wheel. The attitude of the boomerang is in this way stabilized, similar to the way in which spin stabilizes the attitude of a gyroscope, so that the wings always present only a small angle to the airflow. This keeps the drag small, but gives the necessary lift.

On a straight flying boomerang the wings are carefully shaped and twisted so that the total lift acts through the centre of gravity of the boomerang. Such boomerangs are then thrown with their plane of rotation almost horizontal and the lift force acts in a direction opposite to that of gravity throughout the flight. The design is such that the spin rate is maintained almost constant throughout the flight, and even when the boomerang loses forward speed, near its extreme range, the rapidly rotating wings can still inflict severe injury. Since they fly only a few feet above the ground they can be aimed with greater accuracy than such high trajectory weapons as the spear, and aboriginal Australians used them for hunting kangaroos and other animals, and in tribal conflicts.

Return boomerangs These are less widespread in Australia than straight flying boomerangs, as they are of less practical importance to the aborigines. Their design differs only in detail from that of straight flying boomerangs, but they are generally smaller and lighter. Typically they measure 18 to 30 inches (45 to 75 cm) tip to tip, with a mass in the range 4 to 10 oz (100 to 280 g), which compares with 24 to 35 inches (60 to 90 cm) tip to tip and a mass in the range 7 to 14 oz (200 to 400 g) for straight flying boomerangs. Apart from their smaller size, return boomerangs are also characterized by their more bent appearance, the angle between the wings being typically 90° to 130°, whereas for straight flying boomerangs this angle may be up to 150°. As their name implies return boomerangs can be thrown so as to return to the thrower, and in skilled hands their trajectory often approximates a large flat circle, some 30 to 60 yards (27 to 54 m) diameter, round which the boomerang flies at a near constant height of 3 to 6 feet (1 to 2 m) above the ground. Though this trajectory is a typical one, many others are possible.

Return boomerangs, unlike straight flyers, must be thrown with their plane of rotation almost vertical. The diagram gives the thrower's view of a return boomerang, just after it has left his hand and is flying away from him. The plane of rotation as shown is about 20° to the right of the vertical (for a right-handed boomerang) and the more curved side of the wings is uppermost. The lift, which acts perpendicular to the plane of the boomerang, can be thought of as being made up of two components, one vertical and one horizontal, as indicated by the dashed lines. The smaller, vertical component helps to keep the boomerang in the air, by opposing the weight. The larger, lateral force makes the boomerang accelerate to the left. The wing design of return boomerangs is less advanced than that of straight flying boomerangs, and the total lift does not act through the centre of gravity. The upper wing cuts through the air more rapidly than the lower wing (just as the upper parts of a car or bicycle wheel move forward more rapidly than the parts close to the ground) and more lift is consequently produced by the upper wing. The total lift from the two wings, therefore, acts through a point above the centre of gravity and this tries to make the boomerang lean over in an anticlockwise direction. Since it is spinning rapidly, however, the boomerang behaves like a gyroscope and instead of leaning over it turns (or more precisely *precesses*) to the left. It does this throughout its flight so, since it is constantly turning and accelerating sideways, its trajectory is approximately a circle and it will return to the thrower.

Although return boomerangs are thrown with their plane

Why a boomerang returns. The thrower's eye view and side view show that whichever wing is uppermost, its greater airspeed produces more lift. This would tend to twist the boomerang upright but for gyroscopic precision, which transforms it into a leftwards movement. This causes the boomerang to fly in a circle, as shown by comparing the top view with that of a gyroscope rotor. A boomerang will return quite well if both wings are identical. But to increase its flight duration, it is made to 'lie down' by shaping one wing, as the lower diagrams show, to give greater lift (here exaggerated). This tries to turn the boomerang leftwards, but precession converts this into a more horizontal attitude.

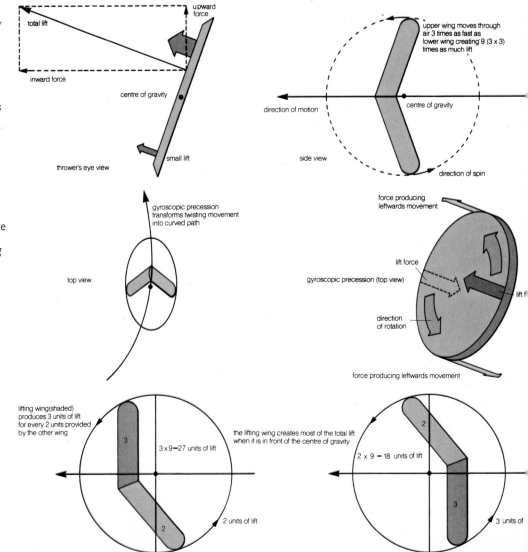

of rotation nearly vertical, near the end of the flight the plane of rotation is usually nearly horizontal. The more curved side of the wings is uppermost and they lean over in this way because the total lift acts through a point which is forward, as well as above, the centre of gravity. (This makes the boomerang precess, or turn, about a horizontal axis at the same time as it is precessing, as already described, about the vertical axis.) This leaning over, or lying down as it is usually called, is essential to a successful return flight, because as the boomerang slows down and loses lift, a greater proportion of the total lift has to be directed upwards, in opposition to gravity. With the boomerang plane horizontal, near the end of its flight, all the lift acts upwards and the boomerang may sometimes hover for a few seconds before sinking slowly to the ground. (This and other variations from a circular flight may be induced by a skilled thrower by slight alterations in the angle and speed at which the boomerang is thrown.) Return boomerangs were sometimes used for hunting birds, principally by frightening them into nets.

History It is commonly believed that boomerangs are peculiar to the Australian aborigine, but this is incorrect. Similar weapons were used until recent times by the native inhabitants of south India and north-west India, and by the Hopi Indians of the south-west USA. Boomerangs were also used by prehistoric Europeans and ancient Egyptians. In later Egyptian times (about 1500 BC) fowling with boomerangs in the Nile marshes became a popular sport with Egyptian nobility.

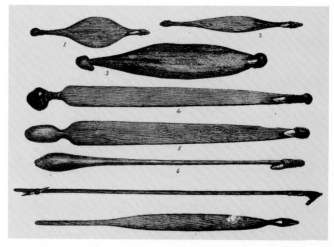

Below: a selection of non-returning throwing sticks from various parts of Australia, drawn in 1870.
Bottom of page: an aborigine throws a returning boomerang. He holds it almost vertical and grips it by the angled end. European enthusiasts generally hold the other end, which makes it easier to throw.

BRONCHOSCOPE & related devices

Bronchoscopes are instruments designed to visually examine the interior of the lungs. They are a type of *endoscope*, the general name for instruments made for internal visual examination of the human body.

Until the 1950s, all endoscopes consisted basically of a rigid tube containing a series of lenses. A miniature light bulb, connected to an external power supply, was mounted on the tip. The endoscope was normally provided with an outer sheath through which saline solution could flow to ensure a clear field of vision. This system had the disadvantages of limited light intensity, and poor definition due to distortion because of the large number of lenses used.

In modern instruments, fibre optics are used for transmitting light to the interior part of the body undergoing examination. In some cases, the image is provided by the same means. Fibre optics operate on the principle that light can be conducted along a curved glass rod by means of internal reflection at the walls. In practice, it has been found that the rod may be reduced in thickness to that of a flexible

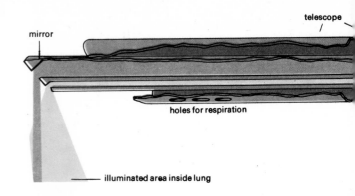

mirror   telescope

holes for respiration

illuminated area inside lung

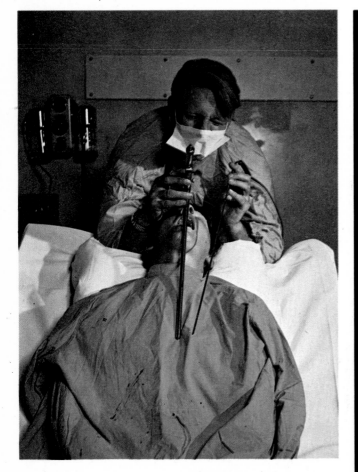

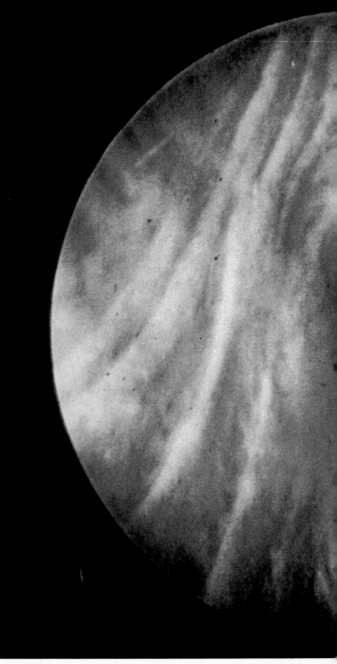

Top of page: section through a bronchoscope. The holes in the side of the outer tube allow the patient to breathe and anaesthetic to be introduced. The viewing system consists of a telescope and light guide contained in a smaller tube which projects through the main one. Bundles of glass fibres are used along which light is transmitted.
Above, he holds the device outside the patient but in the intended position. A sucker in his right hand is used to clear secretions.

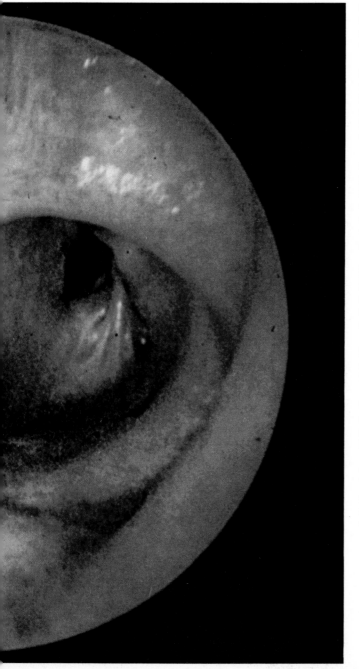

air or anaesthetic inlet

rubber seal

eyepiece

fibre light guide

illumination

glass fibre. Such fibres are manufactured from glass of differing light-bending powers (refractive indices); they have a central core to transmit the light, and an outer layer of lower refractive index to prevent the escape of light. A rod of light-transmitting glass is placed within a tube of glass with a lower refractive index. The two parts are heated in a furnace to soften them before being drawn out into a fibre. Such fibres are grouped into bundles, and used as light guides.

Because the actual light source is outside the body of the patient, lamps of extremely high intensity, such as halogen lamps, can be used without risk of heat damage to the tissues. This makes for greater efficiency in either direct viewing or photography through the endoscope.

When fibres are grouped in *coherent bundles*, with each fibre in the same position at either end of the bundle, they can be used to transmit visual images. Each fibre carries one small element of the picture, allowing reasonable detail to be distinguished without distortion. In contrast, the bundle used to transmit light to the area to be viewed can be *incoherent*, or composed of fibres arranged at random. Endoscopes that have both viewing and illuminating elements composed of fibre optic bundles are extremely flexible, and can be inserted into otherwise inaccessible areas of the body with minimum discomfort to the patient.

Where colour photography has to be undertaken through an endoscope, and good colour definition and fine detail are necessary for diagnostic purposes, a rigid instrument operating on the Hopkins system is used. In the traditional instrument, glass lenses were fitted within a tube filled with air. In the Hopkins system, however, a solid glass rod is used, with the lenses formed by air-spaces. The ends of the rod or rods are ground to an appropriate curvature. This system allows about nine times as much light to be transmitted than previously.

Different types Some more advanced forms of endoscope can be used for certain surgical procedures, such as electric cauterization. Others are used to remove small portions of tissue for biopsy. Such uses of the endoscope can often eliminate the need for more extensive surgery.

The bronchoscope may be a rigid instrument which is inserted into the trachea, viewing the orifices leading to the lungs by means of obliquely places lenses. It is equipped with an anti-fogging device, and is useful when taking a precise sample in biopsies. The flexible fibre optic version has a controllable tip, allowing it to enter the smaller bronchi.

The arthroscope is a specialized type of instrument used for examination of the fluid-filled capsule in the major joints of the body. A large hollow needle, or *trocar*, is inserted into the joint, and the arthroscope inserted through this.

Gastroscopes are flexible instruments, and are passed through the mouth and down the throat for the examination of the stomach and upper intestinal tract. This process entails a minimum of discomfort for the patient, and is widely used in Japan in mass screening campaigns for gastric disease.

Teaching attachments are available to allow simultaneous viewing through an endoscope by more than one individual. These usually consist of a secondary fibre optic bundle and eyepiece. For viewing by a larger number of people, a closed circuit television camera can be attached to the instrument via a fibre optic bundle.

Left: the view shown here is what the doctor sees when he looks through a bronchoscope.

BUGGING DEVICES

Eavesdropping has always been a popular form of spying. Until this century it has been restricted to listening at keyholes or under windows, but modern electronics has changed all that. The importance of surreptitiously gaining information is so high that large amounts of money have been spent developing sophisticated eavesdropping techniques. And the availability of such equipment has, in turn, vastly increased this sort of spying.

Governments regularly listen in on both friends and enemies. In most countries of the world, including the USA and UK, the police use telephone taps and bugs to keep watch on criminals and political dissidents. Many large companies use such devices to find out what their competitors are doing. Indeed, these devices have become so common that private detectives and even ordinary citizens now use them, although it is illegal in both the USA and the UK to do so.

Impetus for development of electronic eavesdropping devices comes primarily from government spy organizations and the military. It is generally considered that the US Central Intelligence Agency and the US National Security Agency lead the world in developing new techniques. A handful of electronics companies that supply military and spy apparatus to the smaller nations also develop new hardware. The US space programme led to the development of much smaller and more reliable electronic components, and this, in turn, led to the development of more sophisticated eavesdropping devices. Electronic devices have reached such a high level that it is now possible to listen in on conversations thousands of miles away.

Telephone tapping One of the first pieces of electronic communications, the telephone, led quite naturally to the first electronic eavesdropping, the telephone tap. In its simplest form, a tap is no more than an extension telephone installed secretly. In a large building it could be in the next flat or in the basement, or anywhere that the telephone lines are exposed and the eavesdropper can make his connection.

But an extension telephone need be nowhere near the main telephone—it can be half way across the city (indeed, this service is used by many businessmen who want to be able to answer business calls at home). If the telephone tapping is done by the police or government with official approval, they can have the secret extension put anywhere they want. Most large cities of the world have secret telephone listening rooms, usually in the central telephone switching office or in the main police station.

1, a briefcase tape recorder.
2, 3, tape recorder and pen microphone. 4, pen transmitter.
5, transmitting microphone.
6, 7, 8, miniature cameras.
9, ashtray with microphone in base. 10, transmitting microphone to attach to windows.
11, microphone watch. 12, telephone transmitter. 13, 15, microphone. 14, cigarette lighter containing microphone.
16, 17, microphone, transmitter and amplifier modules.

Most commonly, the eavesdropper is monitoring only a few telephones and he simply takes notes of any relevant information. This technique is still used in many cases, for example, by the US Federal Bureau of Investigation. Recently, however, there has been a trend to tape record the conversations using a device that turns on the tape recorder only when the telephone is in use. This saves time for the eavesdropper as he need only check the tape periodically. Undoubtedly, the bulk of electronic eavesdropping is government and police tapping of the telephones of suspected criminals and political dissidents.

It is often quite easy for even the unauthorized listener to connect his own extension telephone. For example, he can rent another office in the same building as the person to be listened to, and then rewire the building's central telephone junction box so that his telephone becomes an extension. If that is impossible the usual technique is to connect a small radio transmitter to the telephone which is to be tapped.

Bugging Listening in on conversations in a room has never been as easy as tapping a telephone. With a telephone the telephone company has already supplied the wires and the microphone; the victim speaks directly into the instrument. Most important, the telephone tapper need never enter the room where the telephone being tapped is located.

But the electronic eavesdropping on conversations in a room, commonly called 'bugging', is much harder. First, the spy must gain access to the room to place the microphone, and he must run his own wires. Second, until quite recently any microphone that would pick up all conversation in a room was so large that a simple search would discover it unless it was actually built into the wall.

The second problem has been solved by the development of small and highly sensitive microphones. Now the spy can choose the locations of the microphone based on the ease of running wires.

But wires remain a problem. A newspaper reporter, for example, once stumbled on wires in a park opposite the White House and followed them to a microphone hidden under a bench.

As with telephone taps, the answer is radio transmission. A radio bug is little different from a walkie talkie or a radio microphone of the sort used by entertainers, except that it is placed secretly. Radio transmitters for 'bugs' have a major drawback in that they require power, which means they must have batteries and will not work once the batteries wear out. Nevertheless, most bugs now use radio transmitters rather than wires.

Miniaturization The most significant change in electronic surveillance has been the development of high quality miniature components. The smallest widely available microphone and transmitter are each 0.1 inch × 0.2 inch × 0.3 inch (3 mm × 5 mm × 8 mm). The smallest battery is less than 0.15 inch (4 mm) in diameter. Put together with a small coil aerial, it is clear that the smallest possible bug really could fit in a sugar cube or the proverbial martini olive.

But at this point reality intrudes into the spy movie fantasies. One miniature battery will not give enough power for a bug to transmit very far (in practice the eavesdropper would have to be in the next room) and the battery would be quickly worn out.

The smallest practical bug is the fountain pen transmitter. There exist a number of these, and they actually write because the electronics take up so little space that there is room left for an ink pod. A typical pen would have the miniature transmitter and microphone and at least four miniature batteries (making the electronics already bigger than the sugar cube). Such a pen will pick up speech clearly at 10 ft to 16 ft (3 m to 5 m). But the batteries will last only 7 hours and the transmitter will broadcast only 65 ft to 100 ft (20 m to 30 m) in normal conditions, which is just enough to reach listeners in a nearby parked car, for example.

The pens have another problem. The microphone is sufficiently sensitive to pick up the noise the pen makes when it writes. So the user does not actually write with the pen; instead he leaves it on the table after writing just a word or two.

Seven hours of transmitter life may be enough for the eavesdropper who has legitimate access to a room and can carry the bug with him. But it is rarely sufficient for a person who must take the risk of breaking into a building to plant the bug. Thus common radio bugs are much larger, more powerful, and last longer. A typical unit would be a box 3 inches × 2 inches × 1 inch (75 mm × 50 mm × 25 mm),

half of which is taken up by a 6 volt battery. In addition, it will have a 3 ft (1 m) long aerial wire to increase its broadcast range. Such a device could pick up speech 100 ft (30 m) away (even in an adjoining room if the door between them was open), transmit 1300 ft (400 m), and work for 25 days. Because of its high sensitivity, it could be hidden under a desk or behind the top of a curtain. Indeed, one British firm sells such a unit with a sticky backing already on it so that it can be quickly placed under furniture. Most radio bugs use frequency modulated (FM) signals broadcast at roughly 100 MHz.

When radio transmitters are used for telephone taps, on the other hand, they can take their power from the telephone circuit. No batteries are needed and the device can be very small. If the spy does not want to enter the victim's home or office, he is likely to hide the transmitter in a central terminal box, either in the basement of a large building or on a telephone pole in suburban areas. It is quite common, however, to hide the transmitter in the victim's telephone. One technique is to put the microphone and transmitter inside the telephone's own microphone, using the telephone line as an aerial. This is particularly convenient, because the spy can drop in a doctored microphone in place of an ordinary one in less than a minute.

Lapel microphones, fitted into the buttonhole and hidden by, for instance, a carnation, are connected to a transmitter or tape recorder and can pick up voices within about 5 feet (1.5 m).

Miniaturization has reached other areas of electronic spying as well. It is now possible to get a tape recorder that will record for two hours and which is only 5 inches × 3 inches × 1 inch (125mm x 75mm x 25mm). A voice operated switch, to turn the recorder on only when someone is speaking and thus save the tape, is 4 inches × 2.3 inches × 0.7 inch (100 mm × 60 mm × 20 mm). Both contain their own batteries.

The future Miniaturization and the increasing use of radio make it almost impossible to detect a bug or telephone tap simply by a physical search. But the use of radio has its own problems. A London detective was caught using a bug in a divorce case when a radio ham picked up the transmission and notified the police. Similarly, businessmen who think they are being bugged can use a broad band radio receiver to detect radio transmissions in their offices.

Probably in the future there will be less use of radio. The British military is reported to be working on two alternative devices. One is a tape recorder that transmits only when the tape is full and rewound rapidly. This enables two hours of conversation to be broadcast in a few minutes. On the receiving end, the tape is also put into rewind to record the broadcast and then played forward at normal speed to reproduce the original conversation.

The second technique is to display the speech pattern in digital form on a miniature TV screen and send just one TV picture when the screen is full. Both techniques have the advantage that the short broadcast would be virtually undetectable. But they have the disadvantage of requiring a considerable bandwidth—range of radio frequencies—during the short time.

The laser has been suggested as another answer. Sound makes objects vibrate; if a laser beam is bounced off a vibrating object, then the vibrations can be recorded and decoded to produce a record of the original speed. Such devices do exist, but they are very expensive and do not work very well. Sometimes there is no other choice, however. For example, to overhear people talking in a car, the only options are lip reading using binoculars or a laser bounced off the back of the rear view mirror. (A window will not reflect enough light unless it is very dirty.) To pick up room conversation, an ordinary bug is cheaper and works better, but a laser beam could be bounced off a thin vibrating object such as a calendar page.

The harmonica bug The most dramatic eavesdropping invention of the past few years is a device which uses the long distance telephone as a bugging aid. It is called an 'infinity transmitter' or 'harmonica bug'. The device is a bug, in that it picks up room conversations, but it is installed in a telephone. The spy telephones the victim, then apologizes for getting a wrong number. The victim hangs up but the spy does not. Because the person who originates the call must hang up to break the connection, the line remains open. Next, the spy uses a small whistle to sound a particular note that activates the bug (the device is called a harmonica bug because when it was first used in the US is was reputedly activated by a harmonica note). Conversations in the room are transmitted over the open line until the bug is automatically shut off when the spy hangs up the telephone. There are no time or distance restrictions (the spy can plant the bug and then ring weeks later from another continent), and the bug can be reactivated an unlimited number of times. Telephone systems vary throughout the world, but in some countries the spy need never actually get his 'wrong number'; by sounding the proper tones after he has dialled the number but before the telephone rings, he can open the line without the other telephone ever being answered.

Techniques do exist to detect bugs. One US company has a machine that automatically checks complex office switchboards for taps and infinity transmitters, and a British firm offers a hand-held device to detect radio transmitters.

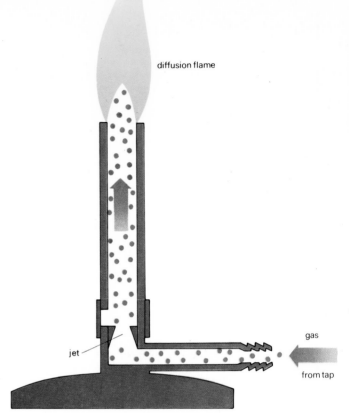

diffusion flame

gas

from tap

jet

BUNSEN BURNER

The Bunsen burner, one of the most useful laboratory tools, is called after the 19th century German chemist Robert Wilhelm Bunsen—though he did not invent it, but only contributed to its development. It is designed to give an extremely hot, nonluminous flame, which does not produce sooty deposits on the articles being heated. The burner consists of a metal tube into which gas is injected under pressure through a narrow jet near the bottom of the burner. As it enters the tube, it causes air to be drawn in through holes level with the jet. The gas and air mix in the barrel of the burner, and are ignited at the top. The proportions of the mixture are usually 3 parts air to 1 part gas: this will burn with a blue flame. If the volume of air is reduced, by rotating a collar which partly covers the air holes, the burner produces a luminous, smoky flame.

When the burner is operated with a 3 to 1 gas mixture, the flame will consist of two distinct zones. The inner cone of the flame, or reducing zone, consists of partly burned gas. This partial ignition results in a mixture of carbon monoxide, hydrogen, carbon dioxide, and nitrogen (whatever the type of gas being used). In the outer cone, or oxidizing zone, this mixture of gases is completely burned, using the oxygen present in the surrounding air. This is the hottest part of the flame. This type of burner can be made to operate on almost any type of inflammable gas, such as natural gas, coal gas, butane or oil gas.

Other designs With simple modifications, the Bunsen burner's efficiency can be considerably increased. In the Tirril burner, both gas and air supply can be adjusted to give optimum burning, and a flame temperature of more than 900°C (1600°F) can be achieved. In the Meker burner, the barrel is much wider and a grid is attached to the top. This produces a number of tiny Bunsen flames, with the outer zones fused to give a 'solid' flame without the cooler central zone. This burner produces temperatures in excess of

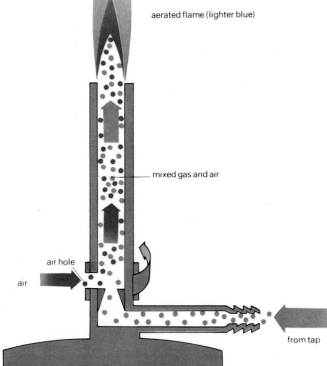

aerated flame (lighter blue)

mixed gas and air

air hole

air

from tap

1000 °C (1832 °F). A further modification is the blast lamp, in which air or oxygen is fed in under pressure. This type of burner is commonly used for glass blowing in the laboratory, and is frequently used for brazing and silver soldering of metals. It produces temperatures of up to 1830 °C (3326 °F).

The burners used in gas cookers and gas fires operate on the principle of the Bunsen burner or of its derivatives. The Welsbach burner is another variant commonly used for domestic purposes, being used in conjunction with a gas mantle to produce incandescent light. This type of burner is very precisely machined, to ensure that a steady flame is produced, which will not cause the light to flicker.

BURGLAR ALARM SYSTEMS

Security systems in recent years have benefited considerably from advances made in other electronic fields. The range of equipment available today includes ultrasonics, microwave, infra-red and television in addition to the more traditional window tube and batten, pressure mats, switches and simple bell systems. Window tubes, for example, are made from aluminium and held in position by wooden battens. Through them is threaded a small delicate cable which is easily broken if the tubes are tampered with—thus causing a break in the electrical sensing circuit.

Most systems are designed to monitor a continuous signal by means of a control panel. When the circuit producing this signal is broken by an intruder or by a failure in the equipment, the signal is interrupted and an alarm is set off. By this means a fail-safe system is provided in the majority of circumstances.

The application of these devices can be divided into two distinct groups: perimeter and space (volume) protection systems. Perimeter protection is concerned with sensing intruders at vulnerable points in a protected area such as fences, walls, windows and doors. Space protection employs sensing devices in a restricted area such as a warehouse, and detects the presence of an intruder within it.

Control panels The nerve centre of any alarm system is the control panel. It is where all the detection circuits terminate and where electronics monitor the circuit for the continuous presence of a voltage. When the detection equipment is disturbed, the circuit is broken and the control panel immediately registers the loss of voltage and activates bells, sirens or telephone dialling equipment, warning of an intrusion.

Control panels range in complexity from a single relay to more complex circuitry monitoring several alarm circuits and providing sequence switching to telephones and audible warning devices, in addition to test facilities, controlled exit routes, and special circuits for the continuous protection of areas with restricted access.

Switches The most commonly used switches for the protection of doors and windows are small encapsulated *magnetic reed switches* inserted in the framework of a door or window and activated by a separate magnet fitted in the door or window itself. When the magnet and reed are within $\frac{1}{4}$ inch (6 mm) of each other the switch remains closed, but once the magnetic field is removed, say when a door or window is opened, it opens the switch and alarms the system.

Vibration transducers and inertia sensitive switches have

Above, diagrams: gas entering a burner passes through a jet near the bottom of the barrel. This causes a partial vacuum in the area around the jet outlet and air is drawn in through holes on the outside of the barrel. The gas mixes with the air and is ignited at the top of the burner. By rotating the collar to cover the holes less air is available and a cooler luminous flame results. Exposing the holes provides more air and the gas/air mixture burns with a hot blue flame. The inner core of the flame consists of partly burned gas.
Left: types of burners. On the right a sectioned burner, at the rear a standard burner, and in the front two modifications to give large, hot flames.

been developed to be operated by high frequency vibrations caused by intrusion or drilling. False alarms caused by traffic rumble or weather conditions can be minimized by careful mechanical design or frequency sensitive circuitry inserted between the switches and the control panel.

Pressure mats are widely used and these are generally constructed from two pieces of foil held apart by a perforated piece of foam. The switch created by this arrangement is open until pressure forces the two pieces of foil together and creates a closed switch across the circuit. This causes a loss of voltage on the circuit to the control panel and an alarm is created.

Space protection Ultrasonics and microwave systems have now become standard for space (or volume) protection and a variety of equipment is manufactured. The majority of systems are designed to use the doppler effect whereby a transmitted ultrasonic or microwave signal is altered in frequency when it is reflected off a moving object back to the detector unit. The change in frequency of the returned signal is detected, amplified, filtered and finally used to operate an alarm. The characteristics of the two systems are determined by the fact that ultrasonics are high frequency sound waves, usually in excess of 20 kHz, while microwave systems use electromagnetic waves. Although the principles and processing of the detected signal are similar, the difference in their characteristics determines their application.

The ultrasonic unit radiates an output frequency from a transducer and the space to be protected is filled with standing waves which form a complex pattern because of reflections from various surfaces in the area. Any movement of an object within the space protected will modify the frequency of the signal received by a second transducer and from this the speed of movement can be calculated.

In a microwave system the frequencies normally transmitted are in either the X or J band, and the most common frequency used in Britain is 10.68 GHz. The most suitable

Below: an alarm system incorporating fibre optics. A break in the light circuit will immediately register on the control board.
Bottom of page: an ultrasonic intrusion detector. Any movement within the monitored area will disturb the standing wave pattern and a modulated signal will arrive at the receiver, setting off the alarm.

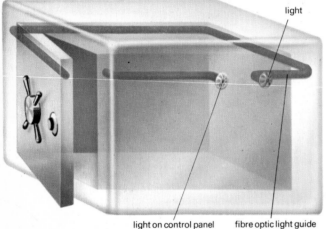

light

light on control panel   fibre optic light guide

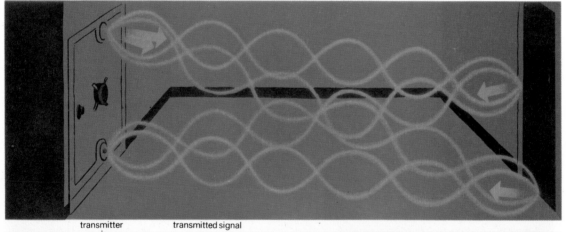

transmitter   transmitted signal

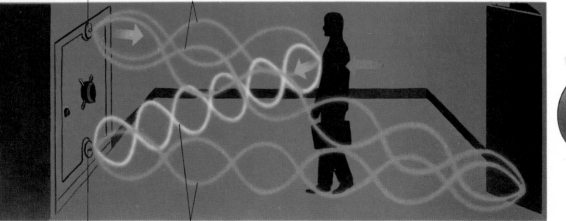

receiver   reflected signal

source of microwave frequencies for intruder alarm systems is the Gunn diode which is mounted in a resonant cavity. The returned signal (from reflections) is mixed with a small portion of the transmitted signal and any Doppler shift in frequency is detected. The signals are then processed in a similar way to the ultrasonic unit relying on the duration of a movement in order to activate an alarm.

Both the ultrasonic and microwave units, however, may be subject to false alarms because of certain conditions. Microwave signals, for example, can penetrate glass and thin partitions and consequently any movement outside a protected area may raise an alarm. Ultrasonics, however, are contained within the protected area but the system can be set off by high winds or additional ultrasonic vibration produced, say, by a telephone ringing. The furnishings also greatly affect the sensitivity of the ultrasonic system, since strong reflections occur from hard surfaces while soft furnishings such as carpets and curtains absorb the ultrasonic signal.

The type of systems described are suitable for applications within a building. Microwave units, however, using a transmitter and a remote receiver have been developed for external use and an alarm is created whenever the narrow microwave beam is interrupted. Such systems find wide application for perimeter protection and can be arranged to form an invisible fence around the area to be protected.

Television systems Surveillance by television cameras is widely used in stores and places where a larger number of people have access to an area. The systems usually require continuous monitoring by an operator but with the additional facilities of video recording equipment any incident can be recorded on magnetic tape and retained as evidence.

The use of television for surveillance has been restricted by the fact that a considerable amount of light is required for the camera tube to produce a television picture but, by using camera tubes with image intensifiers, pictures can be obtained at night.

Infra-red systems The infra-red spectrum extends from the highest microwave frequencies to the longest wave-lengths of visible red light and, because they are invisible to the human eye, they have been used for a considerable time in intruder systems. The normal infra-red beam consists of a transmitter which generates the beam either directly from a gallium arsenide light source or by filtering an incandescent light source, and a receiver which is optically aligned with the transmitter. Providing the receiver continuously receives the infra-red signal the photo diode detector produces a signal which holds the system in a non-alarm condition. If the beam, however, is interrupted an alarm signal is created. More recently passive infra-red devices have been developed which use the material triglycine sulphate. This material is extremely sensitive to infra-red light and can detect the infra-red radiation given off normally from objects because of their temperature. The processing of the signal is complex, as it is designed to sense a change in radiation due to an intruder's movements against a background of infra-red radiation from other sources.

Central stations The alarm is raised either by an alarm bell on the premises or, more probably, by alerting an alarm company's central station or the police. In the latter instances, the intruder can be unaware that he has been detected and consequently the police are more likely to make an arrest. The central stations themselves are high security premises. Signals which indicate the state of the alarm system are constantly transmitted on ordinary telephone lines from the protected premises. If an alarm is activated this is registered at the central station where the operator immediately informs the police. Similar systems can also operate directly into a police station.

Automatic dialling machines are also widely used which, when operated, dial the police and other interested parties and transmit a prerecorded alarm message.

Digital techniques Intruder systems are also benefiting from circuitry and techniques used by the computer industry. Systems are being used for monitoring distant alarms, operating entire systems, and for gaining access to premises by the use of codes. The coding of signals provides greater security than a normal DC circuit because a code is so much more difficult to fake. Not only must the intruder determine the correct pulse code for that particular sensing unit, but also the correct pulse shape and magnitude. Such systems are, however, highly expensive, and used only in the most vital high security areas.

CAM

A cam is a device for converting rotary motion into linear motion. The simplest form of cam is a rotating disc with a variable radius, so that its profile is not circular but oval or egg-shaped. When the disc rotates, its edge pushes against a *cam follower*, which may be a small wheel at the end of a lever or the end of the lever or rod itself. The cam follower will thus rise and fall by exactly the same amount as the variation in radius.

By profiling a cam appropriately, any desired cyclic pattern of linear or straight line motion can be produced. With most cams some form of spring action ensures that the cam follower remains in contact with the cam. In practice, cams are not necessarily rotary in action. The same form of linear movement can be obtained from a cam profile that oscillates back and forth. In both cases the output movement will be at right angles to the initiating forces.

The idea of a cam goes back to Hero of Alexandria and it forms one of the basic devices in engineering. It has found innumerable uses throughout history to provide the oscillating action of bellows, the timing action of valves as in steam engines, and in the control of windmills.

Today it finds uses in the internal combustion engine,

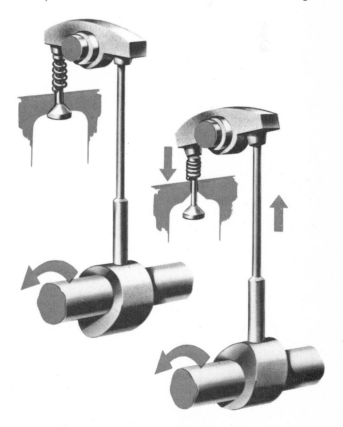

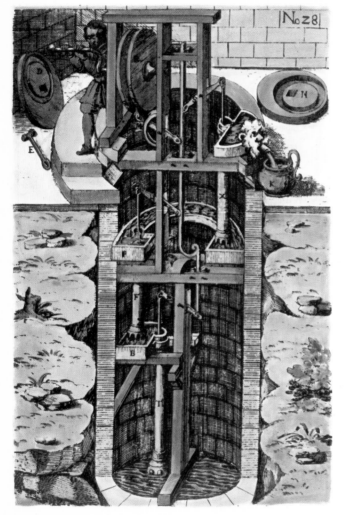

An illustration from a book by Agostino Ramelli, published in 1620. A large cam, R, shown in detail at D and N, is used to raise and lower the pistons of a water pump. The cam converts the rotary motion of the crank into the linear motion of the pumps.

Because the cam can be made with any desired profile and in almost any material, it provides a fine controlling device for machine tool cutting feeds where high stresses are involved. In effect, the cutting tool is made to follow the same path that the cam's profile would follow if it were stretched out in a straight line. It is fairly expensive to make such cams but they can be justified where high production rates are required.

Cams are also used in electromechanical systems to operate electrical switches, providing an ideal means of timing experiments or controlling machine cycles. Such electrical contacts, however, tend to wear out so newer non-contacting systems have been developed where the change in cam profile is detected inductively or optically.

CAMERA and projector

Camera obscura The camera obscura (meaning 'dark room' in Latin) is a device which projects an image of an external object or scene onto the wall of a darkened chamber. The camera obscura works because light travels in straight lines. The light from one point on the subject shines on just one part of the screen, so forming an image consisting of a large number of continuously overlapping images of the pinhole. The sharpness therefore depends on the size of the hole, but the image is in focus at all distances. This property can still occasionally be useful for photographing large buildings or rooms without using an expensive lens.

The first accounts of a true camera obscura appeared in the 10th century. A small hole was used to cast an image on the whitened wall of a dark room. Such a device was described by the Arabian scholar Alhazen, who used it to observe an eclipse of the sun without endangering his eyes by viewing it directly. Much later, in notebooks written about 1490, Leonardo da Vinci clearly described a camera obscura. This primitive device, however, was limited by the feeble illumination provided by its small aperture. To make the image brighter the hole could be enlarged, but this destroyed the sharpness of the image. By the middle of the 16th century, the use of lenses was becoming widespread, and it was discovered that a suitable lens could produce a sharp and bright image.

The camera obscura then became a useful aid to sketching, since the image of the scene to be drawn could be cast directly on the paper. Collapsible camera obscuras, in the form of a tent or a huge box, came into use, allowing the artist to select his scene at will. The lens was usually placed at the top of the structure, with an adjustable mirror, like that in a periscope, to reflect the view down onto a table. The great mathematician Johannes Kepler was one of the first people to use a portable apparatus of this kind.

With the improvement of lenses, still brighter images could be produced, and in the 18th century the device was scaled down to a manageable box, with a lens at one end and translucent paper at the other. By shielding his head and paper from direct light, the artist was able to sketch a small, convenient image. Camera obscuras of this type, more correctly called 'camera lucidas', were very popular at this time, being used both for outdoor views and for indoor portraiture and still life.

It was the use of these camera obscuras that led directly to the invention of photography, when experimenters sought ways of fixing the image on the screen by chemical means. Two pioneers of photography, Louis Daguerre and W H Fox Talbot, both began as artists, sketching with the camera obscura.

The original need for the device vanished with the development of photography, but even in an age where colour television is commonplace the fascination of being able to see full-colour images on a screen has prevented it from

machine tools, electromechanical timing gear, and a wide range of machinery from juke boxes to computer equipment. In the internal combustion engine a set of cams on a rotating camshaft control the opening and closing of the inlet and exhaust valves. The shapes and positions of the cams ensure that the valves are opened and closed in the correct sequence and at the correct time intervals. Various systems have been adopted over the years, one of the most common being pushrod action in which the cam acts on a cam follower and from there on a vertical rod in a lubricated sleeve. Movement of this rod is then transmitted in turn through a pivoted rocker arm to the valve's stem. Other systems involve the elimination of the pushrod as in the 'overhead' camshaft arrangement, where the camshaft is set on top of the cylinder block above the valves. Internal combustion engines operate at high speeds, so the spring force maintaining the cam follower in contact with the cam has to be exceptionally powerful—the faster the engine, the more powerful the spring. If the spiral spring used is insufficiently powerful, the inertia forces in the valve system will prevent the cam follower from maintaining intimate contact with the cam and an inefficient condition known as 'valve bounce' will occur.

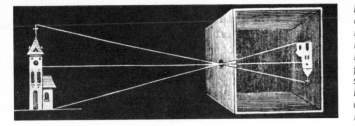

Fig. 3.

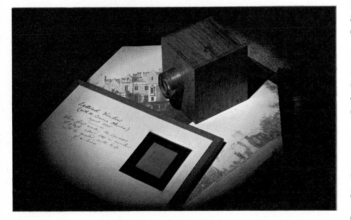

Left: the principle of the lensless
camera obscura, showing how
light rays travelling in straight
lines produce an upside down
image. A lens in place of the
pinhole gives a brighter,
sharper image.
Left centre: a typical camera
obscura design. This one has a
rotating roof carrying a fixed
mirror, below which is the lens.
This casts the image onto the
table. The latest camera
obscura at Blackpool, England,
has a rotating sphere inside
which the viewers sit. They are
carried round with the lens,
mirror and screen. The sphere is
40 feet (12 metres) above
ground level.
Left below: Talbot's original
camera, used in 1835 to make
the first photographic negative,
of a latticed window at his
home, Lacock Abbey.

dying out. Early in this century, many seaside resorts had a tent form of camera obscura. Today there are public camera obscuras in several places around the world. The one at Bristol, England, was built as long ago as 1829, and work is in progress on new ones. A typical camera obscura has a mirror in the roof, reflecting light downwards to a lens from 4 to 6 inches (10 to 15 cm) in diameter, with a focal ratio of about *f*15 to *f*25. This projects an image of the surroundings onto a white table; the view can be altered by moving the mirror with pulleys or electric motors.

Modern cameras The earliest photographic cameras were basically light-tight boxes with a lens at the front to form an inverted image on a flat light-sensitive plate at the back. There was no shutter, but instead a lens cap was removed for a period to make the exposure, which had to be several seconds or even minutes in duration because of the poor sensitivity of the plates.

Plate cameras are still made, though today sheet film (also called cut film) is used rather than plates. The plate is held in a flat container which can be made light-tight by a sliding sheath at the front. The whole plate holder, or *dark slide*, can be removed from the camera for loading of the plate in the dark. It can be replaced by a ground glass screen, on which the image can be focused. This is done by moving the lens assembly in and out, with leather bellows to exclude light.

As the plates became more sensitive, refinements became necessary to limit the exposures. A variable lens aperture was provided—that is, the effective diameter of the lens could be changed. This was done either by providing a range of masks with different sized holes for insertion near the lens, or by an iris diaphragm—a system of metal leaves arranged to close together so that their edges form a roughly circular aperture of any chosen size.

A shutter soon became necessary, to give brief exposures lasting for seconds or less. This usually had a few metal leaves, like the iris diaphragm, arranged to open or close completely, and operated by air pressure from a rubber bulb —hence the B symbol on modern camera shutters, for giving exposures longer than a second.

Plate cameras are mainly used for high quality work, with plates up to 10 by 8 inches (25 × 20 cm). Plates are fragile, heavy and awkward to use, each one requiring separate loading. Although they are still used for scientific purposes, where the extremely flat and non-shrinking properties of glass are invaluable, the popularity of the camera is due to the introduction of flexible film to carry the light-sensitive emulsion.

Many attempts were made to find a suitable transparent base, the first solution being stripping film where the emulsion was peeled off its backing after processing. By 1889, however, roll film had been introduced.

The first Kodak cameras, produced in 1888, gave a great stimulus to amateur photography since they were the first to use roll film. In 1895, Eastman produced the first Brownie, using daylight loading roll film. This type of camera marked the beginning of modern snapshot photography.

Today, the most widely used cameras are of the 'instant load' variety, in which the film is carried in a light-tight cassette. Before the introduction of this system, 'box' cameras were popular, being simple light-tight boxes with spools for the film, a fixed aperture and a single shutter speed, provided by a spring operated moving leaf. The capabilities of these cameras are very limited, and for good quality photographs a wide range of cameras has evolved, with refinements at every stage.

Taking pictures The technical skill of the photographer lies in selecting the correct shutter speed, focusing and focal ratio for the picture concerned. The focal ratio is the lens's

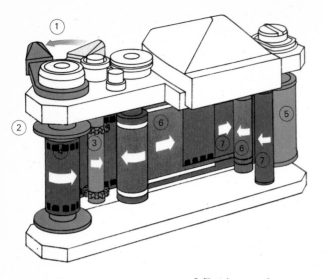

A typical film transport and shutter system with the film in dark green, the transport in light green and the shutter blinds in dark blue and brown:
1. film wind and shutter cocking lever, shown in wound on position

2. film take up spool
3. film transport sprocket
4. 35 mm film
5. film cassette
6. 1st shutter blind
7. 2nd shutter blind

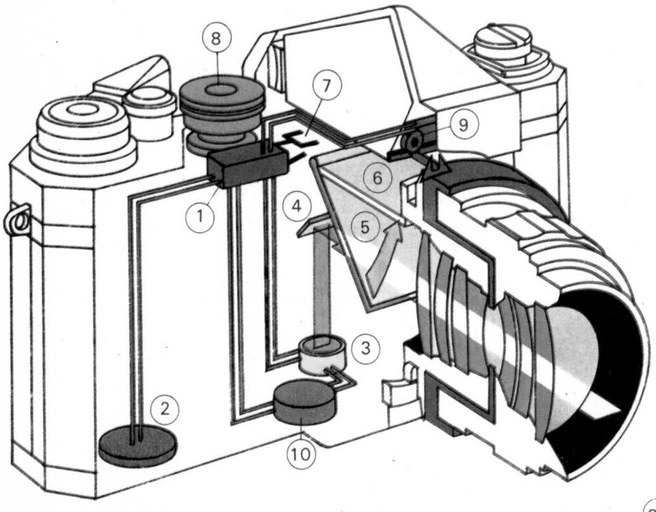

A 'match needle' through the lens light metering system on an SLR:
Heart of the system is the meter (1) which can be biased by the setting of the film speed scale (2) and moves the needle according to the light falling on the photocell (3). The light reaches it via the small mirror (4) and through the semi-opaque centre of the main reflex

mirror (5), shown retracted at (6). The needle can be seen in the viewfinder, and has to be centred in a claw (7), controlled by the shutter speed dial (8) and the aperture of the lens using the connected variable resistor (9). The whole system is powered by the (10) mercury battery in the base plate.

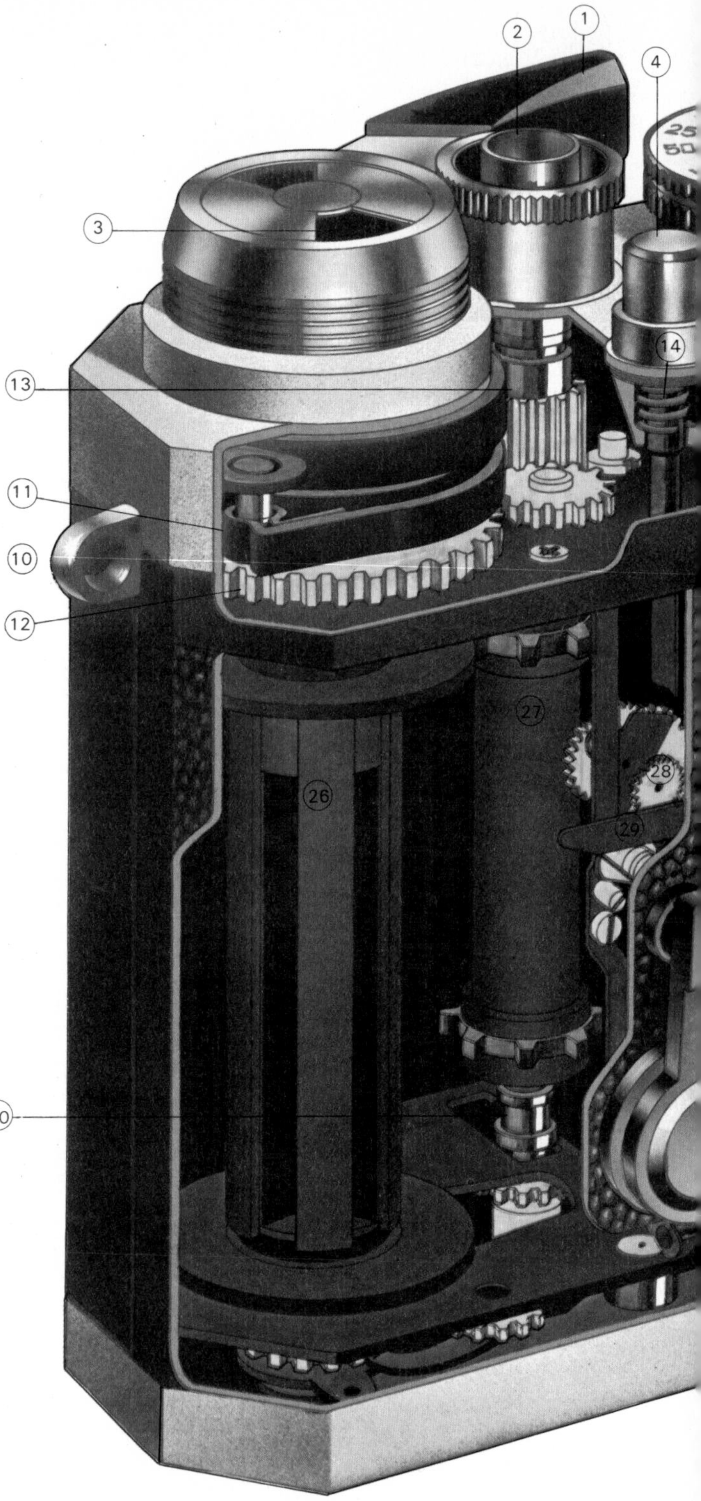

How the reflex viewing system works:
1. components of the camera lens
2. reflex mirror
3. focusing screen
4. fresnel or field lens
5. pentaprism
6. eyepiece optics

1. rewind lever
2. shutter release button
3. frame counter
4. depth of field preview button
5. shutter speed dial
6. eyepiece optics
7. viewfinder
8. pentaprism
9. rewind lever
9A. accessory shoe
10. strap lug
11. main spring
12. film transport gear train
13. shutter release axle

14. return spring
15. shutter speed ratchet
16. meter
17. meter inputs
18. meter inputs
19. meter claw and needle
20. field lens
21. focusing screen
22. diaphragm variable resistance
23. coupling prong
24. max. aperture setting (here f.2)
25. synchronization coupling
26. film take-up spool

27. filmtransport sprocket
28. self timer movement
29. self timer escape lever
30. shutter release lever
31. self timer lever
32. self timer release button
33. diaphragm stop-down linkage
34. diaphragm stop-down linkage
35. diaphragm stop-down non return unit
36. diaphragm stop-down pin
37. mallory cell
38. cadmium disulphide (CDS) cell
39. lens mounting ring

40. CDS incident light 45 mirror
41. main reflex mirror
42. main reflex mirror axle
43. mirror box
44. ribbed mirror box interior
45. diaphragm coupling
46. diaphragm ring
47. lens mounting
48. focusing ring
49. moveable front element mounting
50. rear lens group (4 elements, 3 groups)
51. front lens group (3 elements, 3 groups)
52. iris diaphragm

The iris diaphragm has a number of thin metal blades. At one end of each is a pivot, while the other end is attached to a pin on a movable ring. When the ring is rotated by means of the projecting lever, the blades move across the aperture. The number of blades varies from five to over twenty.

lens. The most advanced types have speeds from 1 to 1/500 second but simple cameras may have only one or two settings such as 1/90 second for sunny weather and 1/30 second for dull lighting or when using a flash.

Focal plane shutters are placed as near to the film as possible. The exposure is made by a roller blind moving across the film closely followed by a second blind. Light passes through the slit between the blinds and the amount can be varied by altering the width of the slit and the speed of the blind movement. Most focal plane shutters have

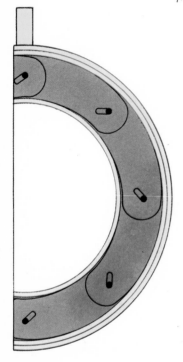
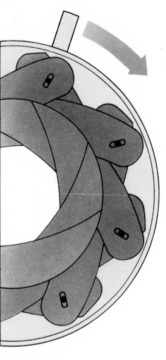

focal length divided by its aperture. A lens at *f*16 or *f*22 is called slow: it has a small aperture compared with the distance to the film, and little light passes through. A fast lens, say *f*2, has a larger aperture and lets through more light. In practice, the lens is 'stopped down' using an iris diaphragm, so that one lens can give, say, *f*2 at full aperture, down to *f*22 at its smallest, the intervals being known as stops. The *f*-ratios of the stops are calculated so that the light gathering area of the lens alters by a factor of two between each one: the series runs 1.4, 2, 2.8, 4, 5.6, 8, 11, 16, 22, 32.

The aperture of the camera also controls the depth of field of the image—the range over which it is in focus. At *f*22, both near and far objects will be in focus, while at *f*2, only a precise distance will be in focus—often a desired effect.

The shutter speed has the property of being able to stop motion. The faster speeds, such as 1/250 sec, will give 'frozen' pictures and will eliminate any camera shake, but do not let as much light through onto the film as longer exposures. The longest hand-held exposures normally possible without camera shake are about 1/25 sec; in dim light, longer exposures are needed, for which the camera must be tripod mounted.

The shutter speeds also vary by factors of 2, the usual values being 1, 1/2, 1/4, 1/8, 1/15, 1/30, 1/60, 1/125, 1/250, 1/500 and 1/1000 second. This makes it easy to compensate for the similar steps of lens aperture: if an exposure of 1/125 second at *f*8 is correct for a particular picture, then so will 1/60 second at *f*11.

Shutters Modern shutters are divided into interlens and focal plane types. Interlens shutters, such as the famous Compur and Prontor, have moving metal blades which open and shut between the lens elements. It is also possible, but unusual, to have the blades behind, or even in front, of the

speeds from 1 to 1/1000 second, although a few models attain 1/2000 second.

Lenses and focusing In the simplest cameras, the focus is fixed on a distance such as 10 ft (3 m) or so, with an aperture chosen so that the focus is acceptably sharp from infinity (in photography this applies to a very distant subject) to as close as five feet (1.5 m). The simplest lenses are moulded of plastic (methyl methacrylate) and have focal ratios as slow as *f*14. Better cameras, however, have achromatic lenses, with two components or elements, working at around *f*9; 3 or 4 element lenses working at around *f*2.8; or 6 element lenses of *f*1.8 and *f*2.

Focusing is carried out by moving the whole lens nearer or further away from the film. This is normally done by twisting the lens, which is mounted in a case with a spiral screw thread so that it moves in or out. A scale around the edge of the lens shows the distance of sharpest focus. A rangefinder is sometimes provided, a common system giving two images in the viewfinder which coincide when the object is in focus. The rangefinder may indicate the object's distance on a scale, or it may be linked directly to the focusing mechanism.

Film sizes Photographic film is made in sheets, sometimes called cut film, but photographers mostly use roll film or 'miniature' 35 mm film. Roll film comprises a strip of film fastened to a longer strip of paper which is black on the film side and bears numbers on the other. The film and paper are wound on a spool with close fitting ends so that the spool can be put into the camera in the light, although preferably not in direct sunlight. The number can be seen through a window in the camera back so the user knows how far to wind on the film and how many pictures have been taken. The most popular film size is 120, which gives 12 negatives, measuring 6 × 6 cm ($2\frac{1}{4}$ × $2\frac{1}{4}$ inch). Other sizes are 220 (a double length 120) and 127 which gives 8 frames measuring 4 × 4 cm (1.6 × 1.6 inch).

Miniature film was originally made for the movie industry. It has no backing paper and comes in 20 or 36 exposure lengths in small cans called cassettes. There are perforations along both sides of the film into which the teeth of sprocket wheels fit to wind on the film. No backing paper is used. The usual picture size is 24 × 36 mm (about 1 × 1½ inch). Another kind of miniature film comes in cartridges which can be put straight into the camera without any loading problems. The 126 size gives frames 28 × 28mm, with one sprocket hole per frame; the newer and increasingly popular 110 size is 12 × 17mm.

35 mm cameras In a cassette-loaded 35 mm camera, the sprocket holes in the film make it fairly easy to wind the film by the correct amount between frames. A sprocketed wheel turns with the film, but is stopped by a ratchet when it has moved far enough. It will not turn again until the shutter has been pressed, and the shutter cannot then be pressed again until the film has been wound on. This makes it impossible to expose the same piece of film twice. The ratchet can be disengaged to allow the film to be rewound when it is fully exposed.

A pressure plate presses the film into contact with a rectangular aperture which defines the image area, so that the film is perfectly flat.

Roll film cameras The old folding roll film camera with bellows is seldom seen today. Most roll film cameras are of the reflex type—that is, they have a reflection viewing system. The twin-lens reflex camera, such as the Rolleiflex, has two lenses, one mounted above the other on a panel. The bottom lens has a shutter and takes the picture while the top lens forms an image on the viewing or focusing screen via a mirror placed so the screen can be horizontal. The two lenses move together on the panel so when the image is sharp on the screen it is also sharp on the film: focusing is carried out by moving the whole lens and shutter assembly in and out. Nearly all twin-lens reflexes take 120 film.

The twin-lens reflex cameras, although excellent for many purposes, are not suited for close-up work since each lens sees a slightly different view; neither can the lenses be changed unless the shutter mechanism is included in each lens, as in the Mamiyaflex. With modern lenses and films, the roll film reflexes are capable of producing work of a high professional standard formerly obtainable only with cameras taking larger sizes such as 5 × 4 inch sheet film.

Single lens reflex A great advantage to the photographer is the single lens reflex, or SLR camera. The viewfinder sees exactly the same view as the lens; it is normal for a focal plane shutter to be used, so that a variety of lenses can be interchanged.

A movable mirror is located behind the lens and in front of the shutter. This reflects the light upwards to a ground glass screen, arranged so that the light travels exactly the same distance to the screen as to the film when the mirror is out of the way. The image on the screen can, therefore, be focused accurately. A *pentaprism* above the screen orients the image the same way that the eye sees it in the viewfinder.

Focusing aids are often included in the screen, such as *microprisms*—small segments which break an unsharp image up into dots.

A device is often incorporated into the shutter lever of an SLR which closes the lens's iris diaphragm down to the pre-set aperture. This makes it possible to focus and compose the picture at maximum aperture, while when the shutter is pressed the lens is stopped down.

When the shutter button is pressed, the reflex mirror flips rapidly upwards and the focal plane shutter operates.

Lenses are rapidly interchangeable, having standard screw thread or bayonet fittings. A range of lenses, from wide

Below: a Mamiyaflex twin lens reflex camera, which takes 120 roll film. Also shown are a cable release, used to operate the shutter without touching the camera, and an interchangeable telephoto lens.

Bottom of page: a Sinar studio plate camera of the monorail type is shown with accessories. Both lens and plate are fully adjustable, to allow for photography of objects at awkward angles and perspectives.

Below, top: a Pentax 35mm single-lens reflex camera with built-in light meter.
Below: two Kodak Instamatics: the 16mm pocket model and the 35mm design.

misled, such as when photographing into the sun.

Cameras can have the light sensitive cells along the front of the top housing or around the lens. Through-the-lens or TTL metering works by having the cells inside the camera so that light must go through the lens before reaching these cells, which are usually placed on each side of the top of the pentaprism. Another meter method is to allow some light to pass through the mirror, the cells being behind the mirror.

Movie camera Motion pictures—movies—are made possible by an optical illusion known as 'persistence of vision', an ability of the eye and brain to fuse a series of still pictures viewed in rapid sequence into a continuous image. Objects that take up progressively different positions in successive pictures will appear to move smoothly provided the images are viewed at rates greater than about twelve per second.

It is the job of a movie camera to record such a series of images onto a flexible, transparent, light-sensitive film analyzing the scene in front of the camera into individual *frames* (the unit picture in a cinematographic film) at a constant repetitive rate. After the film has been chemically processed to give positive images, it can be projected onto a screen and viewed as a reconstruction of the original scene.

Parts of the camera Movie cameras vary in shape and complexity depending on their intended use and the facilities they offer. Certain elements, however, are common to them all: a light-tight container to exclude light from the film except at the point of exposure, which is behind a photographic lens forming an image of the scene being recorded; a transport mechanism that moves the film past the exposure point; a viewfinder to indicate the area being photographed, and a counter to show the amount of film used (or remaining to be exposed).

The light-tight container may be part of the camera body on which various parts of the mechanism are mounted, the film being wound on closely-fitting spools that allow loading in subdued daylight. Before filming can begin, the outer, fogged part of the film has to be run through the camera. Nearly all current amateur cameras have the film coiled in a light-tight cartridge for additional protection and faster loading. Professional motion-picture cameras often house the film in external magazines that are simple to fix on to the camera body, but must be pre-loaded in the dark.

The photographic lenses on professional movie cameras are interchangeable, so that different focal lengths can be used to vary the scale of the image and, as a result, alter composition and apparent prespective from a given viewpoint. Often several lenses are mounted on a rotating turret to allow quick changes to be made. Each lens can be focused separately and incorporates an iris diaphragm to control the amount of light entering the lens. In this way the exposure of the film can be suited to the brightness of the subject and the sensitivity of the film.

In most home movie cameras the lens is built into the body. For this reason the lens is not interchangeable, though in all but the simplest models, a zoom lens is fitted that allows the focal length to be varied, even during filming—the design of the lens makes sure that the focusing stays the same. If desired, zoom lenses can also be fitted to professional cameras. The vast majority of amateur movie cameras incorporate an automatic exposure mechanism that regulates the iris diaphragm to suit the brightness of the scene.

Intermittent movement The analysis of the scene into individual frames is achieved by moving the film intermittently past a rectangular aperture in a 'gate' that locates the film at a fixed distance behind the camera lens. To achieve regular spacing of the images, the film carries a series of perforations outside the frame area. The intermittent motion is produced by a claw mechanism that enters one or two perforations and pulls the film down one frame at a time at regular intervals. In many professional cameras one or two register pins engage with the perforations to hold the film absolutely steady dur-

angle to telephoto, are available in all fittings.

Popular cameras An entire range of cameras is made for instant load 126 film, though the greatest sales are at the cheaper end, replacing the old box camera. It is even possible to obtain an expensive SLR 126 camera, though because the film has no proper pressure plate the quality of photographs is limited. Similarly, 110 cameras are made in a range starting with the most simple and ending with models having automatic exposure control. There are also other subminiatures taking 16 mm film, usually in cartridges. Instant load cameras should not be confused with the polaroid camera, in which finished prints are produced in a few seconds.

Exposure control Many cameras have built-in exposure meters to indicate the correct exposure for the lighting conditions. The simplest types use photoelectric cells which receive the light coming from the direction in which the lens is pointing, indicating the exposure value on a scale. The photographer then adjusts the camera settings accordingly. Some cameras, called *semi-automatic*, have a linkage between the light meter and the scales on the camera lens so that the photographer simply lines up two needles to get the correct exposure. Fully automatic cameras will vary either the camera aperture or the shutter speed, so the other of these must first be set by the photographer. Most photographers prefer to be able to set the values by hand as well if necessary where an automatic meter may be

Left: in the early days of movie-making, cameras were hand-cranked. The normal speed was 16 frames per second, and they were projected at that speed. When shown today at the standard speed of 24 frames per second, they look 'jumpy'. Below: the Kodak Instamatic movie camera and cartridge. Early home-movie cameras had spring motors which needed winding; today they run on batteries.

ing exposure; these are then withdrawn (or the film is lifted off) during pull-down. In less elaborate cameras, including all amateur types, friction from a sprung pressure-plate stops and holds the film during exposure.

In each case it is necessary to allow light to reach the film only while it is stationary and to cut it off during pull-down or else vertical streaks will spread from the lighter parts of the scene. This is normally achieved by a partially cut away disc shutter between the lens and gate, geared to the intermittent mechanism so that the opaque sector cuts off all light from the film when it is moving from frame to frame. A mechanism ensures that the shutter is always closed at the end of a sequence.

In some cameras it is possible to vary the open sector of the shutter, running from a typical maximum of 170° to 180° (sometimes up to 220°) of the full 360° circle down to zero. If this is varied while a scene is being taken, the exposure time for each frame will be reduced from the usual 1/30 second or so to a much shorter time. This has the effect of 'fading' the scene, and by back-winding the film a few frames and carrying out the procedure in reverse, a 'dissolve' of one scene into another is obtained. Alternatively, the shutter speed can be kept constant at, say 1/250 second, and the diaphragm opened to give the correct exposure. This allows rapid movement to be photographed, or the camera to be 'panned'—swept around the scene—without undue blurring. This facility is now mainly found only on the more expensive amateur cameras. Professionals also produced their 'effects' in this

way in the past, but nowadays these are achieved in the processing laboratory.

Film gauges Various widths of film are used in movie cameras. In 1889, Edison pioneered the use of 35 mm wide film. It quickly became the standard for professional use, and remains so today. Originally the picture size was 24 × 18 mm, almost filling the area between the perforations, and the normal shooting speed about 16 frames per second. When sound was introduced, the picture area was reduced to 22.05 × 16.03 mm to make room for a 2.54 mm (1/10 inch) wide soundtrack between the frames and one row of perforations, the standard used today. The shooting rate was increased to 24 frames/second to give better sound quality. This difference in shooting rates accounts for the 'jumpy' look of early films when shown on modern projectors. Typical film capacities of modern studio cameras are 1000 ft (35 m), running for 11 minutes, with 400 ft (122m) loads for portable cameras.

Larger film widths such as 65 mm or 70 mm are sometimes used for shooting and projecting 'spectaculars'—they have the advantages of giving finer grain and brighter pictures when projected. Some popular movies originally made in 35 mm have been transferred to 70 mm simply to give a brighter image in large cinemas.

'Cinerama' and other wide-screen systems now use 65 mm film in the camera and 70 mm film for projection (which includes six sound tracks). Early Cinerama used three 35 mm cameras coupled together, but was never really successful in

Above: common film gauges.
Top, left to right: standard 8mm;
Super 8, with smaller sprocket
holes and larger frame area;
16mm film with optical soundtrack; 35mm CinemaScope or
Panavision film, showing the
image before it is 'unsqueezed'
by a lens or, in older systems,
angled mirrors. Lower:
Cinerama 70mm film, with
multi-track magnetic stripe
soundtracks.

1. matte box
2. front effects stage
3. matte support bar
4. locknut
5. front adjustment
6. rear adjustment
7. front filter rack
8. rear filter rack (revolving)
9. prime lens
10. alternative lens (2nd)
11. mount for lens 3
12. centre turret pivot
13. turret shift tabs
14. lens barrel locks
15. focusing levers
16. 3 lens divergent turret
17. neck strap lug
18. eyepiece cup
19. viewfinder focus
20. 180 mirror shutter at 45 to the film plane
21. rear of prime lens
22. intermittent film claw
23. registration pin
24. film gate
25. film plane
26. ground glass screen
27. field lens
28. prism
29. viewfinder optics (10x)
30. film stock
31. geared sprockets
32. pressure rollers
33. feed film spool
34. take-up film spool

Above, top: a TV news team filming on location. The sound mixer is operated by the man with headphones; the sound is recorded on a magnetic stripe on the film. The colour reversal film is processed rapidly to give a positive rather than negative image.
Above: an Arriflex 35mm movie camera, used for making full-length feature movies. The bellows excludes unwanted reflections from the lens.

eliminating the joins. CinemaScope, with a slightly wider screen than standard, uses 35 mm film and an 'anamorphic' lens, which compresses the image horizontally.

16 mm and 8 mm film gauges were originally designed for amateur use, though 16 mm has long been used in the documentary and scientific fields. Most film for television news work is 16 mm, though attempts are being made to use 8 mm for this. Typical 16 mm loads are 100 ft (30.5 m) spools for hand cameras and external 400 ft (122 m) magazines on larger professional cameras. Where a soundtrack is to be included, the film is perforated on one side only.

8 mm film was originally 16 mm wide, but with double the number of perforations. The film was run through the camera twice, exposing half the width on each pass. After processing, the film was slit into two 8 mm strands which were joined end to end to give a film of twice the original 25 ft (7.6 m) length, running for four minutes.

In 1965, Super 8 was introduced using much narrower perforations, which allowed a 43% greater picture area on the same film width. The film is supplied in light-tight cartridges which are easily slipped into the camera. Single 8 film has the same dimensions but uses a different cartridge. Both hold 50 feet (15.2 m) of film, giving a running time of 3 mins 20 seconds at the normal 18 frames per second running speed.

Soundtracks These are either of the optical or magnetic variety. Some cameras have been designed which record the soundtrack in the camera, but the usual method is to record sound on a separate tape recorder, along with synchronizing signals produced in the camera. These are necessary since even a 0.2% variation in speed of either camera or recorder would mean half second discrepancy after four minutes.

Camera drive Early movie cameras were hand cranked, speed constancy depending on the skill of the cameraman. When sound films were introduced in the late twenties, speed variations became more noticeable, and constant shooting and projection speeds became necessary. Studio cameras were fitted with mains-powered electric motors, and enclosed in sound-absorbing 'blimps'.

Amateur and portable cameras were once mostly spring driven, but developments in compact, powerful electric motors made them both smaller and cheaper than the spring driven type. Today, electric motors are the rule.

The camera drive also turns the take-up spool by means of a slipping clutch, so that the film is always wound on under tension. A back-wind facility is sometimes provided, with a mechanism to withdraw the claws while this is being done.

Viewfinders Various types of viewfinding system have been used to indicate the area of the scene being filmed. These include simple open frames or lens systems which may be coupled with a zoom lens or rotating turret to match the varying focal lengths of lenses.

A more recent type use reflex finders, which have either a beam-splitting prism in the lens or a mirror on the rotating shutter which reflects the scene to a ground glass screen or other focusing aid. This method eliminates the parallax error caused by the different viewing angles of two separate systems, and shows exactly the field of view of whichever lens is being used.

Some film-makers are now using a small television camera to observe what the viewfinder sees, so that the director and production crew can all see what is being filmed. By using videotape to record this, instant 'rushes' of the film can be seen.

Footage counters Footage counters, which show the amount of film used, can be gear driven from the camera mechanism, or operated from a lever gauging the diameter of the feed or take up rolls. Both types have their advantages, and some professional cameras fit both. With Super 8 and Single 8 the film rolls are inaccessible in the cartridge, and the counter is usually coupled to the drive for the take up spool.

A few cameras incorporate counters for the number of single frames shot, mainly for special effects work.

Projectors The projectors that we use today, either for viewing holiday slides or for watching full length feature films, are all developments of the Victorian *magic lantern*, which is the basic projection system. A great amusement in the 19th century was to project *lantern slides*, often simply scenes or figures painted on glass, on to the wall or a screen. Although the light source was a flickering candle, the bright colourful images created apparently out of nothing were a great wonder. Eventually, positive transparent photographs were projected and the lantern slide lecture of some explorer's experiences became a favourite entertainment.

With the invention of the Lumière brothers' cinematograph, movie projectors became a familiar sight. The production of colour reversal films, which made it possible for any amateur with a miniature camera to take high quality colour slides, produced a demand for slide projectors which would operate in the home with the minimum of trouble, nowadays equipped with a magazine for projecting a number of slides in sequence.

The traditional size of lantern slides is $3\frac{1}{4}$ by $3\frac{1}{4}$ inches (8 × 8 cm) in the UK and $3\frac{1}{4}$ by $4\frac{1}{4}$ inches (8 × 11 cm) in the USA. This size became unnecessarily bulky when high definition miniature films were produced, and the standard slide size is now 2 by 2 inches (5 × 5 cm), giving a frame area of 24 × 36 mm. Other sizes such as $2\frac{1}{4}$ by $2\frac{1}{4}$ inches (6 × 6 cm) are sometimes found.

Basic system All projectors, from the first magic lanterns to the latest motion picture models, have the same components: a light source, a condenser, the object plane, and a projection lens. The item to be projected is placed in the object plane so that the light source shines through it to the lens, which projects an enlarged image of it on to the screen.

If there were no condenser, the image would be weakly illuminated and of uneven brightness across its surface. This is because the projection lens would form an out-of-focus image of the light source, which is rarely a point of light or an evenly illuminated surface. Even it it were, the projector is usually sufficiently close to the screen that the distance from the lens to the screen centre is somewhat less than the distance to the screen's corners. This would produce a noticeable

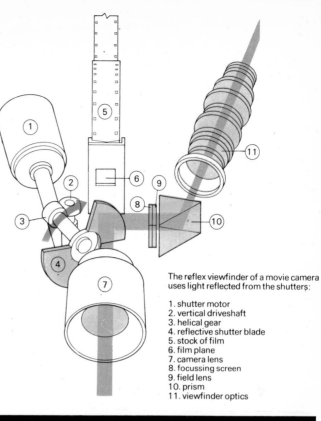

The reflex viewfinder of a movie camera uses light reflected from the shutters:

1. shutter motor
2. vertical driveshaft
3. helical gear
4. reflective shutter blade
5. stock of film
6. film plane
7. camera lens
8. focussing screen
9. field lens
10. prism
11. viewfinder optics

In 1884, this improved magic lantern was made by Rudge, with photographs supplied by Friese-Greene, a well-known photographer. Glass blades dissolve one picture into the next.

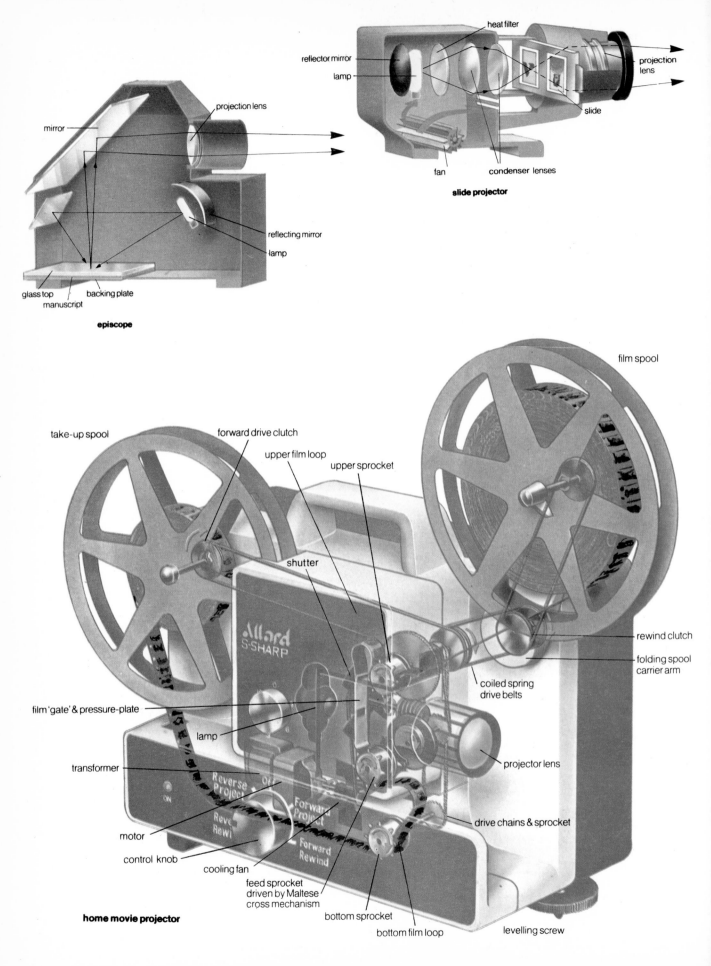

mirror

projection lens

reflecting mirror

lamp

glass top
manuscript
backing plate

episcope

reflector mirror

lamp

heat filter

projection
lens

slide

fan

condenser lenses

slide projector

film spool

take-up spool

forward drive clutch

upper film loop

upper sprocket

shutter

rewind clutch

folding spool
carrier arm

Allard
S-SHARP

coiled spring
drive belts

film 'gate' & pressure-plate

lamp

projector lens

transformer

Reverse
Project

Forward
Project

motor

control knob

Reve
Rewi

ON

Forward
Rewind

drive chains & sprocket

cooling fan

feed sprocket
driven by Maltese
cross mechanism

bottom sprocket

bottom film loop

levelling screw

home movie projector

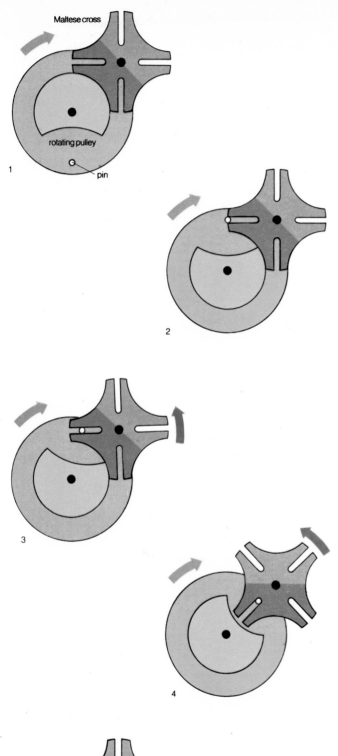

Maltese cross

rotating pulley

1 pin

2

3

4

5 Maltese cross mechanism

Opposite page, top: two projection systems. The episcope, on the left, projects an image of a flat opaque original, such as a manuscript. This is illuminated by means of a lamp and reflector, and the original is then reflected by a mirror before being projected, to avoid its image being reversed on the screen. On the right is a slide projector or diascope, with heat absorbing glass and condenser. In this case the slide has to be upside down to produce an upright image on the screen; no mirror is needed to reverse the image as the slide itself is turned to reverse the image.

Below these is a home movie projector embodying features to be found in a variety of different designs. The film is carried through at a constant rate by the upper and lower sprockets, driven by the motor through gears and a chain device. The intermittent mechanism here is a Maltese cross device, though a claw is more common on small projectors. Since the film runs jerkily through the gate and round the film sprocket, loops are necessary between these and the sprocket drives to prevent the film being strained or even torn by the jerks.
This page, left: the action of a Maltese cross mechanism.

dropping off of light towards the edges. In addition, only the light which happened to fall on the object's area would be projected on to the screen: most of it would be wasted.

Condensers The condenser system usually has two parts: a spherical mirror behind the light source, and a system of lenses in between the light source and the object plane. The spherical mirror catches light shining away from the object and reflects it back the way it came, so effectively doubling the brightness of the source. The lens system catches a large proportion of the light shining in the general direction of the object and condenses it into the object plane so wasting as little as possible. Indeed, up to one third of the total light output can be concentrated on to the object.

The amount of light and heat radiation involved is often large, and blowers have to be provided in many cases to keep the light source and nearest condenser lenses to a reasonable temperature. The object itself often has to be protected by means of heat-absorbing glass—that is, glass which absorbs infra-red radiation which would heat up the object unnecessarily without producing any light on the screen. But the amount of energy in visible light is even more than in infra-red, so it is inevitable that the object will be heated. In the case of motion picture films, which are only projected briefly, it is possible to project brighter images than if they were single slides kept in the projector for minutes at a time.

By designing the condenser system correctly, the outer edges of the projected image can be made brighter, to compensate for their greater distance from the lens.

Above, top: a 35mm cinema projector with a carbon arc lamp and automatic system for keeping rods aligned.
Above: a 16mm projector lamp housing, including a tungsten halogen lamp and concave reflector. A moveable heat filter protects the film when the projector is running slowly.

Light sources Although the first light sources were candles, these were obviously not very suitable, and early motion picture projectors used either electric lamps, or carbon arc lamps. The latter gave brighter images, though were less convenient. Even so, until very recently the carbon arc was the only means of producing light of sufficient power for a large movie theatre, and the carbon arc is still a standard light source. Within the last few years, however, xenon discharge tubes, which give a steady bright light for up to 1500 hours without attention, have come into use. These are also capable of being pulsed at 48 flashes a second, so doing away with the mechanical shutter needed on most movie projectors.

The carbon arc is widely used, and is essential in large theatres, and drive-in establishments. As the carbon burns away, the position of the *crater* where burning occurs changes, so this has to be allowed for by an automatic feed mechanism. Even so, the very largest arc electrodes, which may be over 0.5 inch (13 mm) in diameter, taking 250 amps of current, do not last for more than an hour.

In small projectors, both movie and still, coiled tungsten lamps are often used at powers of up to 1000 watts. These usually require blowers to cool them, but quartz halogen lamps are now becoming more frequently used. These have the advantage of having a colour more comparable to daylight than tungsten lamps, which produce large amounts of infra-red radiation. Consequently, their heat production is less, so that blowers are not necessary, and their colour rendering of the image is more realistic.

Lenses In the case of portable projectors, it is possible to use interchangeable lenses of different focal lengths, in order to allow for the varying sizes of rooms and different 'throws' —projection distances—necessary. A short focal length lens, for example, will be suitable for use in a small room to give a large picture, while a longer focal length lens will produce the same sized picture at a greater distance—more suitable for a lecture room. Since it is easier to design a lens with long focal length, the curvatures required being less, these are generally more satisfactory from the point of view of giving sharp coverage all over the screen with freedom from false colour and distortions.

The energy transmitted by the projection lens is very high, so the components must be heat resistant. In particular, they cannot be cemented together, as is usual with camera lenses, and the coatings must not discolour with heat.

Slide transport With the popularity of the slide projector, methods of projecting a number of slides from a magazine have come into fashion. The single slide projectors, with a simple carrier which moved back and forth, are now rare and devices with increasing complexity are popular. The slides are pre-loaded into a plastic magazine taking 36 or 50 at a time, which is slid into a carrier alongside the object plane. Each slide in turn is pushed into the projector by means of a motor driven device or solenoid which can be controlled remotely from a switch unit on the end of a cable. A servo motor is also provided to alter the focus of the lens, moving it back and forth.

One design which has become almost standard for lecture and display presentations is the Kodak *Carousel*. This has a circular magazine taking 80 slides, each of which drops into the object plane by gravity.

Motion pictures Although the projection system of movie projectors is essentially the same as in any other type, there is the added complication of having to move the film through the image plane, normally called the *gate*, and to make sure that at the same time each frame of the film is stationary when projected on the screen. This is normally done by using an *intermittent mechanism* of some kind.

The simplest intermittent mechanism, which is also used in simpler cameras, is the *claw*. The pins of the claw engage in the sprocket holes of the film, and a cam mechanism serves to move the claw down, withdraw it from the film, and move it up again by a suitable number of sprocket holes ready for the next movement. In this way, the film is moved jerkily through the gate.

Although the claw device is used on the majority of small projectors, such as the 8 mm gauge variety which is standard for home movies, and for a number of 16 mm types, a more robust mechanism is generally found on the 35 mm and 70 mm gauge projectors for cinema use. This is a *maltese cross* mechanism, so called because it has a four-pointed star wheel with slits in each point, these slits corresponding to the gaps between the arms of a Maltese cross. A wheel with a pin projecting from one side rotates next to the cross in such a way that the pin engages in one of the slits for each rotation of the wheel, moving the cross through 90°. The maltese cross is connected to a sprocket wheel which moves the film through the appropriate number of frames. The 70 mm gauge projectors which are becoming widespread in larger establishments generally have both 70 mm and 35 mm sprockets on the same shaft, so that the projector can show either type of film with the minimum of changeover problems.

In order to shut off the projected image while the film is being moved through the gate, a shutter is provided. Most films are projected at a rate of 24 frames per second, but this is found to give a slight flickering effect when viewed, particularly for those people close to the screen. For this reason, the shutter operates not 24, but 48 times per second. Either a disc which has two sectors cut out of it, or a drum with two gaps cut in it, is used. In the case of home movies which are often projected at 16 or 18 frames per second, a three bladed shutter is needed to produce the 48 frame

viewing. Both intermittent and shutter mechanisms are run from the same speed-controlled motor which drives continuously moving sprocket wheels to transport the film from one spool to another and to provide tension, by means of a friction clutch, to the take up spool.

Soundtrack The film soundtrack consists of a magnetic or optical stripe on the film, outside the picture area, which has to be detected by either a magnetic head, as used in a tape recorder, or a light detecting system which will pick up the light and dark variations of the optical track. It is essential that at this point the film has all its jerky movement smoothed out, otherwise a ripple will be heard on any tone or voice. To manage this, the sound head is located well away from the gate, the soundtrack being recorded a suitable distance away from the picture to which it refers. The film passes through rollers and then round a heavy drum, the sound drum, which by its inertia will smooth out any variations which are left.

Spool capacity While 8 mm home movie projectors generally take a 400 ft (130 m) spool, giving between 20 and 30 minutes' viewing, the much larger frame size of cinema projectors, and the longer films shown, require much larger spools. The early films were particularly short reels, mainly because the film base was a fire risk, and so a complete feature might be a two or three reeler, each reel being 2000 feet (600 m) of 35 mm stock. With the advent of safety film base, much larger reels became possible, limited only by physical size (which can be considerable in the case of a long feature film on 70 mm gauge). Although 6000 ft (1800 m) spools are common, allowing up to an hour's showing per reel, some projectors now take 16,000 ft (5000 m) spools, lasting almost three hours. In the case of 70 mm film, the maximum size is limited to about 3500 ft (1100 m) by its bulk, since the projectionist has to lift it up to the top spool carrier.

In most cases, therefore, two projectors are needed at each cinema. One is spooled up and ready while the other is showing the first reel of the picture, which carries marks printed on the film in the top right hand corner just before the reel ends to tell the projectionist when to start the other projector and to change over. He does this by means of a changeover switch which closes a shutter in front of the first machine and opens one in front of the second which, if he has timed it right, should be showing the second reel just where the first left off.

Other projectors One type of projector generally found in educational establishments is the overhead projector. This has a large object area, illuminated evenly by means of a fresnel lens condenser, which is horizontal instead of vertical as in normal types. This is large enough to allow ordinary sized writing to be projected on to a screen behind the writer by means of a lens and mirror. Prepared transparent overlay sheets can also be projected. A photographic enlarger is also a type of projector, with the same components as a slide projector. The condenser system is usually of slightly different design, however.

Left: a cutaway view of a slide projector for school use. Each slide may be on the screen for a considerable time, so a powerful blower is provided. Both lamp and blower motor operate at 22½ volts so at the back of the projector is a transformer to step down the mains voltage. Either individual slides or a film strip can be shown, using different attachments in the object plane between lens and lamp.

CASH REGISTER

The first cash register, invented in 1879 by an American café owner, James Ritty, was inspired by a ship's propeller-revolution counter. It had a clocklike face, the hands showing dollars and cents, with tamper-proof adding wheels inside the case to accumulate totals, and was designed to stop assistants dipping into the till.

Ritty's 'thief catcher' was not a commercial success. Before he sold the patent for a mere $1000, however, he had built a Mark IV model which had the still-familiar pop-up figures at the top of the case and a paper audit roll which, pricked by pins in special sequence when the keyboard was operated, gave the proprietor the means of calculating his daily take.

Mechanical cash registers These broad principles of design, with additions, remain the basis of the mechanical cash register today. Even the most simple 'press down' mechanical register today is a complicated piece of equipment,

Top: Ritty's original Dial machine, with a clock face. The 'minute hand' records cents and the 'hour' hand dollars; one revolution of the cents hand turned the dollar hand ahead one space.
Above: Ritty's Paper Roll machine recorded sales by pricking a roll of paper.

incorporating some 2500 precision-engineered parts. These largely comprise levers, rods, toothed segments and pinion wheels, cams and linkages.

Basically, the depression and release of any key on the keyboard must perform the following functions: indicate the amount registered, print these figures on the machine's audit roll, progressively add up the totals, advance the special transaction counters when necessary, and open the cash drawer.

The keyboard This has 26 keys grouped in units of pence, tens of pence and pounds in Britain, or cents, tens of cents and dollars in the USA. An interlocking device ensures that only one key within a group can be depressed at a time, though single keys within groups may be depressed together to register, for example, pounds, tens and unit pence in one movement. A key depressed in error may be released if the movement is not completed, but after a certain point the transaction must go through.

The cash indicator The transaction amount figures shown at the top of the machine are carried on rotating drums, printed so that the figures can be seen from either side of the machine. The mechanism operates in two half-cycles; when a key is pressed down it restores the amount previously shown on the drums to zero and the release of the key actuates the mechanism to display the correct amount. The first half-circle disengages the drum pinions and the toothed segments moved by the key lever, and a return roll zeros the drums; in the second cycle the segments re-engage to rotate the drums to the correct figure position. The drums are locked in place by the indicator alignment assembly.

Audit counter and roll printer The counter wheels and the printer are actuated by key cams. Numbers are automatically carried over to the next higher counter wheel, as on a mechanical calculating machine. If necessary the counter can be reset to zero by using a special key, normally held by the management.

The roll printer prints amounts on a roll of paper as they are registered and also allows totals to be printed separately. Here again, toothed segments are in mesh with pinioned typewheels and are positioned according to the shape of a cam on the end of the key lever. Thus, depression of the 9p key, say, moves the typewheel nine positions. A platen roller moves against the typewheels, pressing the paper and a reversible ribbon against the type to make a printed impression. A hand-operated total printing assembly is fitted to allow totals to be taken from the counter wheels on a slip of paper.

There are also 'no sale', 'customer' and 'reset' counters which advance one figure every time the appropriate key is depressed, the cash drawer opened or the reset key used, thus providing useful management information.

The cash drawer The cash drawer, running on nylon rollers, is opened by spring tension, being released just before the end of a registering operation. The release is actuated by a mechanism on the main cam line inside the machine.

Control lock The machine is fitted with a control lock functioning in five positions: resetting, reading (of totals), locking, and open or closed drawer operations. A read and reset key, held only by an authorized person, ensures against tampering. The cabinet door is also interlocked so that it can only be opened by the read and reset key. Finally a seal is placed on the cabinet to deter any attempt to remove the casing.

The 'press down' mechanical register is now regarded as obsolescent, and is being replaced with electro-mechanical and electronic machines that provide greater security and much more management information, including data for computer processing in some cases. But millions of mechanical cash registers are still used throughout the world today.

Electronic cash registers The use of electronic components in the design of cash registers has produced a totally new

Left, up and down: two views of a Sweda Class 5000 electro-mechanical cash register. The mechanism closely resembles that of a mechanical calculating machine.
Below: two views of the latest type of register, the Sweda 7000 Series. This is hardly a cash register at all, but a computer terminal which feeds records of sales and other transactions into a central computer, and extracts credit ratings and similar information from it.

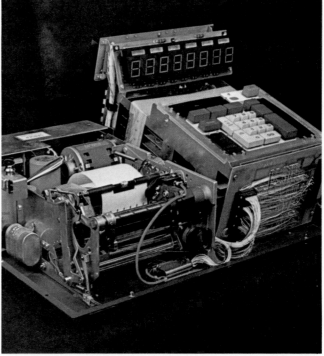

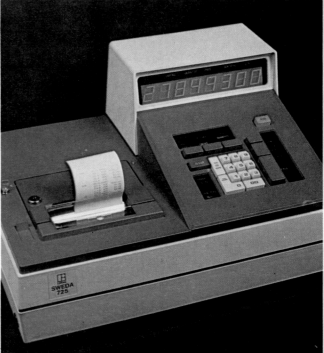

device providing all the traditional functions in a more efficient manner plus additional facilities akin to those supplied by a small electronic computer, including the ability to store and obey a pre-determined set of instructions in the form of computer programs. Such advanced features are possible through the use of integrated circuits and especially the development of metal oxide semiconductors. In addition, neon tubes to display transactions are used, a simplified 10-key keyboard plus a number of function keys are employed for typing in transaction data, and modern printing factilties to produce receipts, audit rolls and so on are available.

Although there are many types of electronic cash registers (ECR) currently available, the main differences lie in the way the cash registers are controlled, the methods used for collecting data for computer input, and the types of tags and tag readers employed where automatic reading of tags is required. As far as the control unit is concerned, the variations range from the completely self-contained cash register which is virtually a miniature computer controlling all its functions independently, known as the 'stand-alone' system, to the other extreme where the cash register has only the basic capabilities and acts on the orders of a computer which may be near or many miles away at the end of a telephone line.

The choice of system largely depends on the number of registers required in each store or shop, the complexity of the data being handled, and the information processing capability of the point of sale system that is required. Most manufacturers have designed their systems so that it is possible to progress from 'stand-alone' systems to elaborate computer controlled networks.

Modern facilities Besides the basic requirements of a cash register (including a cash drawer with security devices) modern ECRs have to cope with the ever changing demands of the financial world. Computation facilities are provided for calculating the percentages of sales tax and discounts. Facilities for recording credit purchases by automatically reading the information on credit cards is available, and, where the register has access to computerized credit files, it is possible to obtain credit authority. Peripheral equipment such as automatic change dispensers can also be attached.

When ECRs form part of a complex computerized system they can be linked to the central computer by standard telephone data lines. It is then possible to store and obey sets of pre-determined instructions as with a computer program. When a computerized product price file is incorporated in the system, individual items can be automatically priced by feeding in the item code.

Information about each transaction, such as product code, sales clerk code, quantities, prices and so on, can either be stored in the central computer or on magnetic tape cassettes attached to each ECR, for future analysis by computer. This information can be fed in manually via the keyboard, but more and more techniques are being developed to do this automatically. Automatic reading devices (character recognition devices) are now available which will 'read' the product information recorded on specially prepared tags or labels attached to each item.

Advantages of ECRs The benefits which are being claimed by the use of ECRs include a reduction in operator training time and increases of up to 20% in checkouts made, which implies a reduction in the number of operators required. Such systems also help to reduce errors.

The major benefits, however, are obtained when ECRs are employed in conjunction with computerized systems. It then becomes possible to ascertain the creditworthiness of each customer immediately. When combined with a stock control system it is possible to keep a continuously updated inventory of available stocks. It also enables automatic control systems used in warehouse storage to be operated more efficiently.

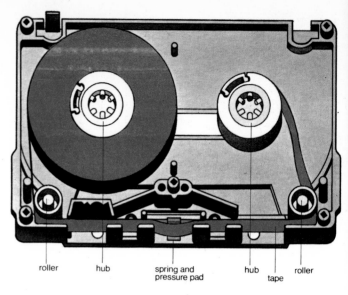

roller   hub   spring and   hub   tape   roller
                       pressure pad

CASSETTE and CARTRIDGE

The use of recording tapes in the home has been greatly simplified by enclosing the tape in a cartridge or a cassette.
Cassettes Cassettes were introduced by Philips in Holland in 1963, and consist of a flat plastic box containing magnetic tape running between two flangeless reels called hubs, the ends of the tape being attached to the hubs by means of a plastic 'leader' tape. Rollers and guides support the tape along the front edge of the cassette, where apertures allow the recording and playback heads and tape driving rollers of the recorder to come into contact with the tape. Two thin foils of low friction material such as PTFE (polytetrafluoroethylene) or graphite coated plastic prevent the tape from jamming against the top or bottom of the box. A pressure pad in the cassette ensures good contact of tape and head. The magnetic tape is shielded from stray magnetic fields from the motors by including in the cassette a piece of metal called a screen. This is made from an alloy, usually of nickel and iron, designed to trap magnetic fields.

A cassette has a number of advantages over the use of reels on an open tape recorder: the tape does not require threading into the recorder, a cassette can be stopped and removed at any time without first rewinding, and contamination by dust and damage to the tape through handling are minimized.

The major advantages, however, are in size and in playing time. Elimination of handling allows the manufacturer to make his tapes very much thinner and thus get much longer length into a small space. Total tape thickness ranges from 0.00037 to 0.00075 inch (0.0095-0.019 mm). In addition, thin tapes give better recording quality at slower tape speeds, so that a standard speed of $1\frac{7}{8}$ inch/sec (4.76 cm/sec) is used on cassette recorders with results as good as at $3\frac{3}{4}$ inch/sec (9.53 cm/sec) on reel to reel recorders.

Playing times are expressed in minutes (designated C60, C90, C120), taking into account that the cassette can be turned over to the other side for further recording on the other half of the tape.

Cassettes can be purchased as blanks, that is filled with unused tape for the user to do his own recording, or containing pre-recorded programmes. In the latter case plastic tabs at the back of the cassette are removed, so that when the cassette is put into the recorder a safety catch engages in the tab-hole and prevents accidental erasure or re-recording. Most programmes are music or other entertainment, but cassettes are also used for computer tapes.
Cartridges Introduced in America in 1965 by Learjet, the cartridge was primarily intended for pre-recorded entertainment tapes, providing continuous play in playback-only

tape   spool

roller   roller

*Far left: the inside of a cassette.
The tape is attached at either
end by leader tape to two hubs
inside the plastic box and runs
freely between them.
Near left: a cartridge opened to
show the endless loop of tape
wound round the single central
hub. The tape is drawn out from
the inner edge and wound in
around the outer, and thus must
be continuously slipping to
accommodate the change in size.
Below left: cassettes are
recorded at 32 times normal
playing speed. Both sides are
recorded at once, one of them in
reverse. On the left is the master
tape, which is in a continuous
loop so that it cannot be stored
on a reel. In the background and
on the right are four recording
machines transferring the signal
on to cassette tape.*

recorders. It has been especially popular in cars. Tapes usually carry 8 tracks of recording, or 4 pairs of stereo tracks.

The cartridge is bigger than the cassette, use $\frac{1}{4}$ inch tape and plays at $3\frac{3}{4}$ inch/sec (9.53 cm/sec). This is twice the speed of cassettes and should therefore provide better high frequency response. It contains an endless loop of tape which is formed by splicing the two ends of the tape together and winding it on to one spool inside the cartridge. The tape winds on to the spool in the usual way, but is taken off the spool from its centre, rather like unwinding string from the centre of a ball of string. The pack of tape is therefore continuously slipping inside its turns because it is being taken on and off at different diameters, and tapes have to have a very slippery back coating applied on the opposite side of the base film to the magnetic coating. At a certain position in the tape a metal switching foil is attached which is capable of operating a switch relay. This causes the player head to move one track down to the next programme. Guides, rollers, pressure pads, and so on are provided, similar to those used with cassettes.

The magnetic tape Tape manufacture for cassettes and cartridges has become a very specialized business. Advanced techniques are needed for mass production of blemish-free plastic films less than a thousandth of an inch (0.025 mm) thick in strong materials such as polyester. The coating on to it of a dispersion or paint of magnetic powder set into plastic resins has to be carried out with accuracies of a few millionths of an inch in thickness and with very smooth surfaces. The methods used are normally proprietary secrets, but it is known that the dispersion is generally applied by a gravure process similar to that used in printing and that after drying the coating is compressed and smoothed between rollers. This achieves the tolerances and high finish required —the process is known as *calendering*. For cartridge tape the additional lubricant coating is then applied. This is usually a black graphite and resin mixture. The graphite lubricant is 'suspended' in a resin base.

The wide rolls of tape are then slit to the correct width and supplied in reels of up to 6000 foot (1800 m) lengths to the cassette or cartridge assembly. Where the tape is to be pre-recorded, the reels are first sent to a duplicator, who specializes in running these reels across high speed recorders at speeds of 60 inch/sec (1.5 m/sec), putting on to them repetitive programme lengths from a master tape, and recording control signals between each programme so that the assembler knows where to cut the section out of the tape for each cassette or cartridge.

CATHODE RAY TUBE

A cathode ray tube or CRT is a device for producing a visual display using electronic circuitry. CRTs are used, for example, in television sets, radar displays and scientific instruments such as oscilloscopes.

A CRT consists of an evacuated (emptied of air) glass tube with a flat screen at one end coated on its inside with a fluorescent material. At the other end of the tube is an electron gun (see below), which projects a beam of electrons down the tube towards the screen. The electrons striking the fluorescent screen excite the atoms of the screen's material and cause it to glow—thus producing the visual display. The intensity of the glow is proportional to the intensity of the electron beam which is controlled by an element within the electron gun.

The components so far mentioned would be unable to make a useful display. The electron beam on leaving the gun is divergent, that is it spreads out, producing a large dim spot on the screen. It must therefore be focused by a *focusing system* which makes the electron beam converge to a fine point at the surface of the screen. Also a *deflection system* is required so that the electron beam (and therefore the spot on the screen) can be moved around.

Electron gun   An electron gun works on the same principle as the electronic valve [vacuum tube]. It has basically three components: a cathode, grid and anode. The anode is maintained at a high positive voltage with respect to the cathode, which is made of the reactive alkali metal caesium. When the cathode is heated by a small heating element, the caesium gives off electrons. The electrons, which are negatively charged particles, move towards the anode. In the centre of this anode is a small hole through which some of the electrons pass, forming a diverging beam in the direction of the screen. The intensity of the beam is controlled by the grid placed between the cathode and the anode. By varying the voltage on the grid the flow of electrons can be controlled.

Focusing system   The diverging electron beam can be focused using either an electrostatic or a magnetic field 'lens' in the same way that a glass lens is used for focusing light. In both cases a carefully shaped field pattern is established which bends the electrons moving through the field into a converging beam.

Deflection systems   Deflection is also achieved using either an electrostatic or magnetic field. Two such systems are required: one to deflect the electron beam horizontally (left or right) and the other to deflect the beam vertically (up or down). In this way, all points on the screen can be covered by the electron beam.

With electrostatic focusing and deflection the required elements must be positioned within the evacuated tube. With magnetic systems, however, it is possible to position them externally, thus making construction simpler and cheaper.

Fluorescent materials   CRTs producing black and white displays commonly use silver activated zinc sulphide and silver activated zinc-cadmium sulphide as fluorescent materials. When mixed in the correct proportions, these produce a bluish white glow under excitation.

In a television receiver the appearance of a continuous picture is created by scanning the screen with the electron beam in a series of horizontal lines—varying the intensity of the beam while scanning. Because of the speed with which this is done, the eye perceives a complete picture with various shades of brightness—the individual lines being too fine and close to distinguish easily.

With TV CRTs the choice of materials is important because of a phenomenon called afterglow or *persistence*, where the material continues to glow for a while after the electron beam has been removed. The persistence of the screen material helps to maintain the picture continuity between scans. Television screens are scanned at a repetitive rate of about 25 times a second. If the persistence or glow time is much greater than 1/25 second blurring will occur when fast

Right: a black and white TV picture tube under construction. Here the face plate is being welded to the cone; the phosphor is then coated on the inside of the face before the electron gun is attached.

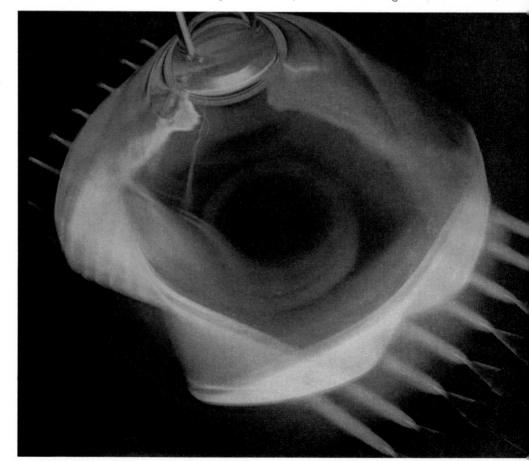

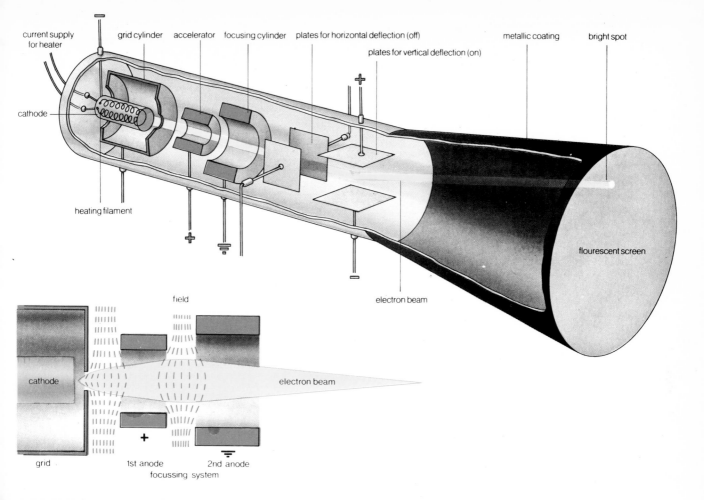

current supply for heater

grid cylinder

accelerator

focusing cylinder

plates for horizontal deflection (off)

plates for vertical deflection (on)

metallic coating

bright spot

cathode

heating filament

flourescent screen

electron beam

field

cathode

electron beam

grid

1st anode

2nd anode

focussing system

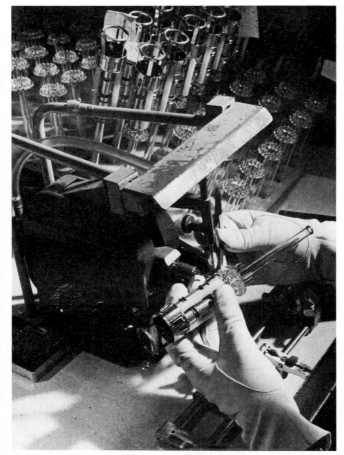

Above: the electrostatic focusing system creates a curved electric field which bends the divergent beam of electrons, emitted from the electron gun, into a convergent beam which comes to a point at the surface of the screen. Deflection is achieved by the deflection plates.

Left: welding the base, with its connector pins, onto the electron gun assembly of a colour TV tube. Each gun is identical, and is similar to that in a black and white tube. The grid, accelerator, and focusing cylinders can be seen; the wide metal ring near one end of the gun contains parallel magnetic plates which are controlled by a coil around the outside of the neck to make the three beams converge.

Below: these are the centrifuges which are used to deposit the colour phosphors, each one in turn, on the inside of the face plate.
Bottom of page: the 'lighthouses', in which ultra-violet light is beamed onto the phosphor-coated face plate through the shadow mask, producing the dot pattern.

actions are being displayed. On the other hand, a persistence much less than 1/25 second will lead to 'flickering' as the picture fades away before being rescanned.

With oscilloscopes and especially radar displays the persistence required of the fluorescent material can be longer because the rate of change of events to be displayed is much less.

With colour television CRTs the screen is covered with a fine mosaic of three different fluorescent materials which glow with the three primary colours—red, green and blue. Associated with each colour is an electron gun and these three guns are arranged to produce a red, green and blue image on the screen which to the eye merge into one high definition colour picture.

Manufacture Almost every cathode ray tube is basically a blown glass structure with a separate faceplate, carrying the screen, welded on under conditions of either high vacuum or an inert gas atmosphere.

The small tubes used in oscilloscopes are much easier to make than the big-screen tubes used in radars and TV receivers. The most complex tubes are those used in colour TV sets. The screen, instead of having a uniform coating, has to be covered with phosphor dots precisely aligned with small holes in a thin metal shadow mask.

This mask is made by preparing a thin sheet of metal, often a nickel alloy, and giving it light-sensitive coatings on each side. Exposure to powerful lights through templates perforated with a fine screen of dots hardens the coating in a corresponding dot pattern, which, after developing and hardening, is etched with an acid spray to leave a precise pattern of some 400,000 0.3 mm holes. The sheet is then annealed (gently heated in a furnace) to remove internal stress, and press-formed to a curving shape to match the spheroidal shape of the face plate.

The shadow mask is then welded into a strong frame which can be fixed to the glass face plate. The face plate, which is moulded in high-quality glass, has to be extremely accurate and free from internal stress. It is coated with a chemical emulsion, normally whitish grey, but which will glow green when bombarded with cathode rays. This is exposed to bright ultra-violet light through the shadow mask, and then washed. Where the ultra-violet light has hardened the emulsion, a dot is left after washing. This procedure is repeated with phosphors which will glow red and blue, with the ultra-violet light slightly displaced in each case. Each phosphor layer is deposited while the face plate is whirled in a centrifuge, to ensure an even coating. After the exposures and subsequent washing, the result is a glass surface coated with an array of dots of the three types of phosphor.

The cone-shaped moulded glass part of the tube is made separately. This is tested for accuracy, cleaned, given a coating of graphite at the neck and a thinner graphite coat over the interior, and then edge-coated with a coat of 'frit' (a mixture of chemical solvent and powdered glass) for the bonding of the face plate.

Older black and white tubes were heat sealed at a high temperature, but the delicate pattern of phosphor dots and the metal shadow mask of the colour tube make this impossible. The use of frit means that rather lower bonding temperatures can be used.

Finally the plate and shadow mask are frit-bonded to the conical part of the tube, sealed in an oven and the joint tested. The electron gun, with its cathode, and focusing and deflection plates, is then fixed inside the neck of the tube, the whole tube evacuated, and the resulting colour picture tube tested. All that remains to be done is to reinforce the tube against implosion ('exploding inwards' under atmospheric pressure), graphite the exterior and inspect it for delivery.

The television picture tube is an example of a product which is expensive to manufacture, and must be made in large quantities on an automated basis to keep costs down.

CIGARETTE LIGHTER

The basic type of cigarette lighter, using a flint and wheel to create sparks which ignite a fuel-soaked wick, is a modern version of the ancient Greek and Roman methods of making fire with flint and tinder.

Tobacco was introduced into Europe in the sixteenth century and the popularity of smoking gave rise to a demand for a simple means of lighting a pipe, which could take fifteen minutes or more using a tinder box.

The first really effective lighters appeared at the beginning of the twentieth century, but many ingenious and sometimes dangerous devices had been tried during the preceding three hundred years. Some of these used percussion caps similar to those used today in toy guns, and there were others that employed a small bottle of acid to which metal chips were added, producing hydrogen gas that was ignited by a spark. The most useful firemaking device invented during this period was the friction match, first produced by John Walker (1780-1859) at Stockton-on-Tees, England, in 1827.

Flint wheel lighters The flint used in cigarette lighters is not true flint but an alloy of iron and magnesium with the rare earth element cerium. This alloy was invented by Baron Auer von Welsbach in about 1900, and the first flint lighter was produced in Treibach, Austria, where von Welsbach worked. There was no wheel action on this lighter, the sparks being produced by striking the flint on an abrasive surface. The first wheel action lighters appeared in 1909, the sparks being struck from the flint by a serrated steel thumbwheel, and lever action lighters were introduced in the 1960s.

Fuel Cigarette lighters today use either a highly flammable liquid fuel derived from oil, or liquefied butane gas in a pressurized tank. In the first type, the fuel is drawn up by a wick from a refillable tank packed with cotton wool which absorbs and retains the fuel, and is ignited by sparks from a flint. The wick is sealed off by a cap or snuffer when the lighter is not in use, and flame height is controlled by adjusting the length of the wick.

Butane gas lighters were introduced in 1945. They are filled from small pressurized containers holding sufficient gas for several fillings, and give a clean and odourless flame. There are two valves, one in the base or at the side for filling the tank, and an adjustable one at the top fitted with a burner nozzle where the flame is produced. Gas lighters are ignited either by a flint or electronically.

Depending on the design, the burner valve is opened by raising a snuffer cap or depressing a level (thumbwheel action models), or else it is opened automatically by the lever action mechanism.

Electronic ignition There are two basic types of electronic ignition. In one, current from a small, low tension battery charges a capacitor; when the operating button is depressed a circuit is completed causing the capacitor to discharge through a step-up transformer. The secondary winding of the transformer produces a relatively high voltage spark across a gap, igniting the gas flow simultaneously released by the operating button. The flame burns for as long as the button is depressed.

The second method makes use of the piezoelectric effect —a voltage is produced between opposite sides of certain types of crystal when they are struck or twisted. A small block of crystals, usually a crystalline lead compound, is struck by a spring loaded, button released hammer. The resulting voltage is discharged across a gap to ignite the gas in the same manner as the battery operated version. The crystal used is virtually everlasting.

Disposable lighters The 'throw away' lighter is a comparatively recent innovation, taking the form of a transparent plastic tube, filled with liquid butane fuel, with a simple flint and wheel ignition device mounted on the top together with a lever to open the gas valve when the flint is struck. The tube cannot be refilled and the lighter is thrown away when

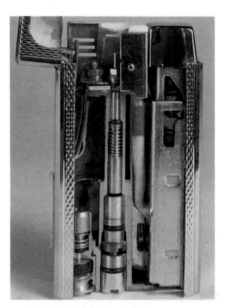

Above: a cutaway view of a gas lighter. Pressing the button opens the gas supply and strikes the crystal, producing a piezoelectric current.
Left: an electrically operated table lighter, from Georges Dary's 'Tout par l'Electricité', Tours, 1883. The lower part of the device is a battery; tilting it closes the circuit which lights the wick.

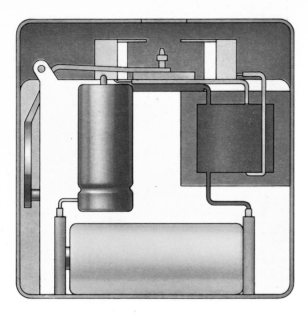

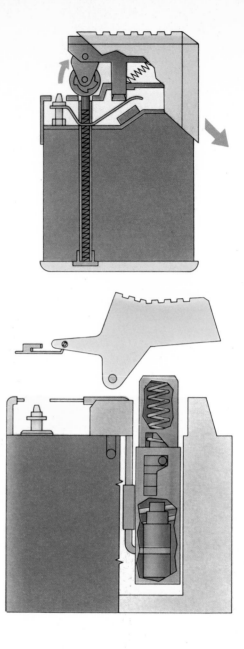

all the gas has been used up.

Car cigarette lighters This type of lighter consists of a push button normally lightly held in a rest or non-inserted position in a receptacle on the car dashboard. An electric element, consisting of a closely wound spiral of nichrome or other electrically resistive metal, is mounted on a heat-retaining ceramic base at the end of the button that fits into the dashboard.

When the button is pushed into its fully inserted position against a spring loaded catch, a circuit is completed between the car's electrical system and the element, which quickly heats up. After the correct temperature is reached a temperature sensitive device releases the catch allowing the button to spring back to its rest position and opening the circuit. It can then be removed, and the brightly glowing element used to light the cigarette. After use, the button is replaced in the receptacle in its rest position.

Above: electronic ignition actuated by the piezoelectric effect. When the crystal pack is struck a voltage is induced and then discharged across the gap above the burner.
Above right: a flint wheel lighter. Depression of the lever raises the burner valve and turns the serrated wheel against the flint, creating a spark.
Centre right: battery operated lighter. When the button is pressed the burner valve lifts and the capacitor discharges. The voltage is boosted by the transformer and sparks across the electrodes.

CLOCK

Clocks are essential to modern life. Apart from acting as timekeepers for everyday matters, their uses extend to the automatic programming of factories, time switches in street lighting and domestic controls, timing in industrial and sporting activities, and for all navigation and space travel.

History Men in the early settlements along the shores of the Nile and Mediterranean first indicated the passage of time by noting the length or position of the shadow of an upright stick in the ground. By 1500 BC, these sticks had been replaced by elaborate obelisks in community centres. Cleopatra's needle (now in London) was one of them. Many elaborate and portable sundials followed over the next millennia, but none was of any value at night or out of the sun. For use at any time, the clepsydra (water clock) was invented about 1500 BC. It was followed by the sand timer, working on the same principle of filling or emptying a vessel at a controlled rate.

The hour was one of the first artificial divisions of time, probably invented by the Egyptians about 4000 BC. Previously man depended only on natural divisions of day, determined by the Sun, and month, determined by the Moon. The 12 daylight hours started at dawn and the 12 night hours at dusk, so they were of different lengths and varied during the year. Early astronomers, however, used hours which were of equal length all the year round. As civilization advanced, more and more human activity was controlled by the *horologium*, as early timekeepers were called. (Horology is the science of time measurement.) A monastery, the centre of learning, would have a horologium operated by water upon which the sexton would depend for sounding the monastery bell to call the monks to rise, work, prayer, and bed.

The first mechanical clock At an unknown point in history, but probably in Burgundy around 1275 AD, an unknown man, probably a monk, devised a new kind of horologium, a mechanical one that struck a single note on a bell at the hour instead of indicating the time visually. It was worked by a falling weight instead of water and became known as a 'clock' after the Latin *clocca*, a bell. The first clocks also had alarms that could be set to ring continuously at a selected hour to warn the sexton, or the watchman of a civil community, when to toll his big bell. Such alarm clocks were quite small, to judge by those shown in early paintings.

The frame was a vertical double strip of iron or brass that stood on a base or was hung on a wall. The wall version had a hoop or stirrup at the top for hanging from a hook and two spurs at the bottom to prop it way from the wall. It was powered by a stone weight on a rope which was wound around a wooden barrel. Small pegs in the barrel allowed it to be turned to wind up the weight by hand. A click (*ratchet*) prevented the weight from falling down again without rotating the barrel in the opposite direction to drive the clock. To the barrel was attached a large toothed wheel, called the *great wheel*, that drove a very small toothed wheel, or a wheel with rods for teeth, called a *pinion*. On the same shaft as the pinion was another large wheel, but of a different kind, called a *crown wheel*, with teeth on one side. (The name arose because of its likeness to a king's crown.) Across the crown wheel was another shaft with two small 'flags' (*pallets*) attached to it, placed to intercept a tooth on each side of the wheel. This vertical shaft was called the *verge*, after the verger who carried a staff with two small flags. As the weight fell, it turned the crown wheel which was checked by a tooth striking a pallet at one side. Movement of the pallet turned the verge and caused the opposite pallet to intercept another tooth. The action was then repeated, so that the verge was oscillated to and fro as the weight descended. At the top of the verge was a horizontal crossbar with a weight at each end, the *foliot*, which had its own period of oscillation, or timekeeping, which it tried to impose upon the clock. Moving the foliot weights inwards increased the rate, and outwards decreased it.

Development Very soon after the invention of the mechanical clock, it became divided into three separate mechanisms. The main one controlled the other two. It incorporated the crown wheel and verge (the *escapement*) and indicated time of day by a moving hand or moving dial. It was known at the time as 'the watch' and is now called the *going train*. The second was the *striking train*, originally called 'the clock', which had its own weight drive. It was released by a lever moved by a pin on a wheel in the going train. The third part was the *alarm*, which was also released by the going train, and had a separate weight to drive it. It was first set by placing a removable pin in one of a radial series of holes representing the hours. Later a friction-held disc with a pin in it was rotated to set the alarm.

On smaller clocks, a balance wheel with one spoke was used instead of a foliot. It was not adjustable, but weight could be added or subtracted from the driving weight to change the rate of drive.

The first improvement to striking clocks was to make them sound the correct number of blows on a bell to represent the hour, perhaps in Italy around 1330. A form of cam wheel was developed to control the number of blows struck. It had

Top: this centuries-old clock is in the clock museum in Vienna. It has 33 dials and more than 40 hands, and indicates the movements of the planets. Above: the courthouse clock at Ulm. This one is elaborately decorated, featuring the signs of the Zodiac.

Right: an anchor escapement of the type designed by Robert Hooke (1633-1703) for use in pendulum clocks.
Below: a clock in a Saigon mosque which indicates hours of prayer.

twelve slots of increasing width around its periphery into which an arm dropped at every blow. When the arm could not drop into a slot, the striking stopped. Count wheel striking of this type is commonly found on English longcase clocks and French ormolu clocks. An improved control called rack striking (or rack-and-snail), invented in 1676 by a parson clockmaker, the Rev Edward Barlow, not only prevented the striking from becoming out of phase with the time shown by the hands, as does the count wheel system, but also made repeating clocks possible. These repeat the last hour struck, and in more advanced clocks the last quarter and even the last minute, when a cord from the side of the clock is pulled. Pull repeaters were valuable at night when there was no gas or electric light, and were popular until matches were invented. Modern clocks use the rack striking system.

The early form of winding was soon replaced by a single rope fitting snugly in a grooved pulley. One end held the large driving weight and the other a small weight to keep the rope in the groove. The clock could be wound by pulling down the smaller weight.

One hand indicating hours on a dial marked in hours with divisions for quarters was sufficiently accurate for some centuries on domestic clocks, although astronomers had used minute hands, even seconds hands, from the sixteenth century. One problem was in setting the clock to the time as shown by the sundial. At first it was necessary to disengage the verge from the crown wheel to allow the hand to turn. Soon an unknown genius thought of providing the hand with a slipping clutch so that it could be turned without affecting the clock.

Winding mechanisms All clocks until about 1475 were weight-driven and were therefore not portable. If moved, they had to be carefully set up again. Someone unknown,

possibly a locksmith or swordsmith, invented a coiled spring to replace the weight, which gave birth not only to the portable clock, but also the personal watch. The first spring-driven clocks were modified weight clocks. Instead of a weight pulling on a line around the driving barrel, another barrel, immediately below the driving barrel, had a coiled spring inside it and wound the line around itself. The earliest known version of such a clock is dated 1480 and is in the Victoria and Albert Museum in London.

The mainspring alone was not an accurate enough power source for the clock because it was strong when wound up and became progressively weaker as the spring ran down, unlike a weight which could provide an absolutely constant source of power. So clockmakers introduced a device known as a *fusee*, to provide a relatively constant source of power from the spring. This was a trumpet-shaped pulley, first illustrated for clockwork by Leonardo de Vinci in 1407, and originally used in his time to make the winding of war catapults easier. It increased the leverage as the spring ran down. The fusee is so efficient that it is still used today in mechanical marine chronometers.

Precision devices The biggest advance in precision, until very recent times, came with the invention of the practical pendulum clock by the Dutch scientist, Christiaan Huygens, also famous for his contributions to dynamics and optics, in 1657-1658. His most effective design was a wall clock with a short pendulum about one foot (0.3 m) long, connected with a horizontal verge acting on a horizontal crown wheel. Huygens hung the pendulum separately and connected it with the clock movement through an arm called the *crutch*. For weight driven clocks he employed an ingenious endless rope or chain that still provided drive while the clock was being wound. It was subsequently used in thousands of English 'grandfather clocks' wound daily, and for modern electrically wound turret clocks. His spring driven clocks incorporated a *going barrel* instead of a fusee. The going barrel was wound in the same direction as it drove the clock, instead of being wound back, and therefore provided power during winding, whereas the fusee, although more accurate, had to have a complicated addition in order to provide this 'maintaining power'. Most clocks were by this time wound with a key.

About 1675, William Clement, an English clockmaker, introduced an escapement designed for the pendulum, known as the anchor or recoil escapement. It employed an ordinary flat escape wheel with teeth on the edge, instead of the crown wheel which had teeth on the side. The arrangement substantially reduced the arc of the pendulum, so that a long one could be used, which dominated the rate of the clock much more effectively. The most satisfactory length was just over three feet (1 m), so that the pendulum swung from one side to the other in one second. This seconds pendulum, also called the Royal pendulum, reduced the error of a clock from about a quarter of an hour a day to only about 20 seconds a day. The minute hand, which had been used on a few astronomical clocks from the sixteenth century became almost universal on clocks with all lengths of pendulum, and the seconds hand became orthodox on longcase clocks.

Better timekeeping by 'artificial clocks', as mechanical clocks were called, made two uncomfortable facts apparent. One was that the days varied in length throughout the year. The other was that sunrise time varied across the country, being gradually earlier eastward and later westward. The first was solved by accepting the 'equal hours' shown by the clock (except in Japan, where complicated arrangements made the clock show hours of different length). The second was not solved until over two centuries later in 1884 when Greenwich Mean Time was adopted in most of the world, adapted by means of local time zones.

It was now apparent that clocks were better timekeepers than the rotating Earth and increasing precision became the

Far left: a wood-framed tower clock in the Science Museum, London. Iron has been the most widely used material for tower clock frames since the end of the thirteenth century, but there is a tradition of wood-framed clocks of which this, from Martley Church, Worcestershire, is a good example. It was probably made between 1680 and 1700. The frame is of oak, with the going train mounted in the upper part, the striking in the lower. Both going and striking trains are wound by a capstan, with ratchets locking on wheel spokes. The weights are of stone.
Left: this pendulum of a baroque clock was an early attempt at more precise time-keeping. The pendulum is made of two different metals, which expand and contract at different rates, thus partially cancelling the effects of atmospheric temperature on the length of the pendulum.

driving force of leading clockmakers from the early eighteenth century. Thomas Tompion, most famous of all English clockmakers, provided the first Astronomer Royal, appointed in 1676, with two high precision clocks for the newly built Greenwich Observatory. In 1715, his friend and successor, George Graham, invented the deadbeat escapement and in 1721 the mercury pendulum, which compensated for temperature changes with a column of mercury on the pendulum that expanded upwards to keep it the same effective length. The deadbeat escapement, by changing the shape of the teeth on the escape wheel, avoided recoil in the mechanism and provided a shorter pendulum swing. The inventions increased the availability of precision clocks. The most remarkable feat of the time was the development of a marine timepiece with a balance wheel and compensated spring by a carpenter, John Harrison. It was tested on a sailing ship voyaging to Jamaica and was only 5.1 seconds out during the 81 days' passage. This was in 1761 and it initiated accurate navigation and the charting and control of the seas by the Royal Navy.

The deadbeat escapement was improved upon, especially for tower clocks, which tend to slow down because of weather conditions. The new idea was to raise weights (by means of the movement of the escape wheel) which when released give an impetus to the pendulum. The most successful device of this type was invented by E B Denison (Lord Grimthorpe), who used it in the Westminster tower clock called Big Ben, which was installed in 1859.

A German clockmaker, S Riefler, built a precision pendulum clock in 1889 that was adopted by many observatories. In 1921, an English engineer, W H Shortt, devised the free pendulum clock, the ultimate form of pendulum clock. The one in Edinburgh Observatory kept time to within 0.1 second a year. It is partly electric and there were two pendulums, one to swing freely and the other to do the work. It was so accurate, it detected a regular wobble in the Earth's rotation, subsequently confirmed and analyzed by the quartz clocks, used later in groups by observatories, that kept time to the equivalent of a second in 30 years. Today the timekeeper from which all time standards are derived is the atomic clock developed by the British scientist L Essen,

which has an accuracy equivalent to one second in 3000 years, and has detected a slowing down in the Earth's rotation.

Manufacture of clocks followed trends set by the inventors. At first all clocks were made individually by single clockmakers and held together by wedges in slots like early wooden furniture, or, after about 1500, increasingly by nuts on threads. Each part was made separately and a wedge would fit only a particular slot and a nut only one particular thread. Some removable pieces were marked to show mating parts. From about 1700 some clockmakers began to make interchangeable parts so that clocks could be produced in batches by workers specializing in one part or another—a process pioneered by Tompion.

In the nineteenth century, after early attempts in France, success in mass production was finally achieved in north America, where in a water powered mill, Eli Terry mass produced wooden grandfather clock movements in thousands. Mass produced clocks of rolled brass also first appeared in North America and flooded Europe's markets at very cheap prices to bankrupt many factories and result in the reorganization of national industries.

Electric clocks The first electric clock was invented in 1840 by a Scotsman, Alexander Bain, and an English clockmaker, Barwise, and worked off earth batteries: coke and zinc buried in the ground. The French manufactured an electric pendulum clock worked by a Leclanché cell, which was small enough to stand on a mantelpiece. An American, H E Warren, invented the electric mains clock in 1918; it is not a true clock as it contains no time standard, merely counting the frequency of the mains and translating this into time of day. The biggest breakthrough was the dry battery clock developed since World War II, particularly in Germany, which is an accurate timekeeper and will run for a year (some versions for five years) on a battery.

The first battery clocks were 'remontoires'. The remontoire overcomes variations in driving power by winding up a small weight (Huygens, 1659) or a small spring (Harrison, 1739) to drive the clock—originally done mechanically. The battery wound a light spring every quarter of an hour or so to drive a normal clock controlled by a balance and spring, a

control system invented for watches by Huygens in 1675 but used also for almost all small clocks in the twentieth century.

In the last ten years a direct drive has been developed. It ignores a long-held horological principle that the less work done by the oscillator, the more accurate the clock. The battery powers an oscillating circuit in which there is a magnetic coil that pulses a permanent magnet fixed to the balance wheel, to keep it oscillating and controlling the rate. The balance turns a wheel to operate the hands. A Swiss invention of the 1950s is an electronic tuning fork movement that further increases the accuracy of the battery clock. The tuning fork replaces the balance wheel. It is driven magnetically at a frequency of 300 Hertz by a coil in an oscillating circuit operating on one of the tines. The hands are turned by a train of gears terminating in a special escape wheel driven magnetically by a small horseshoe magnet on the other tine. The tuning fork is therefore 'free', having no physical connection with other parts.

Most accurate of all is the battery driven quartz clock in which the time standard is a wafer or 'tuning fork' of rock crystal controlling a solid state oscillator that drives the hands. Some are driven by accumulators charged automatically by natural or artificial light.

Alarm clock The alarm clock, now the basic household clock, has always been made in one form or another. Since the end of the nineteenth century, because of the need for it at as low a price as possible, it has followed a rather different direction of development. The mechanical alarm has not been refined much horologically but has been refined to a considerable extent in terms of production engineering so that some factories can turn out every minute of the working day a complete and going alarm regulated automatically to an accuracy of 99.6%. Electric alarms avoid the task of daily winding.

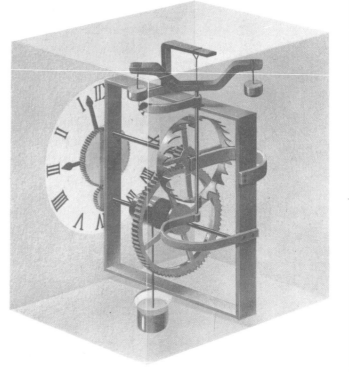

Top of page, left to right:
an all-plastic clock, Shortt's
free pendulum clock (1921), a
chronometer movement. A
chronometer is a timekeeping
device with a spring-detent
escapement, used at sea for
navigation.
Right centre: a diagram of
the earliest type of
mechanical clock.
Right: the oldest surviving
public clock in Britain, the
Salisbury Cathedral clock,
made in 1386.

Left: one of the earliest surviving European compasses, this was made in Italy about 1570. The vellum card is on a central pivot which is on a gimbal mounting bowl, itself inside an ivory case.
Below: an azimuth compass, with sights, made in London about 1770. This would have been used for taking bearings, rather than steering.

COMPASS

The observation that the iron mineral called *lodestone* would align itself in a northerly direction was recorded in China in the first century BC. There is little evidence that this discovery resulted in the development of a floating compass until about 1100 AD. By about 1250 AD the compass was being used by the Arabs, Scandinavians and Europeans as well as the Chinese. When European explorers first penetrated Chinese waters, they found the Chinese compass much inferior to their own. It is impossible to say where the development of the compass first took place, or whether it developed independently in the East and the West.

Today the magnetic compass has reached a high state of development as an invaluable aid to navigation and surveying. Compasses may be classified as direct reading and remote indicating types.

Early development If an iron filing or needle is stroked with a lodestone it will become magnetized. By the thirteenth century magnetized iron 'needles' were used as compasses floated or pivoted in conjunction with a scale (the windrose) showing the direction of the prevailing Mediterranean winds, to indicate north when the sky was cloudy. Later the scales or *cards* were marked with the four *cardinal points*, north, south, east and west, and further subdivided to give a total of 32 points. Subsequently a scale of degrees was made: 0° at N and S; 90° at E and W, making a total of 360°. (The arbitrary number of 360 degrees in a circle is an inheritance from ancient Babylonian astronomers. When measuring time, they had divided the day into six parts which were further subdivided by six. These measurements were originally land distance measurements, which were extended to cover the sky, making a circle. This division of the circle also survives in the time measurements of modern astronomy.)

During the fifteenth century it was discovered that there was a slight difference in the compass reading between *true north* (the direction of the North Pole) and *magnetic north*. The angle of the difference is called *magnetic declination*, or *variation* and it can now be determined for a particular locality from charts. On his second voyage, in 1493, Columbus carried compasses which had been altered in an attempt to allow for variation. The great geographer and map-maker Gerhardus Mercator (1512-1594) was the first to correctly assume the existence of a magnetic pole separate from the North Pole. William Gilbert (1544-1603), in his *De Magnete* (1600), was the first to propose that the Earth itself was a great magnet, but he thought variation was caused by the magnetic attraction of land masses. The first modern charts giving world variations were published in 1701 by Edmund Halley (1656-1742), of comet fame.

A mariner's compass. This is the liquid magnetic type consisting of a float pivoted in a bath of alcohol and water. Attached to the float is a graduated mica card suspended from which are the magnets or magnet—in this case a ring magnet with its magnetic axis set parallel with the north-south diameter of the card.
Below: a binnacle compass of a well-tried design. The mica card floats on alcohol and water, while the soft iron spheres correct for steel in the ship. This compass would be used on open craft like sailing ships.

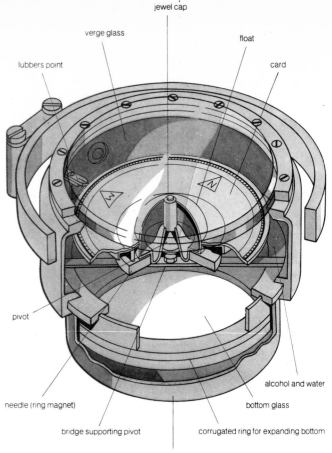

Mariner's direct-reading compass The iron needle originally used in compasses needed frequent remagnetization. In 1766 Dr G Knight patented a compass in England which was the most advanced design of its time. He used a better magnet steel for the needle, a jewelled bearing for the pivot to minimize friction and wear, and suspended the compass on pivots inside a ring (the *gimbal*) to provide insulation from the ship's motion.

Further improvements were made by mounting the compass in a liquid-filled *bowl*, thus damping it from mechanical vibrations. This was complicated on the earliest models by such factors as the leakage or bursting of the bowl due to atmospheric conditions, and the discoloration or corrosion of the compass parts because of impurities of the liquid.

In the meantime other improvements were made to the mounting stand of the compass (the *binnacle*). During the nineteenth century, as shipbuilders began to use more iron in construction, ships were lost on account of magnetic interference with the ship's compass. The solution to the problem, proposed by Britain's Admiralty Compass Committee, was to install in the binnacle a system of compensation for 'hard' and 'soft' iron respectively by separate assembles of magnets and masses of soft iron. 'Hard' iron refers to the iron in the ship's structure and 'soft' iron to changes in the direction of the ship's magnetic field with respect to that of the Earth. (Random magnetic interference is minimized by placing offending objects a pre-determined *safe distance* away from the binnacle.) A ship's magnetic field is determined by trial and error experiments, but once the adjustments are made the compass should only need checking once a year. If a ship is struck by lightning or heavily damaged in military operations, the magnetic field will go awry and may take months to settle down.

The modern mariner's compass A typical 'standard' compass has a 360° card six to nine inches (15 to 22.5 cm) in diameter supported by a jewelled bearing on an osmium-iridium or tungsten carbide pivot in a liquid filled bowl, and carrying either a pair of bar magnets or a single 'ring' magnet with its poles along a diameter. The reference mark or *lubber* is a pointer protruding to the edge of the card from the inner wall of the bowl. A float enables the system to be adjusted to about $\frac{1}{4}$ oz (7 g) minimizing wear and prolonging pivot life. The liquid is commonly an alcohol and water mixture or a light oil. Apart from the system magnet all materials must be strictly non-magnetic. The compass bowl is suspended through gimbal rings at the top of the binnacle, which carries lighting and means for compensating the unwanted fields from the ship. Means are also provided for mounting a prism, an azimuth circle or other device for taking bearings on top of the compass bowl.

COMPUTERS

Ever since man began to count and do simple arithmetic he has tried to make the process easier and faster by the use of machines. From counting on his fingers, man progressed to using pebbles to represent numbers, and this led to the invention of the abacus.

A major advance in mathematics was the system of calculation by logarithms, devised by John Napier at the end of the 16th century and first published in 1614. Following this discovery, the English clergyman William Oughtred invented sliding scales, an early type of slide rule which, by its use of lengths to represent numbers, is in effect a form of *analog* computer. The abacus, which operates by counting rather than measuring, is in comparison a *digital* device.

The first mechanical calculating machines appeared during the 17th century. Blaise Pascal produced a machine that could add and subtract, performing multiplication and division by repeated addition and subtraction. Some years later Leibniz invented a calculator which could perform all these functions individually.

The true ancestor of the modern computer was designed by Charles Babbage in the 1830s. This machine, the Analytical Engine, was never completed, but it was intended to perform any desired calculation automatically by means of a mechanical calculating unit controlled by punched cards. These punched cards, originally developed for the Jacquard loom, formed the basis of the card handling machines developed by Dr Herman Hollerith in the 1880s.

Electronic computers In 1944 Professor Howard Aiken of Harvard, in association with the International Business Machines Corporation, completed his Automatic Sequence Controlled Calculator (ASCC, or Mark I). This machine, a huge electromechanical calculator incorporating about three thousand telephone relays and controlled by punched paper tape, was over 50 feet (15 m) long and eight feet (2.4 m) high: it took 0.3 second to add or subtract, 4.0 seconds to multiply, and 12.0 seconds to divide.

Two years later, at the University of Pennsylvania, Dr John Mauchly and J Presper Eckert completed the first electronic digital computer, the Electronic Numerical Integrator and Calculator (ENIAC). It contained over 18,000 valves [vacuum tubes] which produced a great deal of heat and developed frequent faults, but it could perform as much work in one hour as ASCC could in a week. Although instructions were stored internally, it took hours of manual rewiring to change programs.

Binary arithmetic The early machines used the ordinary decimal system in their calculations, but this was soon superseded by computers using binary arithmetic, which is calculated to the base of two, rather than ten as in the decimal system. In binary, all numbers are represented by combinations of the digits 1 and 0: for example 2 becomes 10 and 3 becomes 11, and 15 becomes 1111. In an electronic circuit a switch, or a switching device such as a transistor or valve, can be in one of two states: it can be 'on' (passing current) or 'off' (not passing current). If a '1' is represented in a circuit by a device in its 'on' state, and a '0' by a device in its 'off' state, any binary number can be represented by a combination of devices which are 'on' or 'off' to correspond with the binary digits of the number. In the logic circuits of a computer, computation takes place through acceptance of data in binary form and processing it by means of on/off electronic pulses or signals. In computer terms, the name 'binary digit' is abbreviated to 'bit', and this name also applies to any device or signal that represents the digit in the circuitry. The bits are arranged into groups known as *bytes* and *words*. The number of bits in a group varies from one system to another, but for example there may be eight bits in a byte, and two bytes in a word.

Stored programs After the development of ENIAC the next significant step in computer technology came in 1949

Part of an early German computer, the Gamma 3. Made by Bull Electric in Cologne and Berlin in 1952, it was the first machine to use germanium diodes in place of the thermionic vacuum diodes normally used at that time.

with the Electronic Delay Storage Automatic Calculator (EDSAC) built at Cambridge University, England, and the machine designed by John von Neumann's team at Pennsylvania, called the Electronic Discrete Variable Automatic Computer (EDVAC). These used binary logic, but more important was their use of a *stored program* or set of operating instructions to control the sequence of operations, instead of the instructions being routed through wires plugged into a perforated circuit board.

At the beginning of the 1950s many commercial organizations were working on the development of computers. The first of these machines to appear on the market were the Leo, made in Britain by a subsidiary of J Lyons, the catering company, and the UNIVAC I (Universal Automatic Computer), built in America by Eckert and Mauchly for Remington Rand and delivered to the US Census Bureau in 1952.

As transistors became more reliable and widely available during the 1950s the computer makers began to incorporate them into their designs. In 1960 the Control Data Corporation marketed the first fully transistorized computer, and in the following years integrated circuits became an important feature of computer design. Improvements in computer *peripherals*, the input, output and storage devices such as

Right: part of the electronics of a large computer system. The rapid development of semi-conductor technology has led to the production of computers that are faster, smaller and more powerful than was possible using valve (vacuum tube) circuitry.

card, tape and disk machines, have also been significant during the last twenty years.

Analog computers Analog computers measure and work with continuously varying quantities such as temperature, speed and pressure. The first analog computers were mechanical, the numbers involved in the calculations being represented by the amount of rotation of shafts and gears or, as in the slide rule, by the movement of a graduated scale. Some early analog machines were made during the 19th century, but it was not until the 1930s that successful and accurate machines were made at such places as the Massachusetts Institute of Technology and Cambridge University. In the 1940s electronic analog computers were built in the USA.

Electronic analog computers work by converting the varying input numbers and quantities into correspondingly varying voltages, and then performing specified functions on these signals such as addition, multiplication or integration to produce output voltages that represent the results of the computations. Analog computers are used in scientific calculations and engineering design research, and in these cases the output voltages may be applied to pen recorders, which produce the results in the form of a graph or a diagram, or to graphic displays that produce a trace on a cathode ray tube screen.

They are also used in industrial process control, and in navigation equipment such as that in space vehicles: in these cases the output signals can be used to control the operation of other mechanisms.

Digital computers The digital computer works with numbers, or numbers that represent alphabetical characters, as opposed to the analog computer which works with varying quantities and measurements. It is essentially a very large and fast calculating machine, but it also has the ability to sort and compare information, and to analyze and store it for future reference. Its main advantage is its speed of operation, processing speeds being measured in terms of nanoseconds (thousand millionths of a second).

The first generation of machines, beginning with ASCC and ENIAC, were those based on valve [tube] circuits. The development of the transistor and other semiconducting devices led to the second generation of computers which used these in place of the valves; about 1965 the introduction of miniaturized integrated circuits to replace the transistors resulted in the third generation computers. With each advance in systems technology computers have become smaller and more powerful, a factor which has led to great increases in the ratios of cost to performance and which brings the prospect of computers in the home much nearer to reality.

Machine organization A computer configuration consists of five main sections: input, storage, control unit, arithmetic and logic unit, and output. The actual machinery is known as the *hardware*; the information that is to be processed is the *data*, and the programs that instruct the machine in its operation are known as the *software*.

For reasons of size and cost it is not possible to store files of data permanently in the main storage (the memory) of the system, so a backing store in the form of punched cards, magnetic tape, magnetic disks or magnetic drums is used for data storage. These devices also act as input and output units, transferring data to and from the central processing unit.

The cpu The control unit, arithmetic and logic unit and the main memory unit are physically combined to make up the central processing unit, although in very large systems there may be additional cabinets to house sections of the memory. All the program instructions are loaded into, and retained in, the main memory and all data to be processed is read into the memory from the input devices and held there while it is being processed. The memory may be in the form of magnetic core storage, thin film storage, or solid state devices such as integrated circuits.

The control unit reads the program instructions, and issues commands to the other parts of the system in order to carry out these instructions.

The arithmetic and logic unit performs computations as instructed by the control unit, adding, subtracting, multiplying, dividing and making logical decisions. During calculation numbers are stored in areas of the arithmetic and logic unit called *registers*. Information can be moved from one register to another. A simple ADD instruction will, for example, take a number from its register and add it to another in the *accumulator*, in this sense a register where the addition takes place. The result of the addition may be put into the memory to await further computation, passed to an output device such as a printer or display screen, or recorded in the backing store.

Input devices Information is fed into the system in coded form through input devices such as card readers, magnetic tape, disk, and drum units, optical or magnetic character readers, paper tape readers, or by direct entry from a keyboard.

The paper tape and punched card machines are relatively slow devices compared with the magnetic units, which are fast access storage devices that the control unit can call on at any time, extracting data from them and recording more onto them. Most modern tape and disk units can simultaneously

'read' and 'write' (record) data, which greatly increases their effective speed, and the latest drum systems can transfer data at a rate of 1440 thousand characters per second.

The transfer of information from source documents to the input media (data preparation) has to be done manually, which makes it slow and costly. Most computers can produce a printed output, but the acceptance of printed input presents a problem, as different sizes and types of print, unusual characters and unnecessary information make the direct input of non-standardized documents impractical. For this reason systems using optical character recognition require the data to be typed or printed on standardized forms using a specially-designed typeface such as E13B or OCR-B, which is styled to avoid possible confusion between similar characters such as zero and capital O.

Output devices The purpose of an output device is to communicate the results of the computations to the operator in a usable form. Output devices currently in use include high speed line printers, graphic display units, typewriters, and paper tape and card punches. The paper tape and punched cards may be used for subsequent calculations and as methods of storing data in a convenient form that can be fed back into a machine without further preparation. As these machines are electromechanical, they are slow in operation compared with the speed of the main processor. For this reason, the cpu may put the results of a set of computations into a storage device called a buffer store, freeing its main store for the next set. The output devices will then read from the buffer store at their own speed, leaving the processor free for other work.

Line printers, so called because they print a complete line at a time, work at speeds up to 2000 lines per minute. The paper is in the form of a continuous perforated sheet with interleaved carbon paper, and after printing the sheets may be separated in a machine called a *decollator*.

Time sharing Output can also be generated in the form of graphs, diagrams and maps by automatic plotters. Input-output devices such as typewriter terminals and graphic display units incorporating keyboards are used mainly to enable an operator to communicate with the machine, either to give it instructions or to ask it questions. These may be part of the main computer installations, or in a completely separate location and linked to the system by telephone or radio circuits. A typical example of the use of these remote on-line terminals (directly connected to the main system) is in banking, where a terminal installed in a branch office can be used to check immediately on the status of a customer's account or to record deposits and withdrawals. A recent development in this field is the on-line cash dispensing machine.

Such a system often involves *time sharing*, where the cpu handles more than one task at a time, splitting each one into a series of operations and then dealing with all these operations in a predetermined sequence until the tasks are completed. It may also use *multiprogramming*, loading the machine with several programs that are run concurrently so that the system can handle more than one type of problem at a time. These operations are carried out in *real time*, that is, computation of the input data, revision of the data files, and display of the result follow almost immediately from the original enquiry, so if for example a customer has withdrawn some money from his account, the details of the transaction will be recorded, the new balance computed and displayed, and the account records updated, all in a matter of seconds. Other computing operations such as payroll calculations, where the data is accumulated over a period of time before processing, are called *batch processing*.

Programming To enable the user to instruct and communicate with the machines it has been necessary to invent new languages. As machine language is based on the binary arithmetic system the first program languages tended to be mathematically based. In FORTRAN (FORmula TRANslation) for example, more commonly used for scientific and mathematical applications, an instruction might say 'ADD X to Y at Z', where X and Y are numbers and Z is an accumulator or register. Another common scientific language is ALGOL (ALGOrithmic Language), taking its name from *algorithm*, a set of instructions for solving a specific problem.

COBOL (COmmon Business Oriented Language) was developed at a later stage as a more alphabetically based language, so that instructions written in this language came nearer to ordinary language. It has therefore come to be one of the most widely used commercial programming languages, enabling people who are not trained programmers to participate in the writing of programs. A typical COBOL statement may simply be 'SUBTRACT TAX FROM GROSS GIVING NET'. Computer languages are often referred to as being 'high level' (near to man) or 'low level' (near to machine).

The basic software of a system comprises the machine language programs, *compilers* and *assemblers* to translate high level languages to machine code, and a program library of commonly used basic programs (*routines* and *subroutines*) that are frequently employed by a particular user. The main *applications programs*, written to instruct the machine to do specific tasks, are sometimes referred to as *middleware*, and are not generally classed as software. The programs, with the exception of the main *supervisory program*, are usually kept in the backing store. The main supervisory program is loaded permanently into the main store, and it controls the other programs, bringing them temporarily from the backing store into the main store when they are needed.

A suite or group of computer programs combine to perform one specific job, for instance a payroll. A software system is a combination of various suites of programs to perform a broader task. For example, it will take several programs to evolve a production control system for a factory, include reports on stock movement, work in progress, purchasing, production line workloads and availability of raw materials. A *systems analyst* will examine the total problem, divide it into sub-groups and assign programmers to each sub-group to write the software.

Hybrid computing There are situations when analog data

has to be fed into a digital computer, and others when digital data has to be fed into an analog machine. This is done by feeding the data through analog to digital converters and through digital to analog converters. An example is in electrocardiogram analysis: heartbeat measurements are taken in analog form, converted to digital form and fed into a digital computer for analysis.

When digital machines are used for controlling production processes they are often dealing with continually changing quantities, temperatures and pressures for example, and this analog data is converted into digital for the machine to process it. On the basis of the data it receives the computer makes decisions, which are then converted back into analog form to control the machinery, pumps and actuator of the manufacturing plant.

The hardware used in mixed analog and digital computing may consist of separate analog and digital computers, connected together through a hybrid interface which does the conversions, or a digital computer with analog units incorporated in the cpu.

Small systems There are many small organizations and companies that need some form of automatic data processing system, but do not have the resources or the workload to justify the installation of a full computing system. In the past they could have used punched card tabulating machines, but these have now been superseded by small office data systems that can handle a wide range of tasks such as invoicing, accounting and stock control. They are often based around a central processor with a magnetic core storage memory and disk, drum or tape backing store. The operator's console is a desk unit with a built-in I/O (input and output) device, commonly a form of electric typewriter or a graphic display unit and keyboard. The number of operator stations can be increased to cope with expanding workloads, and the central processor is often built on a modular system so that it too can be increased in size and computing power. The main programming is often done by the manufacturer, who may market a wide range of standardized program packages (such as payroll or sales ledger packages) in addition to meet the requirements of individual users.

A number of I/O devices are used with these systems, including printers, paper and magnetic tape units, and data collection units that record alphanumeric data—both letters and numbers—onto magnetic tape cassettes. Many of these systems are compatible with larger ones and can be linked to them by landline or used to prepare data for subsequent batch processing by the larger systems.

Above, centre: circuit of a rheostat dimmer switch. A sliding contact moves over a wire wound coil resistance from position 1—the off position, to 2 on zero resistance, through 3 to 4—maximum resistance, minimum current for the dimmed light.

Opposite page, below: many ingenious devices were invented for drilling teeth. Here are two early drills which operated on the Archimedean screw principle, like a 'pump' screwdriver, and a clockwork-driven model.

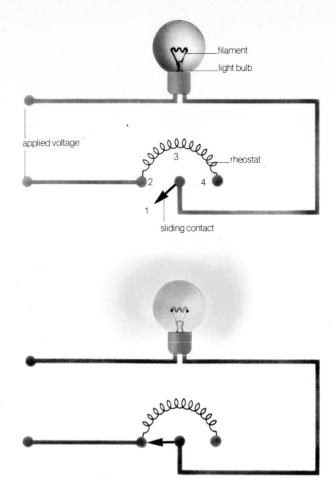

DIMMER SWITCH

A dimmer switch is a device included in an electrical circuit to regulate the amount of current flowing in that circuit. Frequently used in domestic and theatrical lighting, dimmer switches allow the amount of light produced by the circuit to be varied gradually from zero to full. There are three main types of dimming devices: the rheostat, the autotransformer, and the thyristor. (None of these dimming devices can be used in a circuit with fluorescent lights without special circuitry to ensure that the voltage does not fall below the 'striking' voltage between the electrodes in the tube, for if it does, the gas in the tube will not be ionized and no light produced at all.)

Rheostat The earliest type of dimmer to be widely used, especially in the theatre, was the rheostat. This is simply a variable resistance wired into the circuit. A movable contact, usually with a dial face, enables any amount of resistance to be added to the circuit as desired. The higher the resistance the lower the current available to the filament in the light bulb. The nature of resistance means that some of the energy of the current will be dissipated in the form of heat. This means that a rheostat is not only wasteful of current but makes necessary some allowance for the heat, especially since it is likely to be generated in a confined place. For this reason, this type of dimmer is less widely used today.

Autotransformer A type of single coil transformer called an autotransformer is actually a voltage regulating device whose dimming characteristics are not dependent on load. A movable contact on the 'secondary' winding allows fractions of the mains current applied to the 'primary' to be tapped. This type of dimmer is used mainly for sensitive laboratory work.

An autotransformer dimmer switch has the added advantage of being able to produce a voltage on its secondary

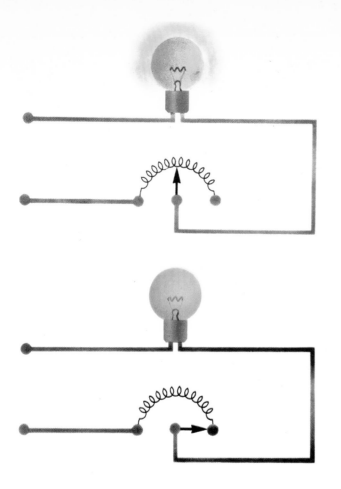

DRILL, dental

The dental drill is an instrument for deep penetration of tooth structure. It is used to remove infected and softened parts of decayed tooth substance and also to shape and prepare the remaining sound tooth to receive the filling or restoration. Rotary instruments are also used to cut, grind and polish in many other dental procedures.

History Originally, cavities in the teeth were prepared with hand instruments such as tiny spoon-shaped excavators, chisels and files. Some of the files were rounded and were used by rotating their pencil-shaped handles between the operator's thumb and fingers while the instrument was steadied in the palm of the hand. T-shaped crutch handles and finger-rings with cups on them were also employed to give the operator greater control of the file. Interchangeable file-tips or burs could be attached to the handles as they became blunted with use. One immediate disadvantage of these straight drills was that the operating tip could not be carried into some of the more inaccessible places at the back of the mouth.

By the middle of the nineteenth century a two-handled Merry's drill-stock was introduced to the dental profession. This drove a revolving bit through a flexible shaft and enabled the drill to be applied at any desired angle, while the operator retained complete control. Even so drilling was still tedious and tired the operator's hands quickly.

The efficiency of cutting was further improved by using mechanical means to rotate the drill more quickly and powerfully yet still allowing it to reach points inaccessible to the straight instrument. Many drill-stocks were invented, for example McDowall's, which used the principle of the Archimedean screw (like a spiral ratchet, or pump screwdriver), and Chevalier's, with a hand-cranked gear system, both of which could produce about 300 revolutions per minute. Clockwork motors were also tried.

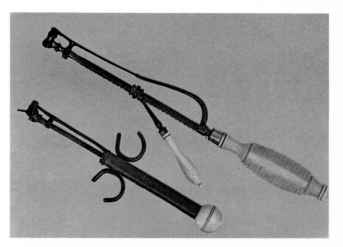

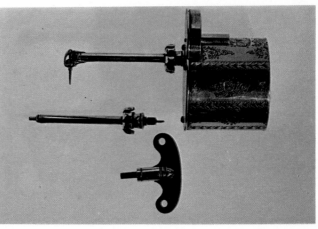

winding that is greater than that on its primary. Like an ordinary transformer, it is very efficient. It produces dimming in a similar way to the rheostat in that it passes a reduced current all the time, but self-induction entails nothing like the energy losses of the resistance in the rheostat.

Thyristor dimmer The development of the thyristor in the early 1950s revolutionized the technology of dimmer switches, among many others. The thyristor is an extension of the transistor, being composed of four layers of crystal instead of the transistor's three.

As well as the common property of semiconductors of allowing current to pass through it in only one direction, the thyristor has the additional property of not allowing any current at all to pass unless a separate 'biasing' voltage is applied between two of its crystal layers.

The dimmer switch that exploits this characteristic consists of a circuit which includes two thyristors. The thyristors are wired to conduct current in opposite directions. In any half cycle of an alternating current, therefore, one thyristor will be passing current (provided the bias voltage is applied) and the other will not. In the other half of the cycle they change roles. The dimmer includes circuitry which draws the small bias voltage (about 5 volts) from the mains current, and includes a variable capacitance and resistance. Thus the point (or 'phase') at which the bias allows current to pass through the thyristor can be varied from the whole to zero. The part of the incoming half cycle applied to the thyristor before the required bias voltage is attained produces no current at all through the thyristor; this means that it is a far more efficient type of dimmer than the others, since it blocks the voltage rather than throwing part of it away. This also means that it can be designed very compactly, and the thyristor dimmer is widely used today, especially in domestic wiring.

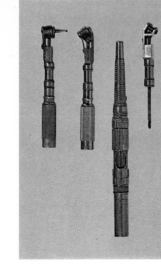

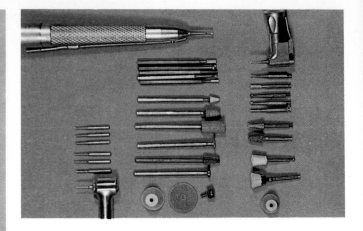

Far right: various modern drill handpieces with a selection of burs and polishing devices. In the top left hand corner is a handpiece with directed water jet cooler attached.
Near right: different heads can be attached to the headpieces, providing a number of drill angles and extensions. These drills are cable and belt driven by an electric motor.

In the 1870s the dental treadle foot drill was introduced where the rotary power was transmitted via a cable arm to a handpiece which held the drill bit or bur and instrument speeds were advanced to 700 rev/min. As electricity became more commonplace over the next 50 years the electric 'engine', first available in the late 1870s, was only gradually accepted into general practice as a safe and reliable innovation and provided operating speeds of up to 1000 rev/min. The speed of the engine was varied by a foot-operated control, and the rotary power of the motor was transmitted to the hand-piece by an endless belt or cord running over pulleys on a three-part extension cord arm. This method can still be seen in use today in modified form. The handpieces were interchangeable and enabled the burs to be set at any desired angle with respect to the handle of the instrument.

A wide range of shapes and sizes of burs were made and these could be sharpened again after use. Other instruments were also developed, including carborundum moulded into round cylindrical shapes mounted upon steel shafts, and discs and brushes. By 1920 some advanced electric dental engines were capable of up to 4000 rev/min, but these were rare and the steel burs then available quickly became blunt at such speeds.

At this stage the dental drill was still inefficient when judged by its ability to remove tooth substance quickly and without effort. Dental drilling was noisy and still time-consuming and tiring. Even with perfect local anaesthesia ensuring that the operation can be carried out painlessly, the vibration and noise of inefficient drilling on the teeth is conducted via the bones of the jaws directly to the hearing apparatus set in the bone of the skull and seemingly amplified. Burs, which are milling devices, remove chips of tooth substance, and bounce and chatter at low speeds. The dental drill, which now gave the dental surgeon the technical capacity to carry out restorations of the teeth far beyond that possible without it, was still the feared and detested symbol of dentistry. People's reactions to dental treatment have always been directly and negatively influenced by the length of time that has to be spent using the dental drill to prepare their teeth.

In 1942 the first diamond instruments were produced. These were precision steel shafts and discs to which diamond abrasive powder was firmly attached. They were found to perform better at speeds in excess of 2000 rev/min, and at 6000 the vibration and noise were reduced and a lighter touch was possible with reduction of tension for both the operator and his patient. Treatment became speedier and more extensive tooth preparations could be considered.

By 1947 much tougher tungsten carbide burs were available and now speeds of up to 12,000 rev/min were used.

Handpieces had to be made with even greater precision and with adequate provision for lubrication of the fast-moving metal-to-metal parts. At the same time the problem of heat production in the tooth was overcome by arranging that a stream of cooling water should play over the surface being cut.

The next step forward was the introduction of the ball-bearing handpiece in 1953 with speeds of 25,000 rev/min. Ducts were now being incorporated into some handpiece heads to convey coolant sprays more accurately to the cut surfaces. Even higher speeds can now be obtained with gear systems included in the drive mechanisms to the handpieces.

The air-turbine drill In 1957, after over a decade of research and development, the first air-driven turbine handpieces became available and they transformed modern operative dentistry by providing a running speed of 300,000 rev/min. In 1962 this was even further improved by the use of air-bearings to 800,000. It was found that operating speeds in excess of 250,000 rev/min were above the level of vibratory perception of the patient, who was only aware of a whistling noise as his teeth were drilled.

The principle of the dental air-turbine is that a foot-controlled flow of compressed air is carried via a flexible hose to the handpiece where it is directed against the blade of a miniature air-turbine, causing it to spin. The air is exhausted partially from the head of the handpiece, returning through a tube back to the control box. An oil mist is introduced into the compressed air to provide constant lubrication of the turbine motor. An air and water bur coolant is also conducted by a concentric tube to be blown accurately upon the bur from jets in the head of the headpiece. The bur, of diamond abrasive or tungsten carbide, is held in the central shaft of the turbine by a friction grip chuck and is easily interchangeable, although each shape of instrument is more versatile and long-lasting at these ultra-high speeds.

The cutting efficiency of the air-turbine is outstanding. A very light pressure is used and the hard tooth substance gently wiped away. If the pressure on the tooth is in excess of 5 to 6 ounces (140 to 170 grammes) the engine slows and stalls. The ultra-speed air-turbine is used for rapid removal of bulky amounts of tooth substance while the relatively slower high-speed drills are used for other purposes such as finishing the cavity surfaces. Rotary instruments are still used at low speeds for slow removal of decayed tooth in circumstances where the operator's sense of touch is needed, and they are also employed for polishing.

Recent developments Air-motors have now been developed where the turbine is larger and has a greater torque, or turning-power, so that burs can be used at slower and more conventional speeds powered by the same compressed-air source as the air-turbine. Miniature lightweight electric motors held in the hand are obtainable and these have great versatility and increased mechanical efficiency because they do not require a drive transmission to the handpiece.

DRILL, electric

The electric drill is a portable drilling machine, powered by its own electric motor carried in the case. Apart from the motor, the essential components are the *chuck*, which holds the drill, and a simple gear train, for gearing the speed of the motor down to a suitable speed for the drill.

The domestic electric drill usually takes the shape of a pistol. The power cord enters the case at the base of the grip, and the on-off switch is the 'trigger'. Often an interlock is provided so that the 'trigger' need not be held down continuously while using the drill. On larger, heavier models there are other handles on the case as well as the grip, so that both hands can safely be used to bring pressure on the work. The other handle may be a simple bar extending from the case at the top, opposite the grip, or may be a stirrup at the back end of the case, opposite the chuck.

The chuck   The chuck is a three-jawed self-centring device which protrudes from the gearbox end of the case and holds the drill. Turning the outer sleeve of the chuck in an anti-clockwise direction opens the jaws; the other direction closes it, final tightening being achieved with the use of a key supplied.

The motor   Housed in the case is an electric motor of the *series* or *universal* type. The advantages of this type of motor are that it is suitable for use with either DC or AC current, and that it produces a high torque at low speeds: as one pushes harder while drilling, thus increasing the load, the speed decreases but the torque increases. The gear train reduces the speed to about 2500 rev/min.

A disadvantage of this type of motor is that it provides interference with nearby radio and TV reception; this can be overcome by fitting various chokes and capacitors to the motor circuitry.

Electrical safety   For safety, electric power tools are either *earthed* or *double insulated*; tools with plastic bodies are double insulated.

Earthed tools are fitted with three-core power cable. The green-yellow conductor is connected to all the metal parts

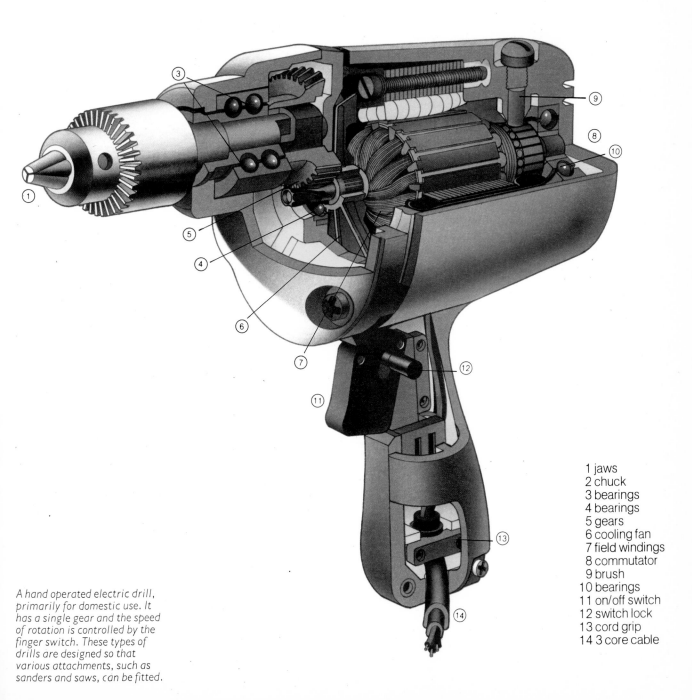

A hand operated electric drill, primarily for domestic use. It has a single gear and the speed of rotation is controlled by the finger switch. These types of drills are designed so that various attachments, such as sanders and saws, can be fitted.

1 jaws
2 chuck
3 bearings
4 bearings
5 gears
6 cooling fan
7 field windings
8 commutator
9 brush
10 bearings
11 on/off switch
12 switch lock
13 cord grip
14 3 core cable

Below: the chuck of the portable electric drill is tightened with the key provided.
Below right: an oil drilling rig showing the kelly, rotary table and drill pipes.

DRILLING RIG, oil

The first producing oil well was drilled in Pennsylvania, in 1859. Since then over two million bore holes have been sunk worldwide. Many of these have failed to find commercial quantities of oil, as opposed to the exceptionally productive few. Drilling is a very expensive business, and a costly gamble too; hence the importance of preliminary geology surveys.

Offshore oil wells are more expensive than those on land, but they are not basically different. Most of the world's oil wells have so far been drilled on land but now, partly because most of the likely land areas have already been explored, drilling at sea is increasingly important. Drilling for oil has been described as analogous to a dentist drilling a tooth with his patient the length of a football field away. This gives an idea of the problem involved in controlling from the surface a drill at the bottom of a well up to 8 km (5 miles) deep.

The drill string and bit Oil drilling is done by rotating a drilling bit to make a hole. The bit may be a fishtailed steel one for soft ground, but it is usually a rotary bit with hardened teeth. In very hard rock, diamond or tungsten carbide teeth are used and it may take an hour to drill 2.5 cm (1 inch). (In softer rock, however, rates of about 100 metres (or yards) per hour are possible.) The bit is fixed to a 'string' of drill pipes which rotate it as it bores the hole. Each length of pipe is normally 9 m (30 ft) long and about 11 cm ($4\frac{1}{2}$ inch) or 14 cm ($5\frac{1}{2}$ inch) in diameter. The pipes are joined by heavy tapered threads. The pipes situated just above the bit are heavier than those in the rest of the string. They are called

which may be touched by the operator and must be fitted at the other end to the earth [ground] pin of a suitable plug. The socket outlet must be earthed to make this protection effective.

Double-insulated tools should be clearly marked as such and are provided with two-core cable. The first (*functional*) insulation is the same as found in any electric tool; the secondary insulation is able to withstand a much higher test voltage than the functional insulation, so that the operator is protected from any breakdown causing leakage of current.

Drills, applications and attachments Drill bits are normally of the twist-drill type, having two helical grooves running from the twin cutting edges for about three-quarters of the length, the remainder being plain shank for inserting into the chuck. The capacity of electric drills ranges from $\frac{1}{4}$ inch (6.35 mm) to $\frac{1}{2}$ inch (12.7 mm). Sizes larger than $\frac{1}{2}$ inch are usually difficult to handle in a portable tool.

An electric drill with a given capacity is designed to handle a drilling job of that size in drilling metal; a drill bit of a larger size, with a cut-down shank to fit the chuck, may be used with caution to drill wood or plastic. Caution is always necessary when drilling metal. Certain types of steel may need a very hard cutting edge, a specially ground angle on the cutting edge, lubrication while drilling, or all three. When drilling a hole all the way through a piece of metal, the pressure brought to bear must be carefully applied as the bit goes through, otherwise the emerging bit may 'grab' the rough edge of the hole, giving the tool a severe wrench which can cause loss of control.

Two-speed and percussion electric drills are also available. The first provides a lower alternative speed of about 900 rev/min; the second gives a percussive effect combined with low speed, and is designed for use on concrete and masonry. Because of the abrasive nature of concrete and masonry, drill bits with specially hardened cutting edges must be used.

The versatility of the electric drill is extended by the availability of attachments designed to make use of the rotary motion, such as rotary files, sanding discs, hole saws, grindstones, and so on. An electric drill is often designed to be fitted, if desired, to a bench or table-mounted machine, to turn it into a drill-press or a lathe attachment for great precision or certain applications, such as use with a jig or fixture which can be attached to the table.

drill collars and are used to put enough weight on the drill to force it into the ground while keeping the rest of the drill string in tension. The whole of the drill string may weigh several hundred tons and if it were allowed to bear on the drill under compression the string could easily break or jam in the hole. In fact most of the weight of the drill string is taken by the drilling equipment on the surface.

Rotary drilling The most obvious part of the equipment on the surface is the derrick, looking rather like an electricity pylon and up to 60 m (200 ft) high. Its height is needed to hoist lengths of drill pipe into place, and to stack lengths of several drill pipes screwed together. The drill string is

rotated in the well through a rotating table at the base of the derrick, driven at about 120 rev/min by a powerful motor. This rotating table has a central hole, through which a length of square or hexagonal pipe known as a *kelly* can slide and by which it can be turned. The kelly is the top section of the drill string and drives the rest of the string as it is turned by the rotary table. The drill string consisting of the kelly, pipes and bit is suspended on a hook from the top of the derrick by cables and pulleys. As the bit cuts into the ground, the kelly slides through the hole in the rotary table. When the bit has descended almost the length of the kelly the drill string is wedged in place, the kelly is disconnected, a new length of

This drilling rig, prospecting for oil and natural gas off the Australian coast, is of the shallow water, fixed type, standing on the bottom. The large flat area on the right is a helicopter landing pad. In front of it is a crane for taking on supplies.

drill pipe is added to the string, the kelly is reconnected, and drilling begins again.

This operation will have to be carried out over 600 times in drilling a 6000 m (20,000 ft) well. Each time it is done a team of men have to carry out hard and exacting physical work in connecting and disconnecting pipes and wedges and taking new pipe out of the stack. Sheer hard work, as well as highly developed operating skill, is still a most essential part of oil drilling. As drilling continues the drill itself becomes blunt, perhaps after only a few hours if it is in hard rock. Then the whole drill string has to be taken out of the hole so that the bit can be removed and a new one put on. This 'round trip' can take up to a day to do. As the drill pipe comes up it is unscrewed in lengths of three, not in single joints, to speed up operations.

During drilling, specially prepared 'mud', a complex colloidal suspension, usually in water, is pumped down the drill pipes through a jet in the bit, and back to the surface in the annular space between the drill pipe and the sides of the hole. This space exists because the diameter of the drill is always larger than the diameter of the drill pipe. The mud circulates through the well quite slowly and cools and lubricates the drill. It also flushes drillings up to the surface, where they are separated from the mud, which is then re-used. In returning to the surface, the mud coats the side of the hole and helps to keep it from caving in. The mud also helps to control any flow of oil or gas from the well. The weight of the column of mud is generally greater than any likely pressure of oil or gas, so that the oil cannot get to the surface until the weight of mud is reduced. In early wells, before mud was used, any oil or gas found under pressure shot at once to the surface, causing a *gusher* which was both difficult to get under control and liable to catch fire.

Another method always used in modern wells to prevent uncontrolled flow is a *blow out preventer*. This is an arrangement of heavy rubber-tipped pistons that can be hydraulically closed to shut off the well entirely. The blow out preventer is firmly fixed to the top of a steel casing that is inserted into the well and cemented in place as the well goes down. Depending on the tendency of the strata to crumble, and the drilling programme, casing may be continued all or only part of the way down the hole..

Right: modern rotary tri-cone jet bits. The bit is rotated under the weight of the drill collar to force the teeth into the rock. This also makes the cones rotate about their own axes and the teeth chip and crush the rock. Beneath the slight projection between each cone it is just possible to see the jet for the drilling mud.
Below right: the first oil well dug by 'Colonel' Drake at Titusville, Pennsylvania, in 1859. The rig was adapted from those used for water, and oil was found at the shallow depth of 69 ft (21 m).
Opposite page: a semi-submersible oil and gas drilling rig, Sedco 135, searching for gas (often found with oil) in the Maui gas field, New Zealand. Gas flaring is a routine part of testing to determine the quality, quantity and pressure of any gas found in a well.

When oil is found the first indication is usually from hydro-carbon analysis of the drilling mud returning to the surface. The oil is tested for quality and flow rate and, if this is satisfactory, production tubing is cemented in and a 'Christmas Tree', so called because of a resemblance in shape of the complex of valves and tubing that make it up, is fixed at the well head.

An alternative to rotary drilling is turbo drilling, where the drill is driven at the bottom of the well by a turbine operated by the drilling mud or (electro drilling) by an electric motor. Rotary drilling, however, is still by far the most useful method.

Offshore drilling This is being done in many parts of the world, but the North Sea is one of the most active areas for exploration, at the present time. It is also the most difficult area so far explored, because of adverse weather conditions, and the distance from the coast of most of the fields. Drilling has been going on for gas and oil in the North Sea since the early 1960s, but this has been in comparatively shallow water. At present intensive oil drilling is being carried out in deeper water, under more difficult circumstances. Fortunately the whole of the North Sea is shallow compared with the oceans; much of it is between 30 m (100 ft) and 200 m (650 ft) deep. This is typical of the so-called continental shelf areas which make up about 10% of the world's under-sea surface.

Types of marine platforms To support the drilling rig, ancillary equipment and crew's quarters, some form of floating platform is needed. The first wells were drilled from converted ships, and these are still in use, but a limiting factor is their tendency to drag even the heaviest anchors during rough weather. Fixed or self-contained platforms are used in shallow water, to a depth of about 30 m (100 ft). Another type of rig is the self-elevating (jack-up) platform, which has an operational limit of 90 m (300 ft), or so, because of bending stress in the leg supports. They can be towed into position and the legs jacked down until they stand on the sea bottom and then further jacked until the platform is well above the sea surface, clear of the heaviest waves. The most recent development in offshore drilling has been the use of *semi-submersible rigs*. These have several large hulls with long legs holding a platform above them, and the hulls are ballasted so as to sink about 20 m (65 ft) below the surface of the water.

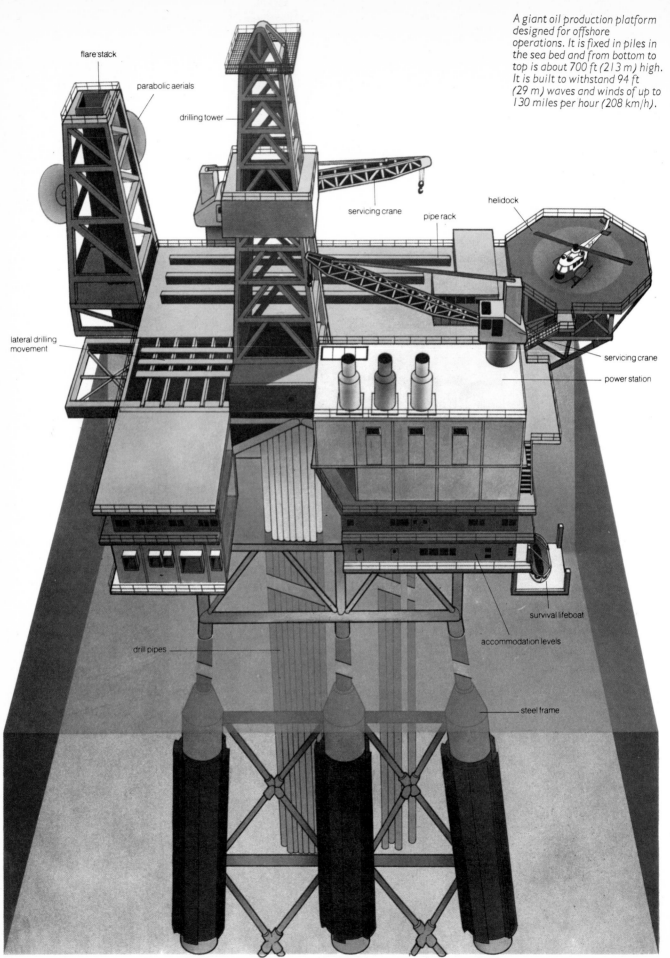

flare stack

parabolic aerials

drilling tower

servicing crane

pipe rack

helidock

A giant oil production platform designed for offshore operations. It is fixed in piles in the sea bed and from bottom to top is about 700 ft (213 m) high. It is built to withstand 94 ft (29 m) waves and winds of up to 130 miles per hour (208 km/h).

lateral drilling movement

servicing crane

power station

survival lifeboat

drill pipes

accommodation levels

steel frame

As with jack-up rigs, the platform is still well above the water and clear of the waves. The rig may be held in place by multiple anchors or it may be dynamically positioned. In this method, multiple propulsion units on the rig respond to signals from a beacon on the sea bottom and keep the rig exactly in position in relation to the beacon, even in the worst weather.

At the hull level a semisubmersible may be about 60 m (200 ft) wide by 76 m (250 ft) long, and its operating draught will be 18 m (60 ft) to 27 m (90 ft). One rig could cost up to £10 million. The biggest semisubmersible in the North Sea at the end of 1973 could drill to a depth of 10,000 m (33,000 ft) in up to 300 m (1000 ft) of water. It could survive in winds of up to 220 km/h (136 mile/h) and in waves of up to 26 m (85 ft). Even larger semisubmersibles are currently being built.

Producing wells After exploration drill rigs have been used to find oil they are moved on to other areas for further exploration. In order to drill producing wells, production platforms are installed. These enormous steel or concrete platforms stand on the sea bottom and, by angling the hole using a technique known as directional drilling, up to 30 producing wells can be drilled from each platform. Oil is treated on the platform to remove gas and water and is brought ashore by pipeline or by tanker. Plans are also being made to drill producing wells and take oil from them without a production platform, working on the sea floor from wellhead cellars serviced by pressure vessels from the surface. A system of this kind has been used to drill and service a well in the Gulf of Mexico in 114 m (375 ft) of water, and a North Sea trial of this system was planned for the 1970s. This undersea technique could show some savings compared to drilling only from platforms, but it will probably supplement rather than replace the conventional platforms.

DRILL, pneumatic

Popularly (and incorrectly) known as a 'pneumatic drill', the pneumatic paving breaker is percussive, that is, it has a hammering action, as opposed to a drill which is either entirely rotary or rotary and percussive in action. As its name implies, the pneumatic breaker is powered by compressed air, which imparts a hammer action to the tool which it holds, enabling it to do a variety of jobs, such as breaking up road surfaces, demolishing buildings, digging trenches, and even hammering down sheet steel piling, using interchangeable tool bits.

The pneumatic paving breaker is a relatively modern invention, replacing other methods of breaking solid surfaces like the pickaxe or the hammer and wedge, which are both time-consuming. The breaker is a T-shaped machine ruggedly constructed in forged steel, in a variety of weights to suit particular job requirements. The main part is a vertical cylinder with handles and throttle control across the top. Compressed air is fed in through a flexible hose from a compressor to the top of the cylinder, below the handle, and is conducted alternately to each end of a sliding piston via a valve. As the compressed air released by the throttle lever first enters the upper cylinder, the piston is forced down on to a cylindrical sliding anvil, beneath which is the working tool (chisel, wedge, asphalt cutter, or spade, for example).

The energy of the piston hitting the anvil forces it on to the ·tool, which is in turn struck down on to the working surface. After striking the anvil, the piston is driven back to the top of the cylinder by compressed air entering the lower end and forcing it upwards. Following both the upward and downward movements of the piston, the compressed air is exhausted to the atmosphere in a downwards direction.

In order to conduct the compressed air to the upper and lower parts of the cylinder, valves and ducts are built into the body, and since minimum wear on the moving anvil and pis-

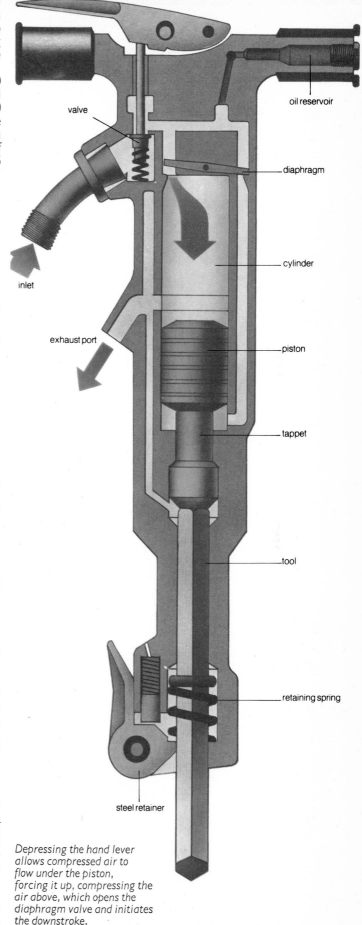

down stroke of pneumatic concrete breaker

Depressing the hand lever allows compressed air to flow under the piston, forcing it up, compressing the air above, which opens the diaphragm valve and initiates the downstroke.

ton is essential, the compressed air is normally lubricated by an oil valve in the throttle mechanism, fed from a reservoir below the handle which holds enough oil for about eight hours' work. Further reduction in wear can be achieved by the design of air 'cushions' at the top and bottom of the cylinder to prevent the piston from hitting the ends of the cylinder during its operation.

Pneumatic paving breakers operate on compressed air at a pressure of about 85 psi (6 bar) and vary in consumption according to size from 40 to 75 cu ft (1.1 to 2.1 m³) of air per minute, producing between 1100 and 1500 blows per minute, again according to size and type.

Noise reduction Vibration and noise are inherent with much pneumatic equipment, not least with pneumatic breakers, but attempts have been made to embody various forms of recoil damping, similar to the telescopic forks of a motorcycle, for the greater comfort of the operators. Many makers now provide integral plastic silencers which considerably reduce the low-frequency exhaust noise which is so irritating to people living or working nearby. The high-frequency 'chatter' of the piston-anvil-tool combination and of the tool hitting the working surface is more difficult to silence, although this problem has been partly solved by fitting a steel-covered rubber collar around the shank of the tool.

DRYCLEANING

Drycleaning, like so many other benefits of modern living, was discovered by accident. The discovery was made by a French dyeworks owner, Jean-Baptiste Jolly, who made the discovery in Paris in 1825, as a result of a simple accident. A maid in the Jolly household upset a paraffin [kerosene] lamp on a tablecloth. Jolly was amazed to discover that the area over which the paraffin had spilled was so clear that it showed up the dirtiness of the rest of the cloth. Operating from his dyeworks, he offered this new discovery as 'dry cleaning' to distinguish it from the soap and water process previously used to clean fabrics.

At the time of the discovery of the drycleaning process, all garments were made from natural fibres such as wool, cotton, and so on which swelled when immersed in water and shrunk on drying. The French public realized the value of drycleaning when they found that garments could be totally immersed in the inert solvent and thoroughly cleaned without distortion through shrinkage. By using the new cleaning process, dirt ingrained over many years was gently floated away. Dry-cleaning spread to other countries where it was at first known as 'French cleaning' because of its origin, and as it developed into an industry the first crude solvent, paraffin [kerosene] was replaced by benzine (an aliphatic petroleum hydrocarbon not to be confused with the aromatic compound, benzene, spelt with an 'e') and later by white spirit, which still survives in many countries as a cleaning solvent. In the USA a controlled quality white spirit called Stoddard solvent is widely used.

Benzine has a flash point of 32°F (0°C) and the fire risk involved reduced its attraction as a cleaning solvent; white spirit in the controlled form as Stoddard solvent has a flash point of 100°F (38°C) and is very much safer. The flash point is described as the temperature at which the solvent gives off a vapour which is flammable or explosive in the air immediately over the solvent.

Solvents used today When the drycleaning moved out of its factory-based environment into local shopping areas a non-flammable solvent was required. The first of these to be established was *trichloroethylene* ($CCl_2:CHCl$), a powerful solvent which is an efficient cleaner. The introduction of clothing made from triacetate rayon, which could be affected by this solvent, caused a general change to *perchloroethylene*

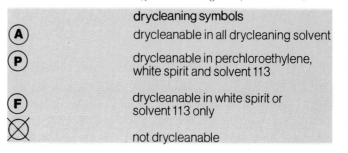

drycleaning symbols

(A) drycleanable in all drycleaning solvent

(P) drycleanable in perchloroethylene, white spirit and solvent 113

(F) drycleanable in white spirit or solvent 113 only

(X) not drycleanable

comparison of drycleaning solvents

property	white spirit	trichloro-ethylene (TRI)	perchloro-ethylene (PERK)	trichlorotri-fluoroethane (solvent 113)	monofluorotri-chloromethane (solvent 11)
boiling temperature distillation	302–392°F 150–200°C vacuum distillation necessary	188.4°F 86.9°C	250°F 121.1°C	117°F 47.57°C	75°F 23.8°C
Kauri-Butanol value (KB) indication of solvency power	31	130	90	30	60
toxicity maximum concentration permissable in parts per million this may vary slightly with local regulations	500	100	100	1000	1000
indication of volatility ease of drying the higher the number, the lower the temperature and time required to dry	difficult	84	39	170	223
flash point temperature at which solvent gives off vapour flammable in air	93°F	non-flam	non-flam	non-flam	non-flam

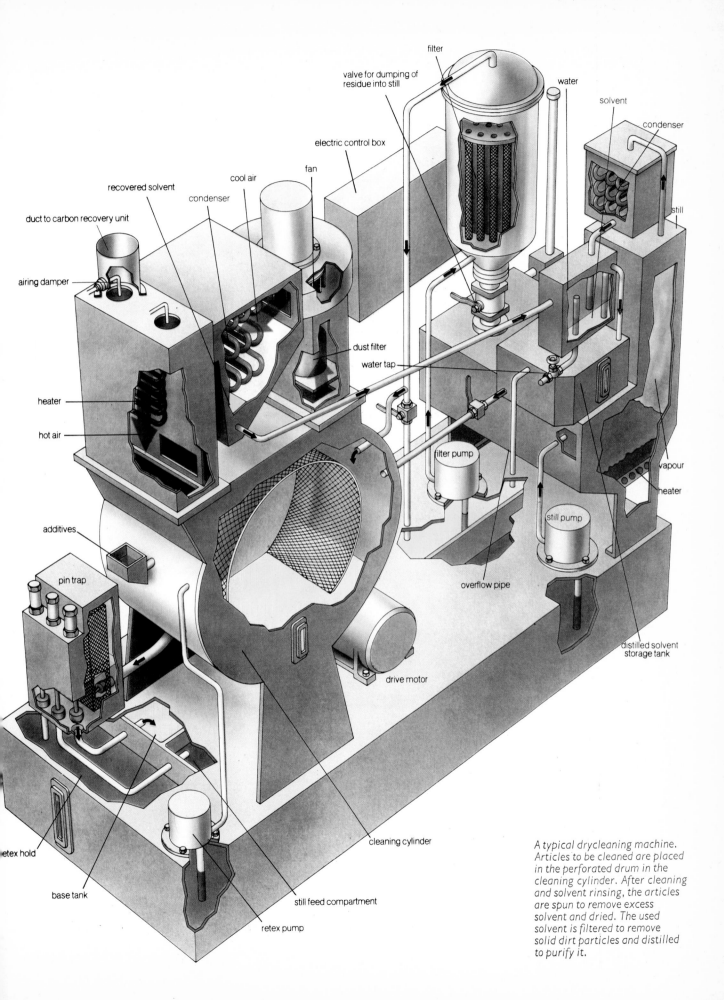

filter

valve for dumping of
residue into still

water

solvent

condenser

electric control box

cool air

fan

still

recovered solvent

condenser

duct to carbon recovery unit

airing damper

dust filter

water tap

heater

hot air

filter pump

vapour

heater

additives

still pump

pin trap

overflow pipe

distilled solvent
storage tank

etex hold

cleaning cylinder

base tank

drive motor

still feed compartment

retex pump

*A typical drycleaning machine.
Articles to be cleaned are placed
in the perforated drum in the
cleaning cylinder. After cleaning
and solvent rinsing, the articles
are spun to remove excess
solvent and dried. The used
solvent is filtered to remove
solid dirt particles and distilled
to purify it.*

(CCl$_2$:CCl$_2$), which is suitable for most garments brought to the drycleaner. Known as 'Perk', it quickly became established and is still the most widely used solvent in automatic drycleaning machines as displayed in 'on the spot' unit cleaners.

Fashion fabrics are subject to continuous change; some of the new fabrics and trimmings on sale are heat sensitive and some are sensitive even to perchloroethylene. This class of fabrics can be handled in a fluorinated solvent in the Freon range (these are used widely as refrigerants), called solvent 113. It will dry at a low safe temperature and being a gentle solvent is suitable for delicate fabrics. There is a further fluorinated hydrocarbon solvent '11' used in some continental European countries, which has a similar low-temperature drying facility as 113 but its solvency power is near to perchloroethylene.

To help the public and drycleaners through the difficulties of identifying fabrics and suitable cleaning treatment, clothes are frequently labelled with advice in the form of symbols.

Drycleaning process In response to the public demand for a conveniently located cleaning service, the machine manufacturers began to develop machines which would complete the whole drycleaning operation as sequential processes (cleaning, drying, aerating, and so on), producing the garments ready for inspection, spotting (stain removal) and finishing.

Perk is a very suitable solvent for use in such automatic machines. The solvency strength of KB 90 is adequate for cleaning without being too severe. The action of the solvent is to dissolve grease, and drycleaning works because most of the soil on the garments is composed of dirt particles associated with oily matter by which the particles become attached to the fabric. The solvent dissolves the grease and thus the dirt particles are loosened. This process is assisted by the agitation of the perforated rotating drum, in which the articles are placed, until all the dirt is removed. Not all the dirt on garments can be removed by the solvent and a small percentage of water-carrying detergents are added to the solvent to remove water-soluble dirt such as food and beverage stains. After cleaning and solvent rinsing, the garments are spun at high speed to extract excess solvent, followed by the drying process in which the clothes are gently tumbled in warm air.

The used solvent is then filtered to remove the solid dirt particles, followed by distillation to remove the soluble contaminants, thus the solvent is continuously purified for re-use. Perk distils easily in a simple vessel similar in operation to a kettle—unlike white spirit which requires distillation in a vacuum. A simple water-cooled condenser transforms the vapour into liquid solvent. Garments (especially woollen ones) hold small amounts of water; this comes off during drying and distillation. The purpose of the water separator is to act as a settling tank and it is constructed to take advantage of the wide difference in specific gravity between the solvent and the water. Because the unwanted water is lighter than the solvent and does not mix with it, it rises to the top and can be drained off before the solvent passes to the storage tank; the solvent from the drying section is similarly passed through the water separator.

After cleaning and drying is complete, the air in the machine has still some solvent content. Before the garments are unloaded, this solvent-laden air is exhausted to a carbon solvent recovery tower, which adsorbs (attracts to the surface of a solid) the solvent vapour in the same manner as a hood over a cooking stove adsorbs smells. Even this adsorbed solvent is recovered.

Additives to provide water repellency, moth proofing, and so on can be included in the process. They are added in small amounts to the solvent as part of a one-stage or, more frequently, two-stage cycle. Here the second solvent wash provides the additive treatment.

DUPLICATING

Duplicating, as opposed to copying, requires the preparation of a master sheet which makes duplicates on a machine. There are two main types of duplicating: *stencil* duplicating and *spirit* or *hectographic* duplicating. Both have been in use for nearly a century.

Stencil duplicating This technique uses a master sheet on to which lettering is impressed as lines of perforations through which ink can be squeezed on to the copy paper. The first experiments with stencils tried different variations on a file plate process, which used a board or a plate with a rough surface on it similar to a file. A blunt stylus was used on a waxed stencil, placed over the file plate, to make a handwritten master. The first major improvement was made in 1882 with the introduction of the Cyclostyle pen by David Gestetner. This was a spiked wheel pen which could be used as for normal handwriting, but made a series of perforations in the stencil. A Japanese paper was imported which had long fibres, providing a firm yet porous base for the stencil. In the 1890s another Japanese paper proved suitable for making stencils on the typewriter, which by then was becoming widely used.

Automatic flatbed machines were patented in the 1890s for making stencil duplicates, and rotary machines were patented in 1896 by Lowe in the USA and Ellams in England. The first successful rotary machines were marketed early in the twentieth century. Gestetner, A B Dick, and Klaber (using the name Rotary Neostyle, later shortened to Roneo) all introduced rotary models, Gestetner using a twin-cylinder design. The materials have changed and the process has become less messy, but the principle has remained the same since then.

The stencil consists of a porous backing sheet and an ink-resisting coating, originally wax. The action of writing or typing on it (using a typewriter without a ribbon) pushed the

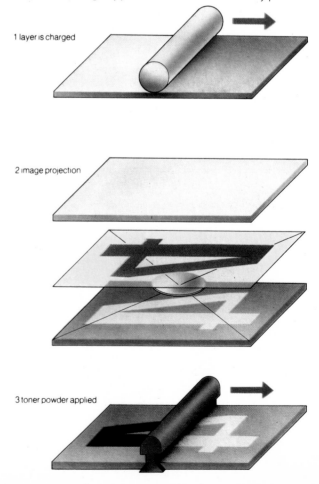

1 layer is charged

2 image projection

3 toner powder applied

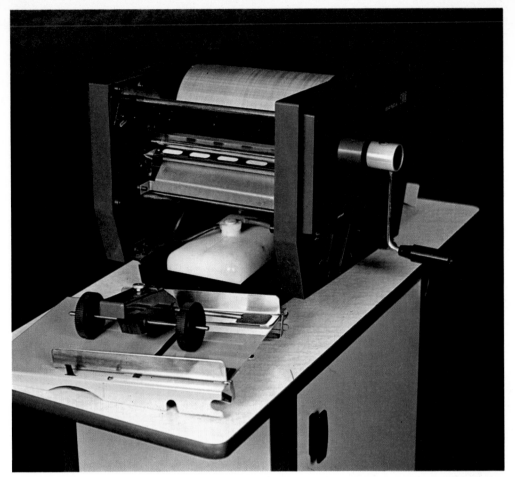

Left: a spirit duplicator, partly dismantled to show the spirit dispenser. The rubber wheels feed paper to the cylinder. Below left: the xerographic process. A semiconducting plate is electrostatically charged. When an image is projected on to the surface only the darkened areas retain the charge and a fine powder dusted on will adhere to these. The 'image' can then be transferred to another piece of paper and fixed.

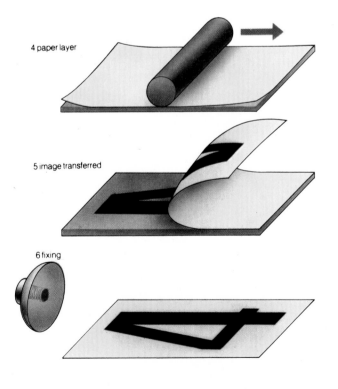

4 paper layer

5 image transferred

6 fixing

wax out of the way, allowing the ink when applied to come through the porous backing on to the copy paper. The fibrous nature of the backing sheet retains the centre of letters like 'o' and 'a' so that their centres do not fill up with ink.

Early stencils required several sheets of different types to make a good typewritten impression. The process was messy and the machine had to be cranked by hand. The stencil could be used for only one short run. Modern stencils use a variety of plastics and can be used several times, although the quality of reproduction is still best on the first run. Modern machines are usually electric. The ink is applied from inside the drum, and some machines feature patented ink systems which can be removed and replaced so that different colours may be run.

Spirit duplicating This process is also called the hectographic method; the term hectographic came from the Greek word for 100 and it was originally claimed that 100 copies could be made. It is also called the ditto process, especially in the USA, and the ink was originally a purple colour. This colour has become strongly identified with the process although nowadays almost any colour can be used.

The method uses a strong aniline dye. Originally the ink was transferred to a sheet of gelatin by placing the sheet of paper with the dye on it in a shallow tray. The moisture retaining qualities of gelatin kept the ink moist, and the copy was made by pressing an ordinary sheet of paper on to the gelatin. A number of copies could be made before the ink became exhausted and a new master had to be made.

The modern process was developed in 1923 by the Ormig company in Germany. The master is in two parts, the lower one like a sheet of carbon paper with the dye on the top side; the dye is transferred to the back of the top sheet when it is typed or written upon. This sheet is then clipped to a revolving drum, and the sheets to be printed are moistened with a volatile fluid which dissolves a thin layer of dye on the master, thus transferring it to the clean paper. The process can continue until the dye on the master is exhausted. The modern process is not messy because the carbon is not softened until in contact with the fluid.

Azograph   This is a process patented by the A B Dick company, and available commercially in the USA. It is similar to spirit duplicating but uses different chemicals. The master includes two colour compounds which become visible when united in contact with a fluid. The compounds are united when the master is typed or written upon, and the fluid is applied in the machine.

Thermographic method   This is a new method introduced commercially at the Business Efficiency Exhibition in 1973. It was developed by Rapid Data Ltd of Britain. The master is typed, forming a heat-absorbing black image on a white sheet. This is attached to a heated drum, and only the image reaches the critical temperature, the rest of the sheet remaining several degrees cooler. The drum makes contact with a continuous belt of waxed material which is highly pigmented. Some of the pigmented wax is melted by the heat on to the image and is then transferred to the copy sheet. The image is fixed and clean before it leaves the machine. The waxed sheets come in several colours and can be changed readily.

Above, top: a more complex spirit duplicator which uses a system of masks to blank out parts of the original master, so that differing versions of a document can be printed.
Above: a stencil duplicator.

DYNAMOMETER

A dynamometer is an instrument for measuring power, force or energy, such as the horsepower developed by an internal combustion engine or an electric motor, or the current, voltage or power in an electric circuit. For car testing, for example, the driving wheels of the vehicle turn large rollers which drive a dynamometer, usually a Froude brake type. There are many different types of dynamometer, both electrical and mechanical.

In order to measure the horsepower produced at the output shaft of an engine or motor, the dynamometer measures the *torque* (turning force) of the shaft, which is the rotating force of the shaft multiplied by its radius. This torque is converted to horsepower by multiplying it by the speed of the shaft in rev/min and dividing this figure by 63,000.

The term 'horsepower' was devised by James Watt in the late eighteenth century as a means of equating the working power of a steam engine with that of a strong workhorse, but it is in fact about 50% more than the average horse can maintain over a period of time. A horsepower is 33,000 ft lb per minute (746 watts in metric terms), the metric horsepower being 45000 m kg per minute (0.9863 hp).

The *brake horsepower* (bhp) is the horsepower measured by an *absorption* dynamometer, and the *shaft horsepower* (shp) is that measured by a *transmission* dynamometer. The 'horsepower' used for car taxation purposes in some countries is a rather meaningless term based on engine dimensions and bears little resemblance to the actual horsepower produced. *Indicated horsepower* is the power actually developed within an engine, calculated from the pressures produced within the cylinders, but as some of this power is used up within the engine to overcome friction and other forces, it is always higher than the power developed at the output shaft. The mechanical efficiency of an engine is its indicated horsepower divided by its output horsepower, expressed as a percentage (typically 90%).

Absorption dynamometers   The absorption dynamometer is so called because it absorbs and dissipates the power it is measuring. The simplest type is the Prony brake invented by Gaspard de Prony (1755-1839), which can handle speeds of up to 200 rev/min and powers of up to about 50 hp. A flywheel drum, driven by the powered shaft, is encircled by a rope or belt attached at one end to a spring balance and with weights attached to the other. The reading on the spring balance with the shaft stationary is the value of the suspended weights, and when the shaft is running at the required speed the friction between the rope and the drum will tend to lift the weights, reducing the reading on the balance. The torque can be calculated by subtracting the second balance reading from the value of the weights and multiplying this by the radius of the drum, and this figure can then be used to calculate the brake horsepower. This type is also known as a rope brake, and other forms of Prony brake apply the friction load by means of wooden blocks, flexible bands and other friction surfaces. The power is dissipated in the form of the heat generated by the friction, and some means of cooling the brakes is often needed.

The Froude or water brake consists of a rotor, driven by the engine or machine being tested, and fitted inside a water-filled casing that is free to rotate a limited distance. The rotor forces the water against the casing, and thus the torque is transmitted through the water to the casing. By measuring the effort needed to prevent the casing from rotating, the output torque and thus the horsepower can be determined. The power is dissipated by the heating and circulation of the water, and this type of brake can operate at up to 10,000 hp and 25,000 rev/min.

Another type of dynamometer, the fan brake, relies on the fact that the amount of torque needed to drive a fan at a given speed is related to the size, number and pitch (cutting angle) of the blades, so these factors can be used to calculate

Left: a racing engine undergoing tests on a Froude brake. This type of dynanometer brake is named after William Froude (1810-1879) who designed one of the first brakes of this kind. He was a civil engineer who turned his attention to the study of applied hydrodynamics, in particular the design of ships. Below: a Prony brake—a simple absorption dynamometer for measuring power output of machines. A rope encircling a flywheel is attached to weights at one end and a spring balance at the other. When the shaft rotates the wheel applies a frictional force to the rope and lifting the weights. The reduction in the tension of the rope is registered on the scale. The torque is calculated by multiplying the difference in registered tension by the wheel radius.

the power of a machine by using it to drive a large fan. The power is dissipated by the displacement of air by the fan, and can be up to 200 hp at 2000 rev/min.

Electrical dynamometers for measuring the power of a rotating shaft are usually in the form of a DC generator or an electromagnetic brake. The generator type is driven by the shaft under test, and the horsepower may be calculated from the power output of the generator, or by using a generator whose stator is free to turn. The strong electromagnetic field created by the rotor produces a reaction torque which tends to turn the stator. The amount of force required to restrain the stator (usually in the form of weights hung on an arm projecting from the stator) is proportional to the power developed by the shaft, which may be as much as 30,000 hp at up to 4000 rev/min.

The electromagnetic brake (also called the eddy current or magnetic drag brake) consists of a toothed metal disc driven by the output shaft, rotating within a stator. The current in the windings of the stator creates eddy currents in the rotor, which produce an electromagnetic force that tends to pull the stator round in the same direction as the rotor, and the shaft torque is calculated from the amount of force needed to prevent the stator from turning. Eddy current brakes can handle speeds of up to 6000 rev/min and powers of about 3000 hp.

Transmission dynamometers The transmission type of dynamometer may be fitted between a power source and its normal load, so that it both transmits and measures the power produced. Where a dynamometer is needed to measure the power input requirements of a machine, it may be in the form of an electric motor which drives the machine in addition to give an indication of the power produced. This can be calculated by measuring the current and voltage needed to drive the motor and converting these measurements into horsepower. Alternatively the force needed to prevent the stator from turning with the rotor can be

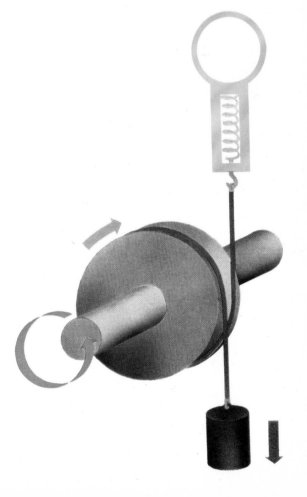

measured, using a bearing-mounted stator in the same way as on a generator type dynamometer.

Dynamometers for measuring the power transmitted by a shaft which is driving its normal load, known as torsion dynamometers, use transducers such as sensing coils or strain gauges bonded to the shaft to measure the amount that it twists under load, which is proportional to the amount of power being transmitted, perhaps as much as 50,000 hp at up to 300 rev/min. Another method is to install a differential gear on the shaft and measure the torque acting on the cage carrying the planetary pinions.

Electrical measurements   The measurement of electrical current, voltage and power is often carried out using dynamometer type (*electrodynamic*) instruments, which work by the interaction of the magnetic field created in two coils, one fixed, with the other mounted on a spindle, and connected to a pointer and a spiral hairspring to control its movement and return it to zero when no current is flowing.

In practice, electrodynamic ammeters and voltmeters are not widely used, as they are more expensive and less sensitive than other types, but the electrodynamic wattmeter is commonly used for AC power measurements.

ELEVATOR (lift)

An elevator is basically a means of vertical transportation between the floors of a building, comprising an enclosed car balanced by a counterweight and moved by a wire rope driven by an electric motor.

The earliest elevators were man-powered, using pulleys and a control rope passing through the car, and this system is still found occasionally in some small hoists. The first power elevators were developed in the mid-19th century, using hydraulic power in the form of an extending ram carrying the car and operated by water pressure. Until the introduction of the telescopic ram, it was necessary to excavate a pit beneath the elevator shaft to accommodate the full length of the ram, which equalled the operating height of the elevator. In industry, where steam was commonly generated to provide power, elevators were frequently driven by belts from the factory's main power shafts.

The first electric elevators were sturdy and simple, and designed for use with direct current (DC) power. The introduction of alternating current (AC) supplies increased design problems, and until the mid-1920s electric elevators used high speed motors which turned the main drive wheel through a worm reduction gear.

The first practical variable voltage control system comprised a DC hoist motor supplied with power by a motor generator running off the mains AC supply. Today geared motors provide the power for the majority of elevators running at speeds of up to 400 to 500 feet (122 to 152 m) per minute. At higher speeds the slow speed gearless motor has many advantages in both speed of travel and in running costs.

The limiting factor in the design of installations is the weight of the motors. In Britain's tallest building, the Post Office Tower in London, the two motors weigh eight tons each, and even larger and more powerful motors have been built for the new National Westminster Bank headquarters in the City of London. These elevators will travel around 600 feet (183 m) at 1400 ft/min (427 m/min).

A picture from the 'Illustrated London News', 1889, of the car on the Eiffel Tower in Paris.

Controls   Although the constant introduction of new materials such as laminated plastics and stainless steel have changed the appearance of elevator cars, the basic travel system has changed little during the last fifty years. By contrast, the control system has changed out of all recognition since the early days, when a pull on a rope actuated a pressure valve or moved a sliding bar across the control panel contacts.

Push button and touch button controls in the car and at landing stations, automatic acceleration and deceleration and the demand for greater travel speeds has led to today's unified control systems which provide the fastest possible service from a minimum number of cars.

One of the objects of modern design is to provide a minimum average travel time. The simplest system provides service when any landing call button is depressed. The first call registered is the first one answered; on entering the car the passenger presses a button to register the destination floor, and the car is then despatched to that floor. Other calls are registered in simple 'first come, first served' sequence.

Automatic control   The simplest automatic control system requires only one button on each floor, no matter how many cars there may be in the service bank. The approach of a car is signalled by an electrically lit up or down arrow above the appropriate landing station.

All landing calls are divided into sectors, each comprising a number of adjacent floors. The number of sectors equals the number of cars and a car becomes available to answer a demand for sector service when the doors have closed and there are no assigned calls to be cleared.

As the car becomes available it is allocated to the despatching system, in which the nearest car to a priority sector is allocated to that sector. The nearest car is chosen by an electronic assembly which constantly compares the location of all the cars in the bank with priority demand.

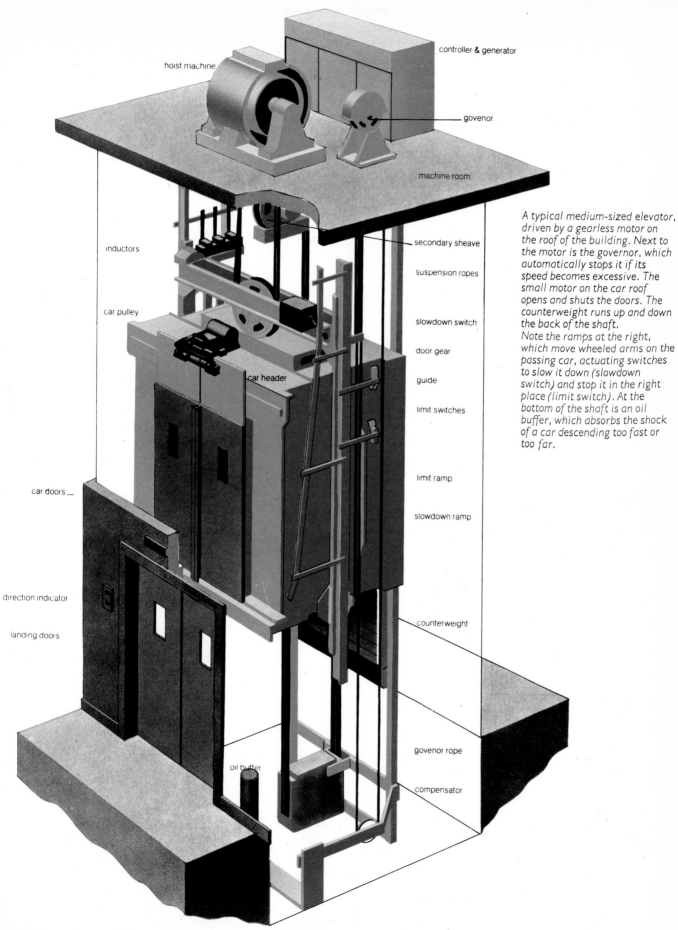

controller & generator

hoist machine

govenor

machine room

secondary sheave

inductors

suspension ropes

car pulley

slowdown switch

door gear

car header

guide

limit switches

car doors

limit ramp

slowdown ramp

direction indicator

landing doors

counterweight

govenor rope

oil buffer

compensator

A typical medium-sized elevator, driven by a gearless motor on the roof of the building. Next to the motor is the governor, which automatically stops it if its speed becomes excessive. The small motor on the car roof opens and shuts the doors. The counterweight runs up and down the back of the shaft.
Note the ramps at the right, which move wheeled arms on the passing car, actuating switches to slow it down (slowdown switch) and stop it in the right place (limit switch). At the bottom of the shaft is an oil buffer, which absorbs the shock of a car descending too fast or too far.

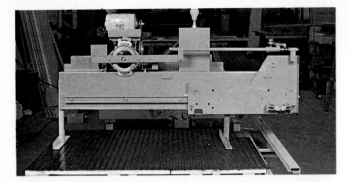

Above: three pictures of an elevator door-opening mechanism showing, from top to bottom, the fully open, half closed, and fully closed positions respectively. This assembly is mounted on top of the elevator car, and the door is hung from the sliding part of the mechanism. The mechanism is belt-driven by the electric motor at the top of the assembly, whose operation is controlled by a system of safety interlocks and the command buttons in the car and at the landing stations. The interlocks reopen the doors if they cannot be closed fully because of an obstruction, prevent the doors opening while the car is moving, and prevent it from moving unless the doors are fully closed.

In many systems, time is saved by initiating the despatch sequence immediately passengers have stopped entering or leaving the car, a situation detected by photoelectric cells fitted into the door edge.

Most control systems are electromechanical, but an increasing number of installations are now controlled by solid state systems, which use a complex arrangement of logic circuits to control car movements.

Counterweights The cars of passenger elevators are usually counterbalanced by a heavy counterweight equal to the weight of the empty car plus about 40% of its maximum load. The effect of this counterweight is to reduce the amount of power needed to raise the car, and to provide a certain amount of deceleration to help control the speed of the descending car.

Safety Under normal circumstances car speed is controlled by governor switches acting on the motor and brake circuits. In the event of the speed of a descending car exceeding a predetermined limit. powerful braking arms (activated by a cable connected to a governor unit on the winding machine) are brought into contact with the guide rails to stop the car smoothly and safely. Hydraulic or spring operated buffers are situated at the bottom of all elevator shafts so that any falling car or counterweight will be stopped safely at governor tripping speed.

In many modern installations there are devices to sense the weight of the loaded car, and when the car is fully loaded it will bypass all landing calls and service only those calls registered within the car. If the weighing mechanism detects an overload, the starting circuits will not function and the car will remain at rest.

Door interlocks are fitted so that the car will not move until all the doors are fully closed, to prevent the landing doors opening unless a car is present, and to ensure that the car doors remain closed until the car has stopped at a landing. Additional devices protect passengers by reopening the doors if they begin to close as a passenger is entering or leaving the car. Many cars are provided with a trapdoor at the top so that passengers can escape if the elevator gets stuck between floors.

belts turned through 180°

step

bottom return wheel

ESCALATOR

The modern escalator provides the most efficient method of moving large numbers of people from one level to another at a controlled and even rate. In its simplest form it consists of a series of individual steps mounted between two endless chains which move upwards or downwards within a rigid steel truss frame.

The pioneers of the escalator were Jesse Reno and Charles Seeberger, whose designs were produced independently in the early 1890s, and in 1900 escalators were installed in Paris and New York. The first modern type of escalator, however, did not appear until 1921, incorporating the best features of both the Reno and Seeberger designs, and today escalators are commonplace in railway stations, airports, department stores and anywhere that a speedy service is required.

Most standard escalators are factory prefabricated and

A section through an escalator, which consists of a continuously moving series of individual steps on an endless chain. Each step moves on small wheels along rails. These are positioned so that the steps fold to give a horizontal surface at the top and bottom.

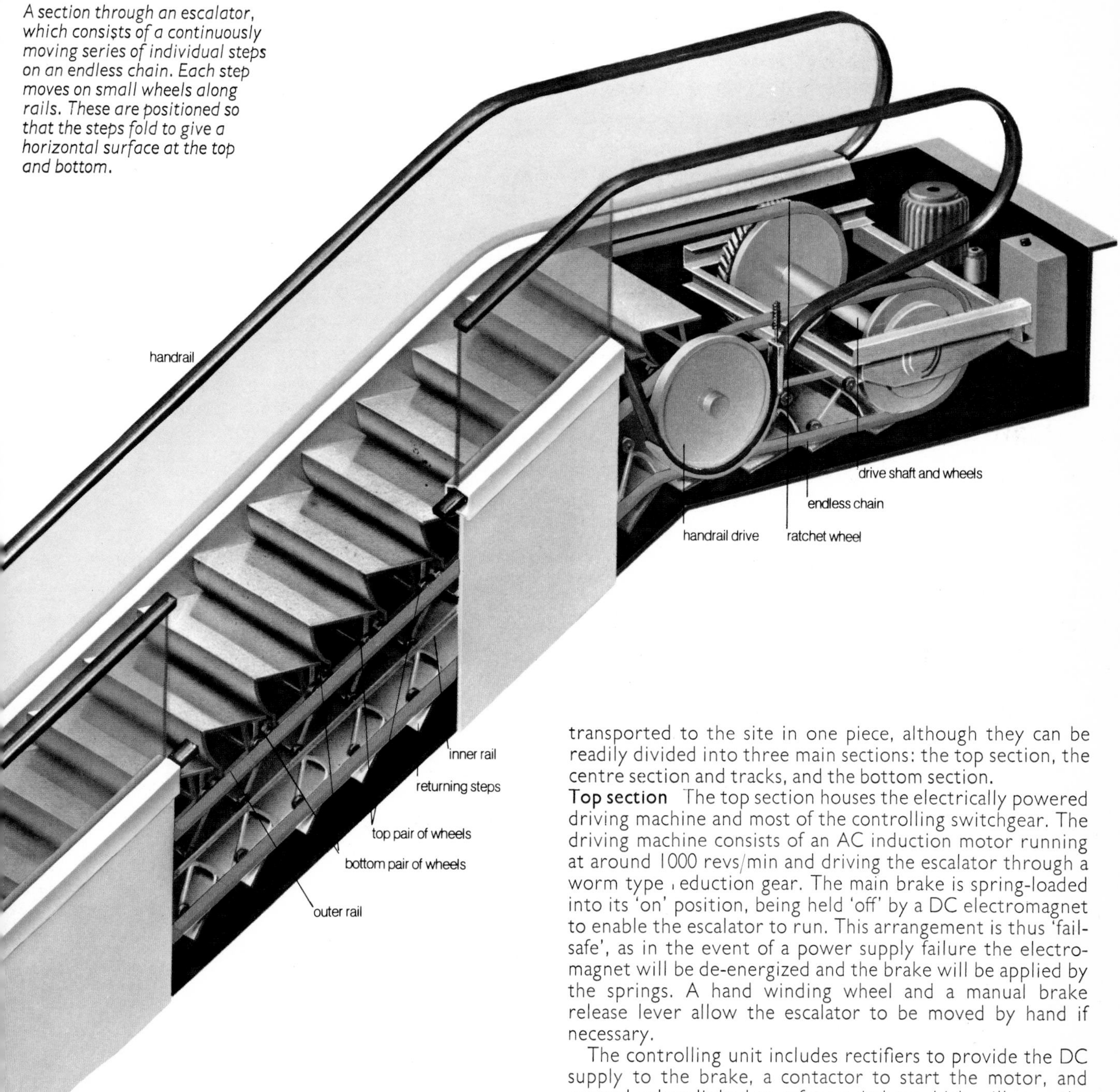

handrail

drive shaft and wheels

endless chain

handrail drive   ratchet wheel

inner rail

returning steps

top pair of wheels

bottom pair of wheels

outer rail

transported to the site in one piece, although they can be readily divided into three main sections: the top section, the centre section and tracks, and the bottom section.

Top section The top section houses the electrically powered driving machine and most of the controlling switchgear. The driving machine consists of an AC induction motor running at around 1000 revs/min and driving the escalator through a worm type reduction gear. The main brake is spring-loaded into its 'on' position, being held 'off' by a DC electromagnet to enable the escalator to run. This arrangement is thus 'fail-safe', as in the event of a power supply failure the electro-magnet will be de-energized and the brake will be applied by the springs. A hand winding wheel and a manual brake release lever allow the escalator to be moved by hand if necessary.

The controlling unit includes rectifiers to provide the DC supply to the brake, a contactor to start the motor, and control relays linked to safety switches which will stop the machine in the event of an overload, drive chain breakage, or an obstruction to the steps or handrail. The controller is also linked to the key-operated starting switches and the emergency stop buttons, and contains a device to prevent an up-travelling escalator from reversing its direction of travel in the event of drive mechanism failure. To reduce wear and running costs some escalators are fitted with speed control devices that run the machine at half speed when no passengers are using it. Photoelectric sensors are fitted at each end to switch the escalator to full speed when a passenger steps on and to return it to half speed.

Centre and bottom sections The bottom section carries the step return idler sprockets (toothed wheels around which the chain turns), step chain safety switches and curved track sections to carry the motion of the escalator from the horizontal plane into the angle of climb, which is usually 30 to 35°. Between the top and bottom sections runs a welded box-type structure which carries the straight track sections.

The steps on which the passengers stand are assembled from aluminium pressure die castings and steel pressings mounted on a frame, usually of cast aluminium, which runs on rollers on the main track sections and is driven by the two main chains, one on each side. Foothold on the step surface is provided by a cleated board faced with aluminium or rubber.

The moving handrails are made from layers of canvas covered with a rubber or plastic moulding. They run in continuous loops in T-shaped guides along the tops of the balustrades, at a speed closely linked with that of the steps. The balustrades and their skirtings are designed to allow a smooth passage for the steps, and all joints are securely masked. The running clearance of the steps has to comply with strict safety standards and the *combplates* (the metal teeth which project at the top and bottom of the fixed escalator base and provide the link between the moving treads and floor level) incorporate safety switches that will stop the escalator if an object becomes caught between the steps and the combplate.

The tread width may be from two to four feet (0.6 to 1.2 m), and the speed of the operation from 90 to 180 ft/min (27 to 54 m/min). Running at 145 ft/min (44 m/min) a single escalator can carry up to 10,000 passengers an hour, and will be powered by a 100 hp motor. A variation of the escalator, the travellator, is a horizontal continuous rubber belt used for moving passengers around such places as airport terminals and railway stations.

Below: an early escalator, based on the design patented by Reno in 1891. As there were no actual steps on this type of escalator, the angle of incline had to be kept relatively small to make it easier to use.

EXPLOSIVES

The first pyrotechnic material to be used in war was Greek fire, invented by Kallenikos in 673 AD. Although not strictly an explosive, it was used with devastating success in several battles in the Mediterranean, and was a primitive but highly effective incendiary material probably based on petroleum oils and sulphur. Its secret was lost with the fall of the Byzantine Empire in 1453 AD.

The origins of gunpowder (or black powder) are obscure, but the Chinese were probably aware of the properties of saltpetre (potassium nitrate) as early as the Chin Dynasty (221 to 207 BC), although they did not develop its use beyond the firework stage until about the 12th century. In the west this material was known as 'Chinese snow', and the first recorded Western experimenter to establish the formula for gunpowder was an English monk, Roger Bacon. In 1245 AD he recorded it as a Latin anagram in his *De Secretis Operibus Artis et Naturae* (Secret Works of Art and Nature).

Its first use as a gun propellant followed in about 1320, and English troops used cannon against the French at Crecy in 1346. The formula has changed somewhat since then, and today the proportions used are 75% saltpetre (potassium nitrate), 15% charcoal (carbon) and 10% sulphur. It is not now used as a gun propellant as it burns too quickly, about 400 metres (1312 ft) per second, its residue fouls the bore of the gun, and it produces too much smoke. On the other hand its combustion is too slow to produce the shattering effect of a high explosive. It is, however, widely used in fireworks and in blank cartridges, and it was used by the Russians in the retro-rockets of the planetary surface probe sent to Mars.

Explosion and detonation When an explosive substance is set off it undergoes rapid decomposition and releases large quantities of gas and heat. *Explosion* is a fast combustion, the burning spreading layer by layer through the material at the comparatively slow velocity of up to 400 metres (1312 ft) per second, and although its rate increases with increasing pressure, it can be controlled. This reaction is often called *deflagration*.

In a *detonation* reaction there is an extremely rapid burning which produces a supersonic shock, or detonating

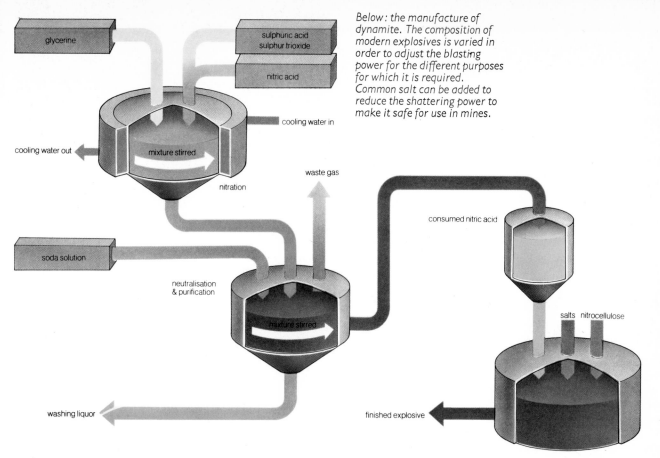

Below: the manufacture of dynamite. The composition of modern explosives is varied in order to adjust the blasting power for the different purposes for which it is required. Common salt can be added to reduce the shattering power to make it safe for use in mines.

wave, in the explosive substance. The detonation velocity is a characteristic of the explosive material itself and is unchanged by changes in pressure. It is usually between 2000 and 9000 metres (6500 and 29,500) feet per second. The detonation wave produces a very high pressure, about 650 tons per square inch (100,000 bar), and this exerts a severe shattering effect on anything in its path. The gases formed travel in the same direction as the detonation wave, so a low pressure region is created behind it. Once detonation has been started it cannot be stopped.

Explosives which react by deflagrating are called 'low explosives' or 'propellants'; they generate gases at a slow enough rate for them to be used to propel rockets or shells. Those which react by detonation have the ability to shatter and are called 'high explosives'. They are used in bombs and explosive shells, and in blasting operations like quarrying and mining.

Classification Explosives can be classified according to their properties and uses. In Britain, for example, there are seven classes: class 1, gunpowder; class 2, nitrate mixtures; class 3, nitro-compounds; class 4, chlorate mixtures; class 5, fulminates; class 6, ammunition; and class 7, fireworks. There are other important forms of classification to define transport, storage and firefighting hazards.

The essential properties of explosives are the velocity of burning or detonation, the explosion temperature, the sensitivity, and the power. For the first three classes above, an absolute measurement is possible, but for the others it is usual to compare the explosive with a standard such as *picric acid*. Picric acid is taken to have a 'value' of 100, with more sensitive or less powerful explosives having values lower than 100. The more sensitive explosives with values of around 20 are used as detonators to initiate explosives. Their impulses can set off the intermediary charges (those of moderate sensitivity, about 60) which in turn will initiate the reactions in the main charges, which are the least sensitive.

Modern explosives The use of gunpowder as an explosive declined in the 19th century, and it was replaced by three main types of composition: those based on unstable molecules, such as the fulminates and azides; ammonium nitrate and the organic esters nitrocellulose, nitroglycerine and PETN; and the nitro-compounds, a large group which includes picric acid, TNT, tetryl and RDX.

In 1845 Schönbein nitrated cotton with a mixture of nitric and sulphuric acids to give nitrocellulose. Until 1875, when Sir Frederick Abel devised a method of pulping the cotton to give a more stable product, there were many accidental explosions associated with its manufacture. Its properties depend on the degree of nitration, which is hindered by the fibrous nature of the cotton, and it is usually a mixture of the di- and tri-nitrates. It is easily gelled by solvents and can be then pressed into the required shape, for example cord, flake or tube. It is very sensitive when dry (with a rating of about 23), but less so when wet (about 120) yet can still be detonated. The velocities of detonation are 7300 (dry) and 5500 (wet) metres/second (23,950 and 18,000 feet/second respectively).

The highly dangerous explosive liquid *nitroglycerine* (sensitivity 13) was first prepared in 1846 by Sobrero in Turin by nitrating glycerol with a mixture of nitric and sulphuric acids. In 1865 Alfred Nobel found that the liquid could be used safely if it was first absorbed into kieselguhr, a form of *diatomaceous* earth (formed by the fossil remains of a type of single-celled *algae*). He also succeeded in solidifying it by adding 8% nitrocellulose to form a gel. He named these famous explosives dynamite and blasting gelatine. Nitroglycerine detonates at 7750 metres (25,426 feet) per second, has a power rating of 160, and an explosion temperature of 4427 °C (8000 °F).

PETN (*pentaerythritol tetranitrate*) is a sensitive explosive (40) with a high power rating (166) and a detonation velocity of 8100 m (26,500 ft) per second. It can be used as an inter-

Right: a composite metal plate, made of a sheet of carbon steel laminated with a cladding sheet of aluminium bronze. The sheets are placed a little way apart, and explosive is spread across the cladding and detonated to bond the two together.
Below: a series of four photographs showing the development of the detonation wave in a block of solid explosive, in this case a mixture of RDX and TNT. The sequence shows frames 2, 3, 4 and 7 of a high speed sequence that was photographed with a time interval of two micro-seconds (millionths of a second) between each of the frames.

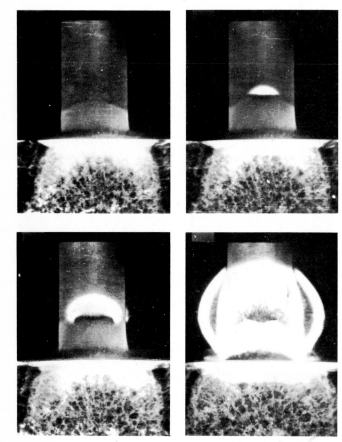

mediary charge but it is extensively used in detonating cords where its small critical diameter enables it to maintain a detonating impulse over a great distance, despite low filling densities.

Picric acid was first prepared by Woulfe in 1771, a hundred years before Sprengel demonstrated that it could be detonated by a mercury fulminate cap. Prepared by the nitration of phenol, it was the first of a large series of aromatic nitro-compounds to be discovered. As it is much less sensitive than nitrocellulose or nitroglycerine, but just as powerful, it safely withstands the shock of discharge from a gun, and in 1888 it replaced gunpowder as a shell filling. Apart from its value as a standard, picric acid is little used today.

TNT, *trinitrotoluene*, is made by reacting toluene with a nitrating mixture of nitric and sulphuric acids. Its sensitivity is 110, power rating 95, and its velocity of detonation 7000 m/sec (22,965 ft/sec). It is cheap and easy to make and widely used to fill shells and bombs, often being mixed with ammonium nitrate for such applications.

Tetryl is formed when dimethylaniline is nitrated. It requires careful extraction and preparation for use as an explosive, as it is powdery and toxic. It has a detonation velocity of 7300 m/sec (23,950 ft/sec), a power value of 120, and a sensitivity of 70, which makes it a good intermediary charge to transfer a detonating shock wave from a detonator to a main charge.

RDX, also called *Hexogen* or *Cyclonite*, was discovered by the German Henning in 1899. It is the product of the nitration of *hexamethylenetetramine*, and is a very powerful explosive (167) with a high velocity of detonation, 8400 m/sec (27,560 ft/sec), and a moderate sensitivity of 55. It has many uses in the military and civil fields.

Applications Inevitably explosives find their principal applications in war, and modern propellants stem from developments made in the 1880s. In 1884 the French engineer Vieille gelled nitrocellulose with an ether-alcohol solvent, then cut the sheets into flakes, and named it 'Poudre B' after General Boulonger. Nobel made *Ballistite* in 1888 by gelling nitrocellulose with nitroglycerine, and the English *Cordite* of 1889 was produced by Nobel's methods. As gun propellants these substances burn quite slowly, yet if the constituents are suitably ignited they detonate at many thousands of metres per second.

Shaping the explosive into a hollow cone and lining it with metal produces a 'hollow charge', which focuses the explosive effect into a jet powerful enough to penetrate substantial thicknesses of steel. The impulse produced on detonating an explosive can easily be used to bond metals together, or to press them into the shape of a former laid below. This technique is used to make such things as dental plates and radomes (the protective domes often fitted over radar antennas).

Explosives are widely used in mining; in the UK the annual use is about 50,000 tons for coal mining, each pound

(0.45 kg) of explosive winning about four tons of coal. Canal excavation, harbour deepening and demolition are other fields in which large quantities of explosives are used. In seismic prospecting, for oil for example, shock waves generated by the detonation of a charge travel into the strata below and are reflected or refracted by the geological features in the area. By detecting these reflected waves a picture of the underlying geological formation of the site can be built up.

An unusual application is that of quenching oil and gas well fires. The largest ever, at Gassi Touil in the Sahara, which had burned for five months, was put out by 'Red' Adair of Texas using 550 lb (249 kg) of dynamite, for a fee of $1,000,000.

EXPOSURE METER

When taking a photograph, the photographer has to make sure that the correct amount of light passes through the lens aperture and shutter of the camera on to the light-sensitive film. If there is too little, the picture will be too dark, or *underexposed*; if there is too much, it will be too light or *overexposed*.

To accurately record his subject on to film, the photographer therefore uses an exposure meter, an instrument which measures the intensity of the light and either indicates or automatically produces the correct exposure for the picture to be taken.

In the early days of photography, this 'correct exposure' was found by trial and error, or by the photographer's own experience. Printed charts and tables were available which gave a rough guide to exposure in average lighting conditions (like the tables packed with modern films today).

A more accurate measuring system came with the invention of the *actinometer*. A light sensitive paper was placed in the meter and put in front of the subject. By comparing the tone of discoloration caused by the light falling on the paper with a separate, pre-printed tint card, timing the process, and referring to tables, an exposure was found for the subject itself.

This rather time consuming method was superseded by the *extinction meter*, basically a tube with a dark interior and an aperture at each end. A transparent strip graded in blackness, and overprinted with black numerals, was placed between the ends of the meter. The subject was viewed through the tube and a numeral found that, in relation to the illumination level, merged with its background tone. This number was transferred by calculation to give an approximate camera setting.

The photoelectric exposure meter was introduced in America in 1932, and is a far more advanced and accurate instrument. It measures the light level falling on a light sensitive surface in electrical terms. In separate (off camera) meters, the photocell activates a needle over a scale of *exposure values*, which can be read out as a range of shutter and aperture settings. The photographer then adjusts the camera accordingly. When the meter is built into the camera itself, the needle can be designed to appear within the viewfinder. In this case in semi-automatic cameras, the cell is linked electrically or mechanically to either the aperture or shutter, leaving the operator one adjustment to complete the setting. On fully automated cameras, it is linked to both.

Types of cell In the *selenium* cell, the light falls on a photocell which has two electrically conductive layers, sandwiching a layer of selenium. The light creates an electrical potential between the two layers in proportion to its intensity and, this current is registered on a sensitive ammeter by a needle. The *cadmium sulphide* (CdS) cell is more sensitive to light than the selenium type, and can be used in duller lighting conditions. This cell is not electrically self generating, however; the electrical resistance of the cell varies with the

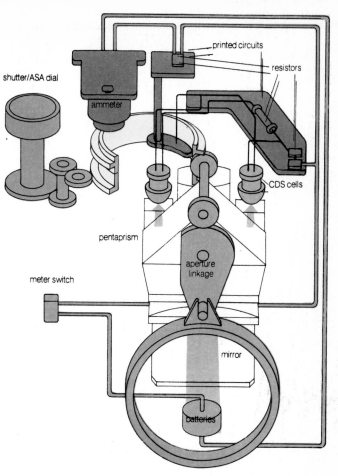

The built-in exposure meter system of the Nikon F2 uses two CdS cells arranged to weight the reading to the centre of the lens' field of view. The meter reading, visible in the viewfinder, varies with the brightness of the scene. The meter is fed a reading which is adjusted electrically to take lens, shutter and film speed settings into account. The aperture ring is linked by gears to a ring of carbon resistance material. The combined shutter and film (ASA) speed is linked to another ring carrying a 'brush' which also tracks along the resistance ring. The current in the cell circuit is passed through the resistance between this brush and a fixed one on top of the assembly. The amount of carbon track between the two brushes thus depends on all three variable settings. Resistors and printed circuits are used to trim the battery output to give constant results and to balance the CdS cells.

intensity of the light, and a battery (usually a long-life mercury cell) is needed to power the circuit.

Using a meter An exposure meter is not infallible, and unless used correctly, will give inaccurate results. The most common method of using the meter is to gauge a general or overall level of brightness. This will take into account the darker and lighter parts of the subject, and give an average reference for exposure. Having been set for the *speed* (sensitivity) of the type of film in use, the meter is pointed towards the subject and an average reflected light value is read off.

Another method is to use an *invercone*, a translucent attachment which is placed over the photocell. This reduces the reading somewhat but instead of measuring the reflected light; the *incident* light falling on the subject is used. This method overcomes problems caused by one part of the picture being much brighter or darker than the subject, thus giving a false reflected light reading.

The exposure value (EV) is a combination of shutter speed and aperture, and its choice depends on the depth of field and movement of a particular subject. For example the camera may be marked in apertures (*f* numbers) between *f*3.5 and *f*22, and have shutter speeds of between one second and one thousandth of a second. The meter will indicate a number of these combinations, all theoretically correct (*f*8 at 1/500 second being equivalent to *f*11 at 1/250, since although the speed has been halved, the aperture has been made small accordingly).

Meters read a particular area of the subject in front of them. This is the *acceptance angle*, and will differ between models. More advanced meters may read a general area, and also have the facility to take a spot reading.

An exposure meter is used to check reflected light by pointing it at the subject, or incident light by holding it near the subject but facing the other way.

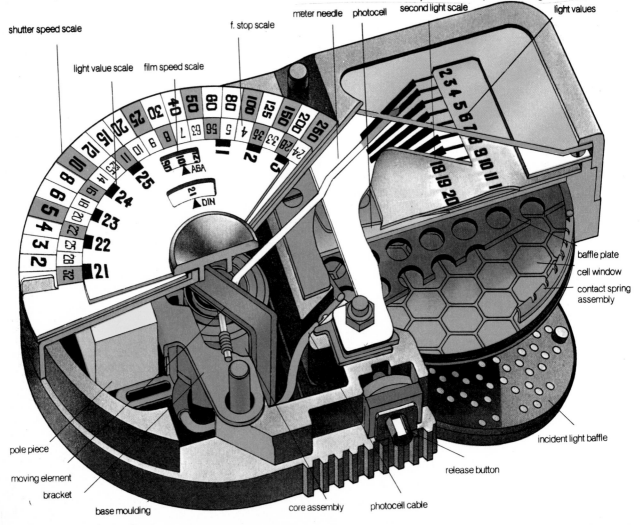

Left: this is Sir Hiram Maxim's Flying Machine, built in 1903 and still in daily use at Blackpool Pleasure Beach, England. Two DC motors are linked by rope drives to 12 foot diameter wheels, one of which can be seen on the right of the picture. These in turn drive the bevel gears which turn the crown wheel attached to the central column. On this column can be seen some of the slip rings which transfer electric current to the ride for lighting. Below: the moving chain drive which pulls cars up the slope of a gravity ride.

FAIRGROUND MACHINES

Fairs were originally annual gatherings of people living in scattered rural communities, offering an opportunity for them to meet, barter and buy goods that would not normally reach them. These events became the occasions for various entertainments, among which would have been small roundabouts or merry-go-rounds in which the children sat on donkeys moving in a circle. With the coming of steam power, these were replaced by faster and larger mechanical devices, called carousels or merry-go-rounds.

The first carousels were operated by steam traction engines, which used steam to drive a large flywheel. This was connected by a belt to the centre of the merry-go-round, while some of the steam was used to play a steam organ.

Instead of real donkeys, the carousel has model horses on poles extending from the base platform to the overhead canopy. These are attached at the top to a crank operated by a bevel gear from the central crown wheel so that as the whole thing rotates the 'animals' move up and down. The movements are arranged to balance each other such that at any one time as many horses are moving upwards as downwards, so as not to put an irregular load on the drive mechanism.

As electricity came into use, the traction engine was made to drive DC generators, first for lighting the ride and then for driving it directly, so that the generator could be kept well away from the ride. Eventually diesel engines replaced the traction engine completely, and are the rule in modern travelling fairgrounds. They drive DC generators, commonly of 110 volt output, since direct current electric motors can be controlled over a large speed range by using a simple rheostat or resistance coil. These have to handle a large load when the motor is running at low speed, so in their simplest form consist of several feet of wound copper bands with a sliding contact.

A typical fairground ride may need 15 to 25 hp (11 to 19 kW) for the rotary motion. Because of the heavy load placed on the generator when the ride starts, it is usual to run the lighting and sound systems off another generator so that the lights do not dim and the music does not slow down or fade away.

Types of ride Travelling fairground rides usually provide simple variations on the merry-go-round principle. The 'caterpillar', for example, has cars running round a circular undulating track, moved by arms extending from a central shaft. Around this shaft is a sleeve with cables attached to it: these cables connect to an awning which can be pulled over the cars completely. When the ride is in motion, the sleeve is stopped relative to the central pivot so that the cables wind up on it, pulling the awning over the cars to the amusement of their occupants.

Another common ride is the 'whip'. In this, cars are carried round a track on the end of arms and are free to pivot on the ends of these arms. The track is elliptical or oval, and the cars obey Kepler's laws just as the planets of the solar system do. They have to move faster around the sharp curves than they do round the straight parts of the track, so they swing violently back and forth on their pivots, which travel at a constant rate.

Right: the underneath of a gravity ride car, showing the dog which engages the chain and (right) the safety pawl.
Below: the shock absorbers on the whip prevent violent shocks to the occupants.
Below right: underneath a large carousel, the Derby racer at Blackpool. 56 horses rotate with a 50-ton platform, each horse riding on an undulating track underneath. There is also provision for the horses to rock back-and-forth; the machine has more than 1000 greasing points.

The 'chairoplane' has chairs or cabins hanging on the ends of cables, so that they are free to swing out by centrifugal force as the machine turns. A modern variation on this has rocket-shaped cabins on arms which rotate from a central hub. These arms can be raised or lowered by compressed air cylinders controlled from the cabin by the passengers. The air comes from a separate compressor and is carried by a rotating knuckle joint with 'O' rings at the centre of the hub. When the ride is over, the operator lets the air out of the system by a central valve, so that all the cabins are brought down to ground level. If air pressure fails for some reason, the same thing will happen; the tubing used in the system is of sufficiently small bore to let the cabins down gently.

A wide variation of rides can be produced using combinations of these techniques, with cranks, cams and eccentric arms to produce the irregular motions. Fairground engineers prefer mechanical rather than electronic solutions to problems, since mechanical devices are more robust and easily maintained.

Other rides One amusement which does not use rotary motion is the 'dodgem cars'. These derive electric power from the overhead wire mesh through a sliding contact strip; their operating area has a metal floor, completing the circuit. This floor is maintained at earth potential, so that people do not get electric shocks as they step on to it. The cars have $\frac{1}{2}$ hp (375 watt) motors, using just a few amps of current at 110 volts DC. This is not enough to cause much heating of the wire mesh or the contact strip, but by gearing of the motor it provides enough power for the motion. The motor runs

with the steering; it drives the front wheel of the three wheeled vehicles. The steering can turn through over 180°, to allow great manoeuvrability. Because dodgem cars often get stuck, an automatic clutch mechanism disconnects the motor drive from the wheel if the load on it is too great, so protecting the motor.

The 'roller coaster' is known as a *gravity ride*: after the cars have been hauled to the top of their track, the rest of the ride takes place by the force of gravity. The tracks are designed with undulations so that the cars just have sufficient force after swooping downwards to make it to the top of the next hill.

There are a number of ways of hauling the cars to the top of the hills: one well-tried method has a heavy duty chain drive continuously operating up the slope. A lug or *dog* under the car engages in one link of this chain, there being several dogs for each train of cars so that if one fails, another will take over. As a further safety device, there is a ratchet (set of teeth) on the track up the slope, in which a pawl under the car can engage if anything goes wrong. There is also a flange projecting beneath the cars, which can be gripped by pairs of boards to brake the cars either in an emergency or at the end of the ride.

Roller coasters are more common in permanent amusement parks, in view of their size. The permanent parks usually have a rather greater variety of rides and amusements than the travelling fairgrounds can offer, such as water rides. One such device is adapted from the timber industry's technique of transporting logs on rapid water flows or *log flumes*. Log shaped vessels are carried along by the water, with occasional ramps rather like those of the roller coaster to lift the boats out of the water and let them slide down again to rejoin the water with a splash.

A permanent amusement park will be more likely to use mains AC electricity rather than DC where possible, for convenience of supply. Modern hydraulic motors are much more compact than electric motors, and can produce high speed and good power application. They consist of turbines operating on hydraulic fluid as pressure; the pressure is provided by rotary pumps, and may be as high as 3000 psi (200 bar). These run continuously, and the pressure is applied to the motor progressively by means of a valve to speed up the ride.

FILLING STATION PUMP

A petrol [gasoline] pump is a means of dispensing an accurately measured volume of fuel, and calculating the cost to the customer of this volume. The basic single fuel pump contains a pumping unit to pump the fuel from a storage tank to the dispensing nozzle, a meter unit to measure the volume of fuel, and a 'computer' to calculate the cost.

The motor driven pumping unit creates a partial vacuum at its inlet port, which is connected to an underground storage tank. The atmospheric pressure acting on the surface of the fuel in the tank forces it up to fill this vacuum. Once it is drawn into the pumping unit, it is then pumped under pressure through a filter, to remove dirt particles, and an air separation chamber, to remove air bubbles, before entering the meter unit. This is normally a *positive displacement* type meter, which produces a rotational output directly proportional to the volume which passes through it. From the meter the liquid passes through a hose to the dispensing nozzle.

The rotational output from the meter is now used to drive the computer, which continuously calculates the cost of the fuel being dispensed. This computer may be a mechanical unit, directly driven by the meter, or it may be an electronic computer, in which case the drive is via a transducer, which is a unit for converting the mechanical rotations into electrical impulses. The computer also displays the volume dispensed, the corresponding cost, and the price per gallon (or litre), the setting of the latter being variable.

Blending or mixing pump The function of the blending or mixing pump is to draw a high octane fuel from one storage tank and a low octane from a separate tank, and mix these in a preset ratio so as to produce one or more 'blended' fuels of intermediate octane ratings.

The pump contains two separate pumping and metering systems, one computer and one dispensing nozzle. The high and low octane fuels are pumped through the two meters, and then pass into a blend valve which determines the ratio of high octane fuel to low octane fuel being dispensed. Separate hoses then take the two fuels to the dispensing nozzle, where they are mixed. The rotational outputs produced by the two meters now drive into a common computer, which arithmetically adds these outputs together to give the total volume of fuel being dispensed. It also calculates the corresponding cost, and displays this together with the volume and the price per gallon. An additional function of the blender computer is to control, via the blend valve, the ratio of high octane to low octane being dispensed, to ensure a high degree of accuracy.

Self service pumps There are two types of self service systems: *post-payment*, where the pumps are controlled by a central controller, and the customer pays at the central control point after the fuel has been dispensed; and *pre-payment*, where the pumps are controlled by an integral counting mechanism, and payment must be made into the counting mechanism before fuel is dispensed.

Below: a modern pump with its outer casing removed to show the pump and motor units, and the drive arrangements for the digital counter dials which indicate the quantity of fuel dispensed and its total cost.

A customer using a post-payment system must signal the central controller that he intends using a particular pump. The controller then switches electrical power to the pump motor, which allows the pump to be used. After the customer has dispensed the quantity he requires he must go to the control point to make payment. The volume dispensed and the corresponding cost is automatically relayed to the control point by transducers.

A customer using a pre-payment system must first insert

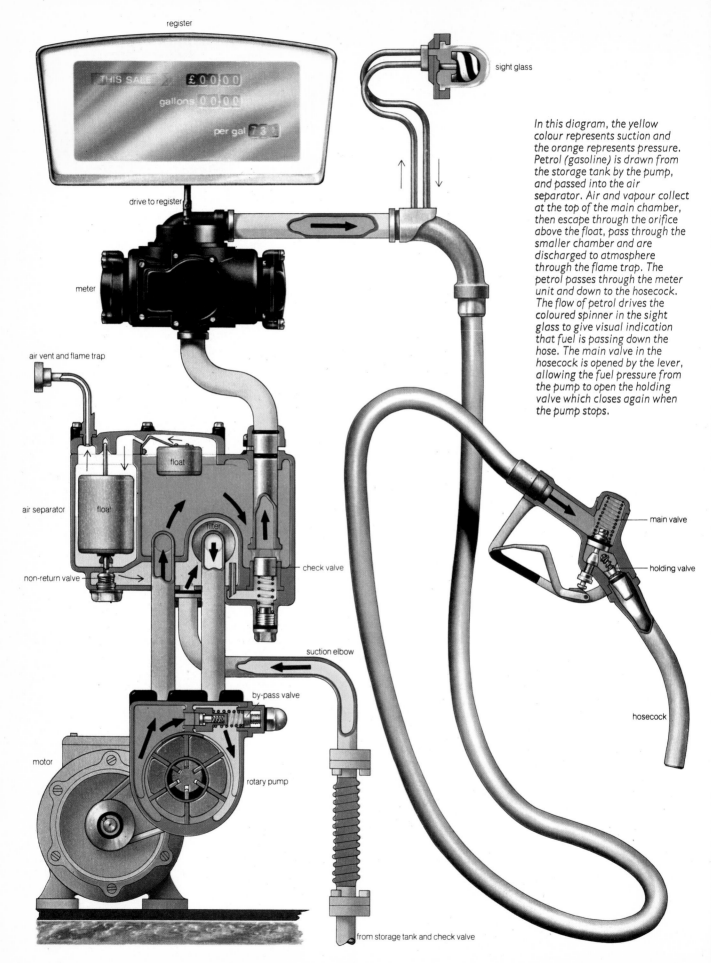

register

THIS SALE £00.00

gallons 00.00

per gal 73

sight glass

drive to register

meter

In this diagram, the yellow colour represents suction and the orange represents pressure. Petrol (gasoline) is drawn from the storage tank by the pump, and passed into the air separator. Air and vapour collect at the top of the main chamber, then escape through the orifice above the float, pass through the smaller chamber and are discharged to atmosphere through the flame trap. The petrol passes through the meter unit and down to the hosecock. The flow of petrol drives the coloured spinner in the sight glass to give visual indication that fuel is passing down the hose. The main valve in the hosecock is opened by the lever, allowing the fuel pressure from the pump to open the holding valve which closes again when the pump stops.

air vent and flame trap

float

float

air separator

main valve

non-return valve

check valve

holding valve

suction elbow

hosecock

by-pass valve

motor

rotary pump

from storage tank and check valve

the appropriate amount of money into the pre-payment counting mechanism. This mechanism then switches electrical power to a specific pump and allows that pump to be used until the corresponding amount of fuel has been dispensed. It then automatically switches the pump off.

Safety Every precaution is taken in the design of the electrical equipment used in pumps, to ensure it cannot produce a spark that could cause an explosion, even if the equipment is damaged or operated incorrectly. Alternatively, each piece of electrical equipment can be built into a separate box that is strong enough to withstand any internal explosion, caused by a spark generated by that equipment, without letting this explosion spread to the rest of the pump and possibly producing a much greater explosion or fire.

FIREWORKS and FLARES

Fireworks probably originated as a consequence of the discovery of gunpowder in China over two thousand years ago, and were possibly used initially to frighten away devils rather than to give enjoyment to the users. The earliest European use of fireworks as an armament was by the Byzantines in the seventh century (greek fire), but the development of fireworks for pleasure use did not begin until about 1500 AD, in Italy, and spread throughout Europe during the sixteenth century. In England, Queen Elizabeth I visited a firework display at Warwick in 1572, and displays gradually became a regular feature of public entertainment on big occasions. Today fireworks are in common use for both public and private displays throughout the world.

Cases and compositions The cases of fireworks are basically laminated paper cylinders or tubes, the thickness and shape depending on the type of firework and the composition (filling). Most cases are plain cylinder shapes, but there are variations on this, such as conical or cubical shapes, and the specialized cases for jumping firecrackers and Catherine wheels [pinwheels]. The jumping cracker consists of a long tube folded back on itself, containing a composition designed to give a sequential series of explosions. The tube of a Catherine wheel is wound spirally around a disc of plastic, cardboard or composite material, and is consumed as the composition burns away.

The basic composition of fireworks contains compounds of potassium, carbon, and sulphur. To produce sparks, salts of lead or barium or finely powdered steel, iron, aluminium or carbon may be added to the composition. Brilliant white

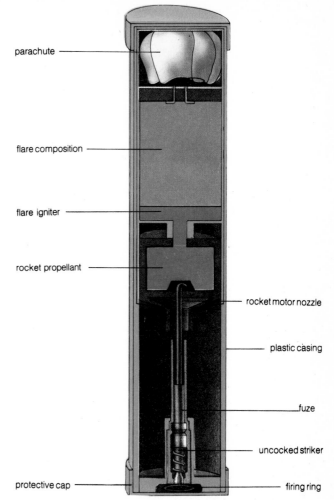

Above: cross-section of a hand-held rocket parachute flare. To fire the rocket the protective caps at each end are removed, and the firing ring pulled downwards to operate the striker, which sets off the fuse. After a two second delay the fuse ignites the rocket propellant which drives the rocket out of the tube and to a height of over 300 m (980 ft), where the propellant sets off the flare igniter and the flare and parachute are ejected. The parachute and burning flare then descend at about 4.5 m (15 ft) per second.
Left: a painting by G M Terrini (1739-1811) showing a spectacular firework display at Florence, Italy.

(labels on diagram:)
parachute
flare composition
flare igniter
rocket propellant
rocket motor nozzle
plastic casing
fuze
uncocked striker
protective cap
firing ring

Right: a miniature signal flare kit comprising eight signal cartridges and a hand held projector.
Below: a hand held orange smoke flare which is used as a daytime distress signal.
Bottom: a hand held distress flare which emits a brilliant red light for about 55 seconds. The case is made of metal, and the flare is operated by twisting an ignition ring.

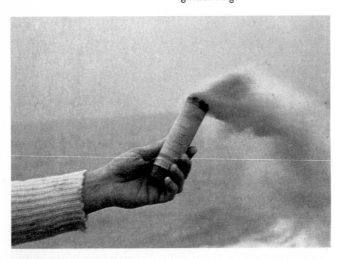

flame is produced by the addition of compounds of potassium, antimony, arsenic and sulphur, or powdered magnesium.

Coloured flame is created by various metallic salts; strontium and lithium salts produce a red flame, green is produced by barium, yellow by sodium, and blue by copper. Coloured stars and similar effects (such as those from a sky-rocket) are made by small pellets of colour composition which are ignited and ejected from the firework by the force composition.

Fireworks can be extremely dangerous if handled carelessly, and even more dangerous, possibly lethal, if made by amateurs. Under no circumstances should attempts be made to produce home-made fireworks, or to use fireworks in any way other than that specified by the manufacturers.

Displays Portraits of personalities, depictions of buildings (sometimes as much as 200 yards, 180 m long) and moving outlines of animals or people have long been a feature of firework displays.

Among the most spectacular fireworks are 'shells', normally round or cylindrical paper or plastic cases projected into the air from a mortar tube sunk into the ground. These vary in size from just under 2 inches (5 cm) to 3 feet (91 cm) in diameter, and burst at a predetermined height giving star, noise and pictorial effects against the night sky. The most recent type, the daylight shell, releases slogans or product dummies on parachutes and is used for publicity and product promotion.

The preferred styles of display vary around the world, the Far East generally specializing in attaining perfect symmetry in shell-burst effects, using small short-burning stars and having a wide variety of noise effects, while in Europe and America the stars burn longer and a wider range of effects is normally produced.

Flares The main uses of flares are for signalling (including distress signals) and for illuminating landing strips or target areas. The flares may produce a brilliant flame, coloured smoke, or fire rockets into the air which release coloured stars or a burning flare carried on a small parachute. The Very pistol, a widely used signalling device, fires coloured stars to a height of 250 to 300 ft (76 to 91 m) from 1 or 1.2 inch (2.5 to 3.05 cm) diameter cartridges.

The main type of parachute flare is the hand held parachute rocket. This consists of a free flight rocket made from an aluminium alloy tube, containing the propellant and the payload, which is a parachute suspended flare. The rocket is fitted into a plastic launching tube and is ignited by means of a striker that fires a percussion cap. The cap in turn ignites a delay fuse and an intermediate charge known as a *quickmatch*, which ignites the propellant. The rocket motor burns for around 3.5 seconds and drives the rocket to over 1000 ft (305 m) before ejecting the parachute flare. The flare burns for over 40 seconds, and is carried on a four string parachute which slows its rate of descent to approximately 15 ft/sec (4.6 m/sec). On a clear night such flares can be seen for a distance of about 28 miles (45 km). Illuminating parachute flares may also be dropped from helicopters or aircraft to light up ground areas.

Handstars are used in lifeboats and by other small craft in distress, and also by mountaineers. They eject two red stars to a height of about 150 ft (46 m), with an interval of 3 to 5 seconds between them. Each star burns for about 5 seconds. Distress *handflares* produce a bright red flame and burn for about 55 seconds.

Distress flares are always coloured red, and contain a mixture of magnesium, strontium nitrate, potassium perchlorate and PVC. Illuminating flares are usually white, and a range of other colours are available for signalling. A recent development is the radar reflecting distress rocket, which carries a payload of two red stars plus about 300,000 tiny pieces of silvered nylon which will reflect radar signals and so enable rescuers to use radar to locate craft in distress.

Smoke flares Smoke flares are used for such purposes as daylight distress signals and ground to air signals. Orange smoke is used for distress signals, and one type, for use by lifeboats, is designed to float and burns for 2 to 4 minutes. The smoke is non-toxic and when burning on oil-covered water it will not ignite the oil. Small hand held smoke flares are available which burn for about 30 seconds.

An important type of smoke flare used on merchant shipping is the lifebuoy marker. One is carried on each side of the ship's bridge, attached by a line to the lifebuoys. When the lifebuoy is thrown overboard, it pulls the marker out of its mounting bracket, igniting the smoke charges. The flotation collar of the marker contains two electric lights powered by sea-water activated batteries. The smoke burns for over 15 minutes, and the lamps stay lit for over 45 minutes.

FORK LIFT TRUCK

Fork lift trucks are established nowadays as the universal means of handling materials in factories and in storage or warehouse operations. The goods are normally placed on *pallets*, which are platforms, usually made of wood, designed to accept the forks of the truck. This provides a convenient means of moving, lifting and stacking loads up to three tons or even more.

Fork trucks fall basically into two categories: *general purpose* vehicles and *high density storage* trucks.

General purpose trucks In this category trucks are called *counterbalanced* fork trucks because the weight of the load is counterbalanced by the weight of the truck itself. They may be battery powered or driven by an internal combustion engine, powered by diesel oil, bottled gas or petrol [gasoline].

On electric trucks, the lead-acid battery may weigh three tons or more, thus providing much of the counterbalance. Advanced systems of electronic solid-state control enable continuous operation for up to 18 hours before re-charging. For round-the-clock operation, two batteries can be used, one being recharged while the other is in use. Electric trucks are especially useful in close quarters indoors, or where food or medical supplies must be handled, because they are quiet and emit no polluting exhaust. Smaller electric trucks may be operated by a pedestrian rather than a driver riding on the vehicle, by means of a handle similar to a lawn mower handle, with the controls built into the handle grips.

Engine powered trucks are used for large warehouse and stock-handling applications, for general yardwork and for loading and unloading of vehicles. Torque converter transmission provides smooth operation; power steering and all-weather cabs are options available.

Fork trucks for general purposes can have pneumatic or solid tyres. The lifting mechanism is hydrulic. An electrically driven hydraulic pump, or mechanical pump driven off the engine, supplies oil pressure through control valves to a hydraulic cylinder, or lift jack. The forks are mounted on a double or triple section mast, which extends upwards as the lift jack raises the forks through a system of chains and rollers. The mast can be tilted forward and backwards by means of smaller tilt jacks. Accessories can be fitted for lifting awkward loads, such as reels of newsprint or barrels of chemicals.

Lift capacities of counterbalanced trucks range from about 1000 lb (454 kg) to 10,000 lb (4540 kg) for electric models, and 4000 lb (1814 kg) to perhaps 100,000 lb (45,400 kg) for the engine powered trucks. Most fork lift trucks have the

Two views of a turret truck. The top view shows the driver's sideways seating, the transistorized controls, and the steering wheels with a gearbox on each. The lower view shows the truck rotating the load within a narrow aisle.

*Left: a fork lift truck with an
attachment for handling paper
reels, which can rotate the reel
and place it on the printing
machine.*

*The drawing on these pages
shows a Deisel-powered truck,
featuring its steered rear wheels
and hydraulic pump driven by
the engine.*

steering on the rear wheels, which allows a small turning radius.

High density storage   The latest development in materials handling vehicles is the high density storage or *narrow aisle* fork lift truck. They are called *reach* trucks and *turret* trucks, and are electric powered. The electric and hydraulic systems are the same as for general purpose trucks, but reach and turret trucks are used in storage systems designed to use the maximum amount of space, with narrow aisles and goods stacked as high as possible.

The reach truck is so called because the whole mast and fork unit reaches out to pick up a load. The wheelbase of the truck is the same length as for an ordinary fork truck of comparable size, but the size of the body of the truck is considerably more compact; 'reach legs' extend from the body of the truck to the front wheels, and the mast travels back and forth on these. Once the load has been picked up it is brought backwards to within the wheelbase of the truck, thus reducing the length of the truck and its load as much as a third of the length of the conventional truck. The truck has an extremely short turning radius, and the driver sits sideways, since in a 'narrow-aisle' situation he is driving backwards as much as forwards. The truck can also be used for ordinary lift truck applications, and can be fitted with attachments for special lifting jobs.

On a turret truck the mast is fixed and does not reach, but the truck is fitted with a 'turret' mechanism which turns the forks through 90° to either side of the truck. Then the forks reach out on a pantograph arrangement to pick up a load. The truck itself does not have to turn to face the load; thus the aisle width required for operation is even less than for the reach truck. The aisle width is governed by the diagonal size of the load, and can be as little as 62 inches (1.6 m).

Because of the narrowness of the aisle, the truck is guided by rails on each side of the gangway at the base of the storage racks; the rails locate with side rollers on the truck. This relieves the operator of the necessity of steering, unless the truck is being operated outside the aisle. Because turret trucks can lift the load to heights in excess of 30 ft (9 m), beyond accurate judgement by eyesight, the driver can pre-select the operating height by pushing a button on a console.

The turret truck makes possible the ultimate utilization of space in large warehouse operations.

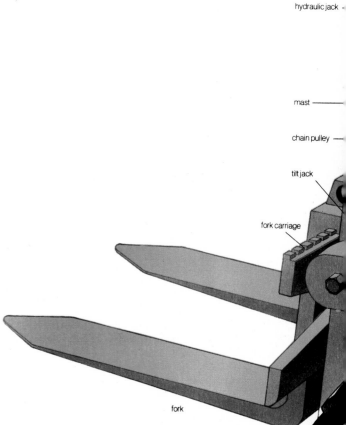

hydraulic jack

mast

chain pulley

tilt jack

fork carriage

fork

drive wheel

protective structure

hydraulic controls

handbrake

control unit

footbrake

air filter

diesel engine

radiator filler

radiator

hydraulic pump

transmission

fuel tank

steering rear axle

FURNACE

A furnace is a closed structure in which heat is applied to a load or charge; the term is derived from the Latin word *fornus* meaning an oven. Inside the furnace the charge may be only melted or may undergo permanent physical, chemical and mechanical changes.

As early as 3000 BC the Sumerians of Mesopotamia were using closed furnaces (kilns) with controlled draught for firing fine pottery at temperatures up to 1200 °C (2192 °F). For metal extraction, the pot furnace, which was buried in the ground, replaced the primitive stone or clay lined hollow. During the Bronze Age, a more advanced type evolved—the shaft furnace—which was widely used for smelting and probably derived from the pottery kiln.

Nowadays the industries using furnaces are those producing iron and steel with blast furnaces, bessemer converters, electric arc and induction furnaces; cement manufacturers using rotary kilns; glass manufacturers with their tank and pot furnaces; the ceramic industries with many types of kilns, which are particular types of furnaces; and the numerous non-ferrous metal producers. *Retorts*, which are used for the production of coke, smokeless fuel and nowadays to a lesser extent, town gas, can also be considered to be a type of furnace.

Furnaces are designed for either intermittent (batch) operation or continuous use. In batch type operation the charge or load is placed in the furnace when cold, fired to the required temperature, cooled and unloaded. The simplest is a box type but there are many variations. A continuous furnace generally consists of a number of zones, preheat, firing, and cooling, and the charge is pushed through the zones on cars using a hydraulic ram which operates from the entrance.

Construction The type of brick used in furnace construction depends on various factors such as the maximum firing temperature, atmospheric conditions, and the reaction of the molten metal, slags and combustion gases with the *refractory* (heat resistant) lining. The furnace structure must remain stable under the prevailing heating and cooling conditions, so that expansion and contraction must be accounted for. Processes involving rapid changes in temperature require refractories that are resistant to *thermal shock*.

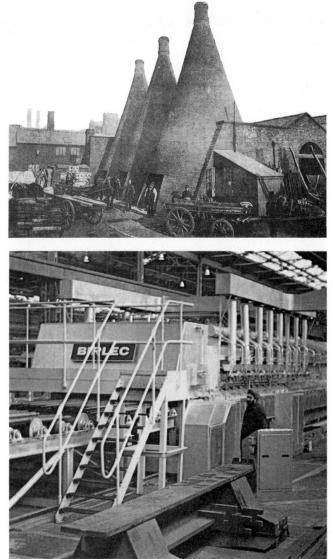

Above top: converter furnaces in Sheffield, about 1880. In this type of furnace, oxygen is blown into and through molten metal to 'convert' it by reducing impurities.
Above: a roller hearth furnace for annealing steel tube. It is 76 m (250 ft) long and divided into three zones.
Left: molten steel from a 130 ton capacity electric arc furnace. The electrodes are lowered into the furnace gradually as they are consumed.
Opposite page: a blast furnace making pig iron at a European Ford factory.

*Above, top: the production
of a molybdenum electrode in
an Austrian steelworks.
Molybdenum has an unusually
high melting point.
Above: a channel induction
furnace used for the bulk
melting of brass.
Opposite page: four types of
furnace: a German glass
pot furnace of around 1750;
an open hearth furnace, in
which the fuel is burnt over
the surface of the pig and
scrap iron to be melted; a
basic oxygen furnace, in
which the oxygen combines
with and eliminates the
impurities in the steel; and
an electric arc furnace, in
which the arc between the
electrodes melts the scrap.*

Construction design is very important, the provision of expansion joints, tie bars and rods being essential to maintain the brickwork's stability. Much use is now being made of cement-like linings which can be sprayed on to give a *monolithic* layer and ceramic fibre blankets which are pinned to the supporting brickwork. China clay, ball clay and fireclay are extremely good refractories, fusing at above 1700 °C (3092 °F). Other materials used include silica, magnesite, dolomite and alumina.

Fuels The traditional fuels were coal and coke, but increasing concern for pollution has accelerated the changes in methods of firing. Shovelling solid fuel on to fire grates was replaced by mechanical stokers, but there were still problems of handling ash and the harmful effects of oxides of sulphur, which, among other things, can corrode metal ducting and fan blades. Many power stations now use pulverized or powdered coal and expensive precautions are taken to minimize the grit and noxious gases which are released to the atmosphere.

Furnaces have, in many cases, been converted from solid fuel to oil and gas, and the use of these fuels has resulted in improved furnace and burner design. Burner design depends on mixing fuel and air in the correct ratio. It is much easier to mix two gases (as in the combustion of methane with air) than a solid and a gas (coal and air). Some fuel burners have now been designed which will handle both gas and oil with very little adjustment. In the past twenty years the same furnace may have been fired by coal, fuel oil and gas—and there is a distinct possibility that there will be a return to solid fuel firing by more refined means in the future.

The combustion process for coal, gas or oil, is one in which the fuel, consisting mainly of carbon and hydrogen, burns to give heat which is passed to the charge by conduction, convection and radiation. Conduction occurs when heat passes through solids, as when the handle of a poker becomes hot. Convection occurs, for example, when the hot gases swirling round the furnace load give up their heat to the cooler charge, and in radiation the heat is transferred in a similar way to that of a conventional electric fire element.

The products of combustion are carbon dioxide and water vapour mixed with the original nitrogen of the combustion

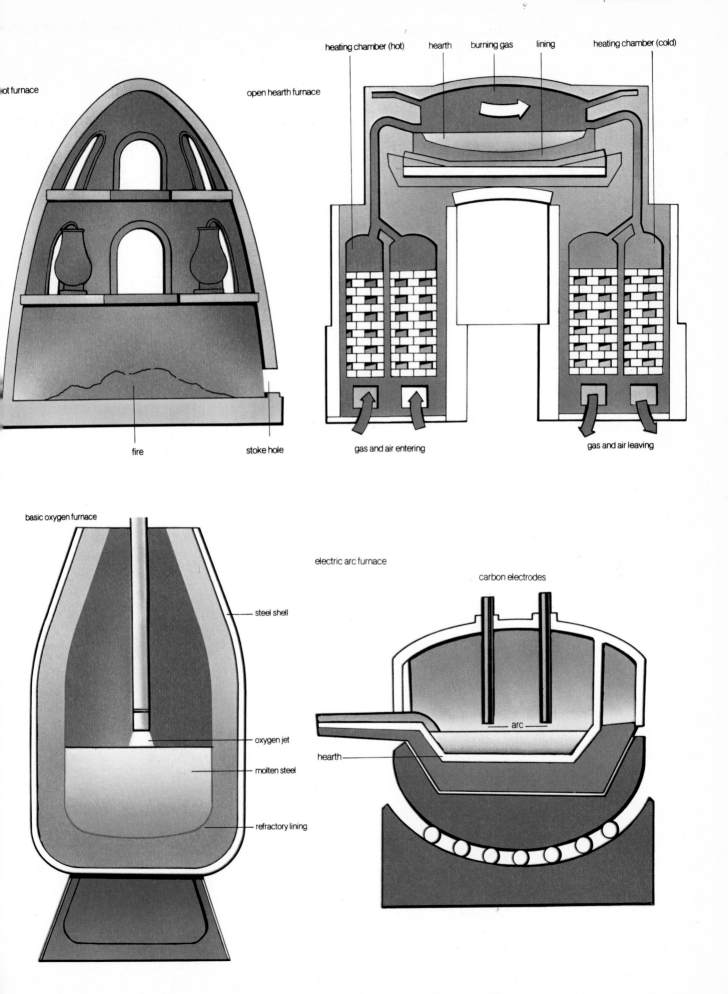

blast furnace

open hearth furnace

heating chamber (hot)　hearth　burning gas　lining　heating chamber (cold)

gas and air entering

gas and air leaving

fire

stoke hole

basic oxygen furnace

steel shell

oxygen jet

molten steel

refractory lining

electric arc furnace

carbon electrodes

arc

hearth

Above: the control panel on
a basic oxygen converter
uses a computer to calculate
oxygen flows and operating
temperatures.
Right: the doors of an open-
hearth furnace.
Opposite page: an electric
arc furnace, showing the
hot electrodes.

air. Control of the furnace is maintained by adjusting the fuel input through a burner, together with the air supply so that the fuel-air ratio is sufficient to give good combustion conditions. Depending on the type of combustion process involved, the atmospheric conditions prevailing in the furnace may be *oxidizing*, when surplus air is present, or *reducing*, when there is a lack of air and unburned gases are present, or *neutral* when air to fuel ratios are perfectly proportioned.

By measuring the percentage oxygen or carbon dioxide content of the combustion gases the furnace operator is able to control the furnaces to give maximum efficiency.

Electric furnaces These furnaces do not depend on a combustion process, electrical energy from the main grid system being converted directly into heat energy. There are three main types of electric furnace: resistance, electric arc and induction. The commonest type is the resistance furnace, which consists of a box with wire wound elements, which heat up when an electric current is passed through. The heat generated is proportional to the electrical resistance of the element and the square of the current. The element can be set into the refractory brickwork which supports them. The maximum operating temperature depends on the wire used for the elements. Non-metallic elements can also be used, such as silicon carbide (1450 °C) and molybdenum disilicide (1700 °C). Such furnaces lend themselves to automatic control.

In the electric arc furnace, the lining is of high quality refractory, and generally the electrodes, made of carbon or graphite, are located in the roof and extend into the furnace chamber and the charge. As the electrodes become consumed they are lowered into the furnace. An arc is struck and the heating effect used to melt metals which can be removed through tapholes. Often the furnace may also be tilted.

In the electric induction furnace, which is used for melting metals and for *heat treatment*, there is no electrical connection between supply and charge. It works on the principle of a transformer with the coil as the primary and the metal charge as the secondary winding. The crucible is surrounded by a water cooled coil and the space between is filled with insulating material. A high frequency power source is applied to the coil and energy induced in the charge appears in the form of heat, capable of giving high temperatures.

Above, top: a furnace operator checking the temperature at a work station.
Above: brass being poured out of an electric induction furnace. The metal itself is the secondary coil of an induction 'transformer'.

FUSES and CIRCUIT BREAKERS

Fuses and circuit breakers are devices for preventing excessive voltages and currents from 'overloading' and damaging an electrical circuit. Fuses are only suitable for light industrial and domestic applications whereas circuit breakers (or CBs for short) are designed mainly to cope with heavy industrial power requirements such as in power station switch gear. Miniature circuit breakers (MCBs) are, however, installed in light industrial and domestic circuits for protection and control. Without these safety devices excessive voltage and current will cause fires, explosions and electrocution.

Fuses A fuse is essentially a piece of wire of controlled dimensions and composition which is designed to melt when the current flowing through it reaches a certain value. The principle employed here is that a current flowing through a conductor generates heat proportional to the square of the current—that is, doubling the current quadruples the heat generated. The wire is chosen to suit the maximum overload current that the circuit can take, at which point the heat can no longer be dissipated quickly enough and the wire melts. Once the wire has melted, the circuit is 'broken' and protection ensured. Fuses are classified by the maximum continuous current they will take without 'blowing' and this should be carefully chosen to suit the appliance to be protected.

Rewirable fuses are the simplest type available in Britain. Replacement after 'blowing' such a fuse entails threading a new piece of wire between the terminals; it is necessary to have the right size wire on hand. The American screw-in plug-type fuse, which is simply thrown away and replaced, is much more convenient.

Cartridge fuses The cartridge fuse offers a solution to these problems because they are manufactured and sealed at the factory. They consist of a hollow, nonconductive body (made of glass or porcelain) which contains a fuse element, the ends of which are connected to metal caps forming the fuse terminals.

Domestic cartridge fuses again use plain wire elements which are relatively slow to melt, but where a fast response is required, especially in heavy duty applications, the element is shaped. The ends, attached to the cartridge caps, are made large to act as heat sinks to remove unwanted heat generated by normal usage. Between these large sections there are one or more narrow sections which will melt first in the event of a current surge. The narrower and shorter these sections are, the faster the response to a current overload. There is, however, a drawback to making this section too short because arcing can occur once the section has melted.

One technique for overcoming this is to include several narrow sections separated by thick sections. In the event of an overload, more than one section will 'blow'. Another technique is to surround the element with a powder—usually silica (silicon dioxide, SiO_2, such as quartz). When the wire melts, metal vapour is produced and an arc is struck through this vapour. The heat generated in this process, however, creates a chemical reaction between the vapour and the powder, fusing them together and producing a high resistance (insulating) substance which extinguishes the arc.

Circuit breakers The theory of circuit interruption is based on the necessity to extinguish the highly conductive arc which is 'drawn' between the circuit breaker contacts as they separate. This is necessary because while the arc continues the circuit is not properly disconnected and current still flows, and also because the arc damages the contact surfaces.

Alternating current (AC) circuit breakers are the most common types found because the mains supply is AC. In AC circuit breakers use is made of the fact that twice every cycle (that is, one hundred times a second where the mains frequency is 50 Hz) the current is zero. Interruption occurring at or near these points will help to prevent arcing, but interruption is only complete when the resistance of the arc gap is high enough to prevent the voltage between the contacts from re-establishing an arcing current.

The fault which initiates the circuit breaker action is detected by protection relays. These react to the fault condition extremely rapidly—usually within half a cycle of the AC mains.

Top: a 150 kV air-blast circuit breaker installed at a substation in Rotterdam, Holland. They are designed to operate in circuits carrying up to 4000 amps. The breakers are mechanically operated by light glass fibre rods and the speed of separation of the contacts enables a breaking time of only $2\frac{1}{2}$ cycles (at 50 Hz mains).
Above: a bulk oil circuit breaker, now replaced by air-blast types.

GEAR

Gears are toothed wheels used to transmit power between components of a machine. Cars, clocks, machine tools, adding machines, cameras, and many other devices essential to our modern way of life contain various kinds of gears. The invention of the toothed wheel is second in technological importance only to that of the wheel itself.

The most important single use of gears is to transmit motion from a power source, such as an internal combustion engine or an electric motor, to a shaft which can do useful work, such as the drive shaft in a car or the spindle of a machine tool. The power must be transmitted at a usable speed. When two gears are running together the larger one is called the gear and the smaller is called a *pinion*. If the pinion drives the gear, the unit is a *speed reducer*; if the gear drives the pinion, it is a *speed increaser*. Gears are more often used as speed reducers than the opposite, because the speed of an electric motor, for example, is normally much too high to be used by a machine without reduction.

The second major function of gears is to provide a usable range of *gear ratios* in a machine: three or four forward gears in a car, for example, or a wide range of cutting speeds in a lathe. The gear ratio is the ratio of the number of teeth on one gear to the number of teeth on the other, and determines the amount of speed reduction or increase which takes place. For example, if a pinion has twenty teeth and the gear has sixty, the ratio is 1:3, and the gear will make one revolution for every three of the pinion. Since the teeth on a pinion each do more work than the teeth on the gear, the pinion is sometimes made of harder material in order to equalize the wear.

When one gear drives another, they turn in opposite directions (unless one of them is an internally-toothed gear). If it is required that they turn in the same direction, a third gear called an *idler* gear is interposed between them. Idler gears are sometimes used in gearbox [transmission] designs to provide the reverse gear.

Configuration If two gears running together are imagined to be two smooth wheels whose surfaces are touching, the diameter of each wheel is the *pitch diameter* or the *pitch circle* of the gear. The part of the gear tooth that extends beyond the pitch circle is called the *addendum*; the *dedendum* is the part of the tooth inside the circle. The *root circle* is the diameter of the gear measured at the base of the tooth. The *pitch* is the distance between a point on a tooth and the corresponding point on the next tooth, measured on the pitch circle. This is known as *circular pitch*. To facilitate calculations, the *diametral pitch* is more commonly used, this being the number of teeth per inch of diameter, measured on the pitch circle.

Backlash is the play between two meshing gears, and can be more specifically defined as the difference between the distance between two teeth and the width of the engaging tooth. Backlash between two gears can be altered by changing the centre distance between them; the correct amount of backlash is designed into a gear system, which means that the distance between the centres must be within tolerance. Incorrect backlash will cause noisy operation, loss of efficiency and excessive wear. In particular, too little backlash may cause generation of heat. In general, the important considerations in determining proper backlash are the operating speed and the space required for a film of lubricant; slowly moving gears need less backlash. Minimum backlash designs are expensive to manufacture and install properly; they are therefore only attempted where timing or accuracy is important, as in certain instruments.

There are several different types of gear systems, which can be described in the following categories: *spur* gears, *bevel* gears, *helical* gears and *worm* gears.

Spur gears Spur gears are the most common type. They have straight teeth and are used to transmit power between

Below: cutting a spur gear. A medium viscosity cutting oil is used. The oil keeps the tool and the work cool, lubricates the contacting surfaces, and flushes away the chips or 'swarf'. Cutting oils contain additives according to the metal cutting job to be done.

two parallel shafts or shafts in the same axis. Their efficiency can be more than 95%.

The sides of the teeth, in profile, describe an *involute* curve. (If a piece of string is wrapped around a cylinder, a point on the piece of string will describe an involute curve as the string is held tautly and unwound.) The sides of the teeth must be curved; otherwise the operation of the gears would be noisy, wear would be excessive, and a great deal of vibration would be generated. The involute curve has been found to be best because when an involutely curved tooth surface transmits power to an involutely curved mating tooth, as much of the power is transmitted as possible even if the centre distance between the shafts varies slightly. The point on the side of the tooth which is also a point on the pitch diameter of the gear is the point at which the power is transmitted most efficiently. The exact curve of the tooth surface is computed from the *base circle* of the gear. The base circle is just below the pitch circle and is the point at which the involute curve from the top of the tooth ends. The tooth terminates in a straight radial flank to the root.

An arrangement of spur gears in which one or more small *planet* gears travel around the outside of a sun gear, while at the same time running around the inside of an enclosing *annulus* gear, is called a planetary or epicyclic system. (An annulus is an internal spur gear; that is, it has teeth on the inside of it instead of the outside.) An epicyclic arrangement allows more than one gear ratio to be selected without

but tapered in length and depth; if extended in length, they would meet at a point ahead of the gear on the axis of the shaft called the *pitch cone apex*.

Hypoid bevel gears have teeth which are straight, but cut on the face of the gear at an angle to the axis of the shaft. They can be used to transmit power between two shafts whose axes cross each other, but not in the same plane.

Helical gears Gears which are shaped like sections of cones, such as bevel gears, or like sections of cylinders, such as spur gears, can have spiral teeth on them, enabling them to be designed to transmit power between shafts at any angle to each other, according to the spiral of the teeth. They are called helical gears. The curved teeth enter the mating teeth while the previously meshing teeth are still in contact; this means that some sliding of the teeth against each other takes place, and that power is transmitted with relative smoothness and silence. Helical gears, like hypoid bevel gears, can be used to transmit power between shafts which are not in the same plane; their combination of running characteristics and design possibilities makes them ideal for pinion and crown gears in the differential of a car. Such a differential runs smoothly and quietly; it also enables the drive shaft of the car to be in a lower plane than the rear axles, making it possible to lower the centre of gravity of the car.

When helical gears are used to transmit power between two parallel shafts, they generate a sideways thrust which may be objectionable. To overcome this, two sets of helical gears can be used, with the thrust in opposite directions, cancelling each other out. For this application, the gears are sometimes machined out of one piece of metal with helical teeth meeting in the centre of the face and spiralling outward from each other; these are called *herringbone* or *double-helical* gears.

Worm gears When a pair of gears having helical teeth is used to transmit power between two shafts whose axes

1. A section of spur gear teeth. The curve of the side of the tooth is an involute curve.
2. Spur gear teeth in diagram.
3. Worm and gear. The ratio shown here is 18:1. If the worm speed were 360 rpm, the speed reduction would be to 20 rpm for the gear. Such a speed reduction might be used to drive a conveyer.
4. Spur gears in mesh.
5. Helical gears meshing. They will generate an axial thrust.
6. Bevel gears connect shafts whose axes intersect.

moving gears in and out of mesh, by locking various components or combinations of components of the system. It is used in bicycle gears, automatic transmissions and other applications.

A flat piece of material, oblong in cross section, with involute teeth cut on one side of it is called a *rack*. A *rack and pinion* gear system, in which a spur gear runs back and forth on a rack, is used in the steering mechanism of some cars, in hydraulic door closers, and for reciprocating motion of tables or workheads of machines.

Bevel gears Bevel gears are shaped like sections of cones. They are used to transmit power between shafts whose axes intersect. The teeth on an ordinary bevel gear are straight

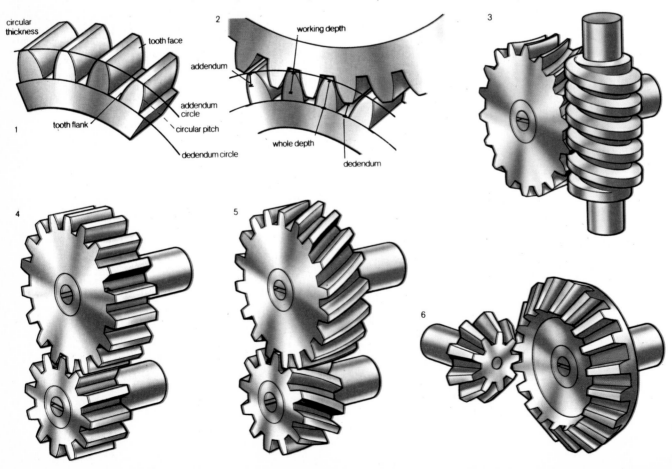

intersect, but not in the same plane, very high gear ratios are possible. In such a case, the gear ratio is commonly as high as 1:100; the pinion may have multiple threads, or only one thread, which will curve all the way around it and resemble a screw thread. Such a pinion is called a worm. The speed reduction of a worm and gear unit is very great, and is used between an electric motor and a slowly moving conveyer, for example.

In such a unit, a great deal of sliding takes place between the teeth, and the efficiency of it is not high. Sometimes the gear is made to partially envelop the pinion in the area where the teeth mesh, in order to increase the load-carrying capacity. Normally, the gear cannot drive the worm, the spiral of the worm teeth being too much greater than that of the gear teeth; if the unit is designed so that the gear can drive the worm, it is more than 50% efficient, and can be used as a speed increaser. Such a unit can be used to drive a supercharger, for example.

Non-metallic gears   Where quiet operation at high speeds is desired, and if the operating load or torque is not too high, one or both gears may be made out of plastic or a fibrous composition material. The timing gear in a car, which drives the camshaft from the crankshaft by means of a chain, is frequently non-metallic. The speed of the timing gear is high; while the timing is important, the load carried is relatively small.

Gear cutting   The machines used for cutting gear teeth comprise three general types. For cutting spur gears, a cutter blade conforms to the size of the space between the teeth. The second type generates the teeth either by a pinion shaped cutter which rotates in unison with the blank and by reciprocation cuts the teeth, or alternatively uses a cutter which is a section of a rack and planes the teeth as it moves across the face. The third type is a *hob*, which is a worm with gashed threads to give a cutting action.

GEIGER COUNTER

The Geiger counter is a device for detecting a charged particle. It was invented by H Geiger and E W Muller in 1928 and, though it has now generally been replaced by much more sophisticated detectors, it was a vital instrument in the early days of investigating the nature of the nucleus of the atom and the behaviour of the minute particles of which the atom is built.

Basic construction   In its usual form the counter consists of a glass tube about $\frac{3}{4}$ inch (2 cm) in diameter enclosing a metal cylinder, often of copper, about 4 inches (10 cm) long along the axis of which runs a thin metal wire, which may be of tungsten. The cylinder and wire are connected through the end wall of the glass tube to a source of electrical voltage. The tube is filled with a gas, usually argon, at a low pressure, equivalent to a pressure of a few centimetres of mercury. A voltage is set up between the cylinder (the negative electrode or cathode) and the wire (the positive electrode or anode) which is just a little less than that needed to create an electrical discharge between the two electrodes. This voltage may be about 1000 volts.

Operating principles   When a charged particle with high energy flies through the glass tube it knocks *electrons* out of

Below: a Geiger-Muller tube. The thin wire anode is suspended along the central axis of a copper cathode. A high speed particle passing through the tube ionizes the air molecules. The free electrons are attracted towards the anode and as they travel release further electrons, building up an 'avalanche'.

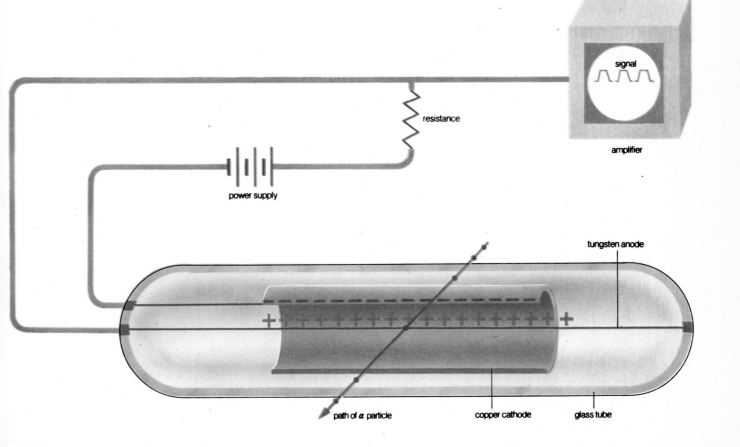

power supply
resistance
signal
amplifier
tungsten anode
path of α particle
copper cathode
glass tube

the atoms of the gas. These electrons, being negatively charged, make for the wire anode and the damaged atoms (which are positively charged argon ions) make for the cathode.

Since there are 1000 volts between the wire and cylinder the electrons are affected by a high voltage gradient—rolling, as it were, down a very steep electrical hill. They pick up enough energy to knock further electrons out of atoms which in turn roll down the electrical hill picking up further energy and liberating further electrons. This is known as *electron avalanche*. At the same time the positive ions hit the cylinder with enough energy to release still more electrons. An avalanche of electrons therefore descends on the wire which can be detected as a pulse of electric current, indicating that a charged particle has passed through the tube.

Before a discharge, the Geiger counter is charged rather like a capacitor, but when a discharge occurs both the stored charge and the corresponding voltage between the cylinder and wire is reduced. This 'kills' the discharge in the tube which would otherwise go on indefinitely—an important feature because while the discharge continues the device is insensitive to further charged particles arriving at the detector.

Another help towards stopping the avalanche is to put about 10% ethyl alcohol into the argon. Because of the way the atoms bump around, it tends to be the alcohol ions rather than the argon ions which reach the cylinder and there they prefer to break up rather than release more electrons. The alcohol vapour also smothers electron production within the gas mixture. It takes about a ten thousandth of a second to snuff out an avalanche and the voltage then builds up again so that the counter is ready to record another particle.

With the addition of an electronic amplifier the pulse of electric current could be made audible in a small loudspeaker. In this form it became famous as part of the equipment of uranium prospectors. It had the virtue of being very easy to operate and the clicks in its loudspeaker were simple evidence of the presence of high energy charged particles.

When a Geiger counter is switched on, irregular clicks will always be heard, originating in cosmic rays (very high energy particles which shower on to the earth from outer space) or in natural radioactivity (spontaneous emission of particles from some forms of chemical element which make up the natural environment).

Limitations The Geiger counter has limitations in pinning down charged particles both as regards their position in space and their time of arrival. The counter can only say that a particle has passed through the volume enclosed by the glass tube. Nowadays scientists want to locate positions to an accuracy of a fraction of a millimetre if possible. The counter can say that a particle has arrived at a particular time to an accuracy of about a millionth of a second. This is usually adequate but the Geiger counter is then 'dead' and cannot detect other particles for a ten thousandths of a second while it recovers from the avalanche. Nowadays scientists want their detectors virtually continuously sensitive to the arrival of particles.

Another limitation is that it cannot give any information about what kind of charged particle has passed through it. This can be improved to a small extent by running the counter with a lower voltage between the cylinder and the wire. The full avalanche that is characteristic of the Geiger counter does not then occur. Instead the number of electrons which reach the wire depends on the number of electrons which the charged particle itself sets free in the tube. This helps to distinguish between particles. The size of the current pulse at the wire is then directly proportional to the liberated number of electrons and, when it is operated in this way, the counter is called a *proportional counter*. In the form of planes of thin, closely spaced wires, the proportional counter is still in vogue for studying particle behaviour at modern particle accelerators.

GENERATOR

A generator is a rotating machine which converts mechanical energy into electrical energy. The term was originally used to cover both alternating current (AC) generators, or alternators, and direct current (DC) generators, but nowadays it usually refers to the DC type only.

Basic principles The generator, like the alternator, makes use of a phenomenon discovered by Michael Faraday (1791-1867). In general terms, when an electrical conductor is moved in the vicinity of a magnet, a voltage is created, or *induced*, in the conductor. If the ends of the conductor are connected in any kind of closed electrical circuit, this induced voltage (electrical pressure) causes a current to flow in the circuit. Mechanical energy has been converted into electrical energy. (In Britain, a generator is called a dynamo.)

For a more detailed explanation it is necessary to consider the principles of electromagnetism. In an electrical circuit it is voltage that causes the current to flow. The voltage, or *electromotive force* (emf) is the cause and current the effect. By analogy, in a magnetic circuit the driving 'pressure' is called the *magnetomotive force* (mmf). This is the cause and *magnetic flux* is the result, or effect. Between the north and south poles of a magnet in the medium surrounding the magnet can be envisaged a set of *flux lines*—the closer these lines are together the greater the *flux density*. Flux density is determined by the mmf of the magnet and the *permeability* of the surrounding medium.

The emf induced in an electrical conductor, moving in a magnetic field, is determined by the rate at which the conductor 'cuts' the lines of flux. The induced emf is therefore related to the speed of the conductor and the flux density. It is also related to the length of the conductor.

A simple alternator When a closed rectangular loop of wire is mounted on a rotating axis (the rotor) and rotated between the north and south poles of a horseshoe magnet (the stator), the following occurs. When the two sides of the

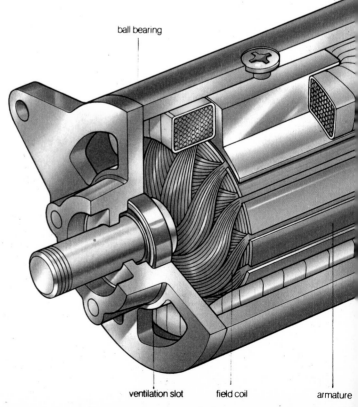

brushes collecting current

ball bearing

ventilation slot   field coil   armature

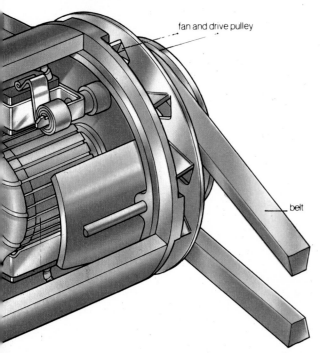

fan and drive pulley

belt

commutator

loop parallel to the axis of the rotor (these are the 'conductors') form a line between the north and south poles, the rate at which these two conductors cut the lines of flux is at a maximum. The inducted emf in the conductors is therefore also at a maximum and the current flowing around the loop at its largest value.

When the rotor has turned through 90°, the instantaneous direction of motion of the conductors is along the lines of flux. No lines of flux are therefore cut, no emf is induced in them and the loop current is zero.

When it has rotated by a further 90°, the rate of flux cutting is again a maximum with maximum emf and loop current. The loop is now, however, 'upside down' compared

Above: a 3350 kW, 660 volt DC generator or dynamo. The principle of the generator is that a wire moving through a magnetic field has a voltage induced in it. Here, the rotor consists of a set of wire coils mounted on a rotating shaft and connected to a commutator. The stator contains electromagnets.
Left: a cutaway drawing of the kind of generator that is used in a car. It is bolted to the engine and driven by the fan belt. The fins on the back of the drive pulley act as a cooling fan, blowing air through the dynamo to cool the windings.

to its position 180° ago and the induced emf and current are a maximum in the opposite direction. A further 90° rotation and the current is again zero.

By breaking the loop at one end near the axis and connecting the ends to two slip rings on the shaft of the rotor, this alternating emf can be tapped using 'brushes' touching the rings to drive an external electrical circuit. This is the basic design of an alternator.

From alternations to generators   To construct a generator, several loops are positioned in sequence around the rotor. This time, instead of taking the loop ends to slip rings, they are connected in the same sequence to individual segments of a commutator (a divided rotating contact). Two brushes are mounted on opposite dies of the commutator such that, as the rotor rotates, the brushes form an electrical contact with the two ends of just one loop at a time.

By positioning the brushes so that they 'tap' that loop which is in the position of maximum flux cutting, then they will tap each loop as it comes into that position during rotation. The induced emf is therefore always in the same direction (that is, a DC voltage) and always with the maximum value possible. This is the simple generator.

Increasing the speed of rotation will, to a certain extent, increase the DC voltage available at the brushes. One other way to improve the generator performance is to increase the number of turns in each 'loop'—thus making a coil—because, as already stated, the length of the conductor also determines the emf that can be induced. Finally, the performance can be further improved by increasing the flux density. Several methods are available for this. First, stronger magnets or electro-magnets can be employed and, second, the magnetic poles pieces can be shaped to concentrate the lines of flux.

When electromagnets are used to provide the magnetic field they draw some of the current generated by the machine. When starting from rest, the current to start the electromagnets working is derived from what little residual magnetism exists in the electromagnets and surrounding magnetic circuit.

GYROSCOPE

The term gyroscope (or gyro for short) is generally applied to a flywheel rotating at high speed about its axis. In a scientific gyro the flywheel is mounted so that its axis can take up any orientation in space, whereas in a toy gyro one end of the flywheel axis is constrained in some way. The Earth, Moon and other planets that rotate about an axis also behave like gyroscopes.

Gyroscopic behaviour   Imagine a flywheel spinning about its axis and supported on a pillar at one of its ends. If the axis is vertical, the flywheel will balance on the supported end of its axis like a spinning top. Imagine now that the axis is moved so that it is horizontal; one would expect the force of gravity to topple the flywheel off its support, but this does not happen. The flywheel axis remains horizontal, apparently resisting the force of gravity, and at the same time moves around its point of support in the horizontal plane. This movement is called *precession*. It is a property of all gyroscopes that when a force (gravity in the above example) is applied at right angles to the spinning axis, this will give rise to movement, not in the expected direction, but in a direction at right angles both to the spinning axis and to

the applied force. This is because the spinning flywheel has *angular momentum*, and any force applied must alter the direction in which the angular momentum applies.

The scientific gyro   In the scientific gyro, a wheel rotating about, let us say, a horizontal axis pointing north and south, is pivoted in an 'inner ring'—literally a ring of metal. The inner ring in such a case might then itself be pivoted about an axis at right angles to the axis of spin, let us say in this case a horizontal axis east and west. The bearings of the inner ring are held in an 'outer ring' and this second ring is yet again freely pivoted about an axis at right angles both to the axis of spin and to the axis of movement of the inner ring. In this example the outer ring axis would therefore be vertical.

The first amazing property of such a gyro is that if the base on which the outer ring pivots are mounted is tilted in any direction, rotated or displaced, the axis of spin of the wheel will remain fixed—fixed that is relative to the framework of the so-called 'fixed star' or 'inertial space' network. Thus, if a gyro mounted in 'gimbal rings'—the name given to the outer and inner rings—is allowed to remain on the ground, or on a bench or table, the axis of spin will *appear* to make a complete revolution in 24 hours. It is of course merely recording the fact that the Earth has turned 360° on its own axis in this period.

The second and even more baffling phenomenon of a gyro as described is that if a twisting force is applied to the outer ring so as to try to turn that ring on its pivots it *resists* such movement—but, the *inner* ring turns about the inner ring pivots so long as the twisting force continues to be applied to the outer. The motion of the inner ring under these conditions is precession. The action can be reversed. A twisting force (torque) applied to the inner ring will move the outer.

There is a simple rule which allows us to predict in which direction precessions will take place. Imagine the twisting

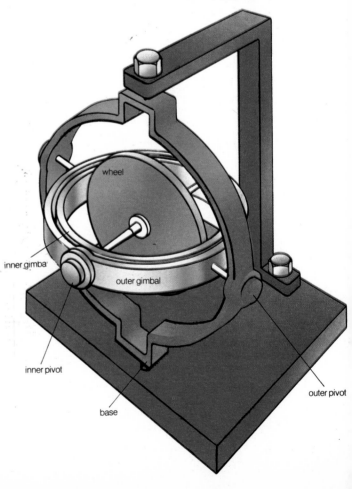

Right: a Sperry 'rotorace' gyro, which is used in many aircraft compass systems.

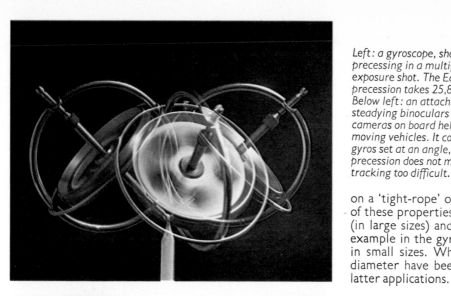

Left: a gyroscope, shown precessing in a multiple exposure shot. The Earth's precession takes 25,800 years. Below left: an attachment for steadying binoculars and small cameras on board helicopters or moving vehicles. It contains two gyros set at an angle, so that precession does not make tracking too difficult.

on a 'tight-rope' of string or wire, or on a sharp blade. Use of these properties are to be found in the stabilizing of ships (in large sizes) and as measuring devices for navigation (for example in the gyrocompass and inertial guidance systems), in small sizes. Wheels of 2 to 3 inches (5.1 to 7.6 cm) in diameter have been rotated at 50,000 rpm or more in the latter applications.

force applied to one ring to consist of a pair of parallel 'straight-line' forces pushing on the edge of the wheel perpendicular to the *faces* of the wheel but at opposite ends of the appropriate diameter of the wheel and in opposition to each other. Then imagine that each of these forces has been carried round 90° (a right angle) by the spinning wheel *in the direction of spin*. The resulting pair of imaginary forces indicate the direction of turning motion that actually occurs.

The toy gyro A toy gyroscope is usually nothing more than a simple wheel with its axle mounted in a metal ring. On the outside of the ring, opposite the two bearing points, there are usually small, nearly spherical knobs, one of which may carry a slit into which a steel ruler or piece of string may fit. The many quite amazing 'tricks' which may be performed with a toy gyro exhibit both types of gyroscopic behaviour—that is, the tendency to maintain a fixed axis of spin and the torque-precession relationship. A small model of the Eiffel Tower is often supplied with the gyro so that an end knob may be supported in a cup-like socket on top of the tower with the wheel axis horizontal. When released the wheel will 'gyrate' in a horizontal plane without falling off or pulling over the tower. A spinning gyro can balance

HEARING AID

The development of the electronic hearing aid was the biggest advance ever made in helping deaf people, and hearing aids are now used by millions of people all over the world. All hearing aids are fundamentally simple acoustic amplifying systems and consist of four basic parts: a microphone to pick up the sound and convert it to a very small electrical signal; an amplifier to increase the size of the electrical signal; an earphone (often called a receiver in hearing aid circles) to turn the electrical signal into an acoustic one which is then fed into the ear through an earmould; and a source of power for the amplifier, usually a small battery. The earmould is not part of the hearing aid but is essential for fitting the aid to the user. It is made of plastic and moulded from an impression taken of the user's ear.

Hearing aids can be divided into three groups: body worn, head worn and educational aids such as group hearing aids and auditory training units. Because of their larger size body worn aids are capable of the widest possible performance and can cater for all types and degrees of deafness. They are, however, large and many people do not like to be seen wearing them, preferring the smaller head worn aids. These can be divided into three main groups: those worn behind the ear, in spectacle frames, and completely in the outer ear itself. Head worn aids are not capable of producing the high output of body worn aids, but otherwise have very similar characteristics. The smaller the aid the more restricted is its performance. Group hearing aid equipment is used in schools for the deaf and partially hearing units where size is unimportant. The large high quality microphones and earphones used ensure the best possible performance and exceed that obtained with aids worn on the body or head.

Performance The performance of a hearing aid is largely controlled by the transducers, that is microphone and earphone. The performance of the transducers can be indicated by means of a frequency response curve which shows graphically how the microphone or earphone responds over a range of frequencies from 100 to 10,000 Hertz (Hz). At low frequencies the limitation is largely due to the microphone, while at high frequencies it is the earphone that restricts the performance.

The amplifier consists of three or more transistors, depending upon the amount of amplification required. The frequency response and power output from the amplifier is such that it places almost no limitation on the performance of the aid. The amount of noise, however, generated by the first transistor stage is important because if it is too high

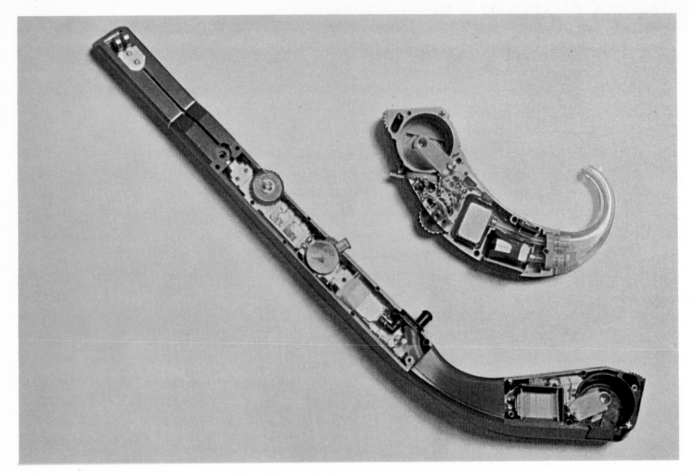

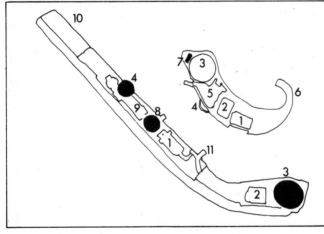

Above: head worn hearing aids.
1 microphone, 2 earphone
(receiver), 3 battery compart-
ment; 4 volume control,
5 circuitry, 6 earhook, 7 on-off
switch, 8 tone control,
9 audiological adjustment,
10 adjustable temple extension,
11 earphone spout.
Right: Beethoven's ear trumpets.

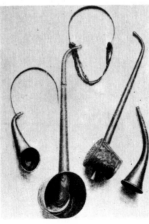

the listener will hear a continuous background rushing noise. The maximum amount of sound available from the aid, called the maximum acoustic output, will depend upon the power handling capability of the output transistor and earphone; in addition the power available will be limited by the current that can be drawn from the battery. The smaller the battery the smaller the current that can be taken while still giving a reasonable battery life, and the lower the maximum amount of sound available from the aid. The amount of acoustic amplification required depends upon the degree of deafness which the aid is required to help. and may vary from 40 decibels (dB) that is 100 times, to 80 dB, that is 10,000 times. Because the transducers have less than 100% efficiency, the electrical amplification necessary is considerably greater than the acoustic amplification. Amplification varies with frequency and most hearing aids amplify high frequencies more than low ones. It has been found that this gives better intelligibility of speech.

The earphone will alter the performance of the aid in terms of frequency response, gain and maximum acoustic output, and on body worn aids this is used as a means of altering the performance and then fitting the hearing aid to suit individual requirements. The main limitation on earphones is at high frequencies.

Frequency response curves The performance of hearing aids can be roughly divided into three groups in terms of the maximum acoustic output or power; that is, low, medium and high power. Low power aids will have maximum outputs of around 110 dB sound pressure level (SPL), medium 120 dB SBL and high power aids 130 dB SPL or more. It should be remembered that the noise from a jet engine close by is only about 130 dB SPL and that some deaf people are not able to hear at this level, but can hear something at 135 dB SPL or more.

On many hearing aids there is a switch which disconnects the microphone (marked M) and switches in an inductive pick up coil (marked T because it was originally designed for telephone use). This coil enables the user to pick up alternating magnetic signals produced by an electric current flowing through a loop of wire around the room, enabling him to move anywhere in the loop and hear the signal without direct connection to the amplifier. Such a system can be used in schools and in the home where it is used for listening to the radio and television. This system has the advantage that speech picked up by a microphone near the speaker's mouth is effectively fed directly into the hearing aid without the interfering effects of distance and background noise.

In order to measure the performance of a hearing aid, a constant acoustic signal over the frequency range 100 to 10,000 Hz is fed in and the output measured.

HEATER

Because man has refused to be restricted in his movements by climate, he has had to protect himself against the cold by clothing and by the heat produced from burning fuel. As housing developed so did domestic heating. The Romans had their floors warmed by the products of combustion from underground fireplaces. In colder climates open fires were superseded by closed stoves with extended flue pipes, and these were the forerunners of the more advanced methods in use today.

Heat transfer There are three ways in which heat is transferred from a hot body to a cooler one: *conduction, convection* and *radiation*. Conduction is the transfer, or flow, of heat through a body when part of it is heated or brought into contact with something hot. The handle of a poker becomes hot by conduction. Convection is the transfer of heat through a liquid or gas by the mixing effect of convection currents which are set up when the gas or liquid is heated. The water in a domestic hot water tank fitted with an electrical immersion heater heats up largely by convection. Radiation is the name given to direct heat transfer from a hot body to a cooler one spaced apart from it. Heating by the sun's rays is an example of radiation.

Domestic heaters generate heat by burning a fuel, such as gas, oil, coal or coke, or by passing an electric current through an electrical element. Such heaters can be broadly divided into two types, those which require a secondary system to distribute the generated heat, as in central heating systems, and those which do not, such as electric 'bar' and fan heaters and free standing paraffin [kerosene] heaters.

Central heating Central heating is the term used for a system which distributes heat throughout a building from a single heat source. The heat is generated by burning a fuel such as coke, oil or gas, in a boiler if the heat transfer medium is water, or a heat exchanger if the medium is air.

In a domestic boiler the heat generated is used to heat (but not boil) water which is then circulated to the various radiators in the system. The radiators are connected to each other and to the boiler by means of piping, nowadays usually small bore copper tubing, so that the heated water flows in a circuit from the boiler through each successive radiator and then back to the boiler to be reheated. Modern systems include electrically driven pumps to circulate the water at the rate required to maintain a suitable temperature difference between the input and output pipes of the boiler. For domestic systems this temperature difference is normally about 10°C (18°F). The system must also allow for the expansion of the water as it is heated. In an *open system* the extra water-volume is accommodated in an open expansion tank. In a *sealed system* the increase is taken up by the compression of air in a closed expansion vessel. Whether the system is open or sealed, it should be connected to a mains water supply so that any loss of water can be made good. The

radiators in a water circulating system emit heat partly by radiation and partly by conduction to the adjacent air. The heat is distributed throughout the room by air convection.

In 'warm air' central heating systems, air is passed through a heat exchanger by an electrically driven fan and is kept separate from the combustion gases which are discharged to the atmosphere. The warmed air is distributed via ducts to the rooms through grilles having regulating dampers, and is returned to the heat exchanger for reheating, usually via grilles in the doors but sometimes through separate ducts. Air for ventilation may be introduced from outside through a fresh air inlet and the rate of recirculation reduced accordingly.

Boilers and heat exchangers In domestic boilers, the water to be heated passes through a series of waterways located above the combustion zone and designed to present a large

Above: a storage radiator. In some countries, off-peak electricity is used to heat bricks and 'store' heat.

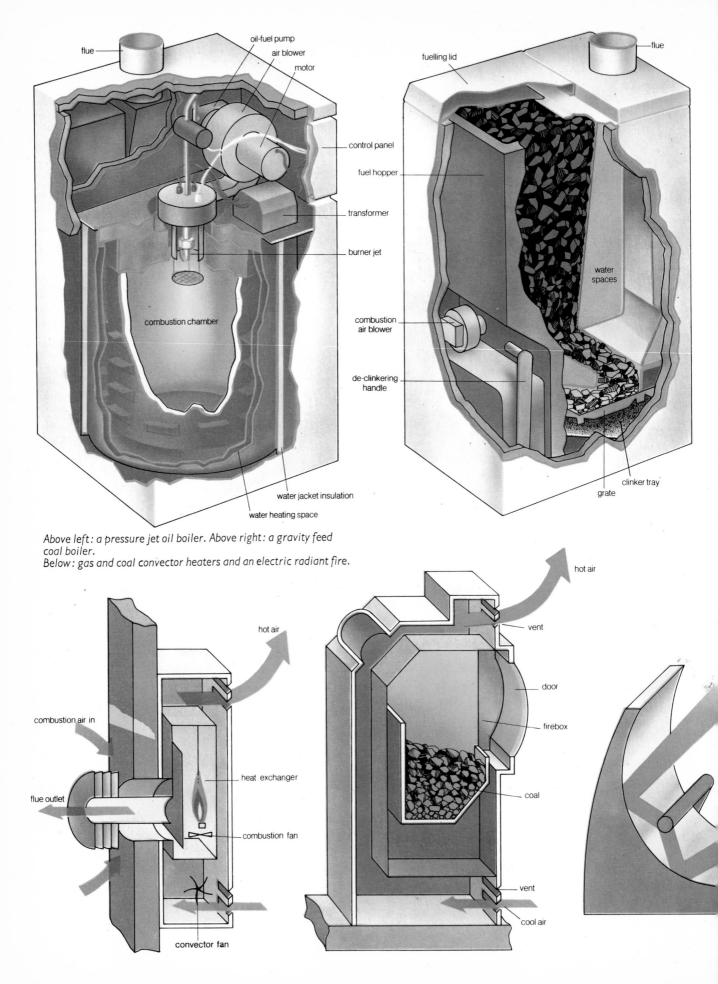

flue

oil-fuel pump

air blower

motor

control panel

transformer

burner jet

combustion chamber

water jacket insulation

water heating space

fuelling lid

flue

fuel hopper

water spaces

combustion air blower

de-clinkering handle

clinker tray

grate

Above left: a pressure jet oil boiler. Above right: a gravity feed coal boiler.
Below: gas and coal convector heaters and an electric radiant fire.

hot air

combustion air in

flue outlet

heat exchanger

combustion fan

convector fan

hot air

vent

door

firebox

coal

vent

cool air

surface area to the hot combustion gases. This ensures that the maximum amount of heat is passed to the circulating water.

The burners in oil-fired boilers can be of the *vaporizing, rotary* or *pressurized* type. In vaporizing burners the oil is fed by gravity along a supply pipe and into a *burner pot* located inside the boiler. Air for the combustion is drawn in or blown by a fan along a *draught tube* positioned around the oil supply pipe. After ignition, the heat from combustion heats up the burner pot causing the incoming oil to be vaporized immediately prior to combustion. In rotary burners the oil is fed on to a spinning *oil cup* situated inside the boiler and air is blown in by a fan. The spinning cup throws droplets of oil outwards on to a burner ring where it is ignited electrically. Once the ring has heated up, the combustion is self-sustaining. Because the flame in a rotary burner is cylindrical in shape and 'wipes' the internal walls of the boiler, rotary burners are sometimes called *wall flame* burners. In the pressurized type of burner, the oil is pumped under pressure to a nozzle inside the boiler, from which it emerges as a very fine spray. As in the vaporizing burner, a draught tube is positioned around the nozzle and air is blown into the combustion chamber by a fan. Combustion is started by means of a spark generated between two ignition electrodes located close to the nozzle.

Gas boilers have a series of gas jets inside the combustion chamber and a pilot jet for ignition. Air for combustion is drawn in from the room in which the boiler is located or from outside through a draught tube. It is not normally necessary to use a fan to blow in the combustion air.

Solid fuel boilers are fed from a hopper above the combustion zone. A fan is sometimes used to blow the combustion air in beneath the grate of the boiler. Solid fuel boilers require a brick chimney or a well-insulated flue pipe carried to above roof level. For small central heating systems having only three or four radiators and where a chimney breast is available, a combined unit comprising a back boiler fronted by an open fire or by a gas or oil-fired incandescent radiant fire can be used.

Domestic heat exchangers are of two general types: *direct* and *indirect*. In the direct type the air to be warmed is blown directly through ducts above the combustion chamber of the heat exchanger whereas in the indirect type a radiator, similar in construction to a car radiator, is fed with hot water from a boiler and air is blown through the radiator. Domestic heat exchangers use the same fuels as boilers and also, sometimes, electricity.

The heat output of boilers and heat exchangers is usually regulated by a thermostat. In the case of oil and gas burners, the thermostat operates a magnetic valve which restricts the fuel supply to the burner. In solid fuel burners the thermostat controls the fan.

Some countries, notably the United Kingdom, offer electricity at a lower price if it is taken 'off-peak', that is during hours of reduced demand. Since these occur mainly at night, off-peak heaters must be able to store the generated heat. The usual method is to enclose the electrical heating elements within blocks of a special high density refractory clay. With water circulating central heating systems the heat transfer can be directly to the water by an integral pipe coil or indirectly using an air-to-water heat exchanger. For warm air systems the heat storage unit operates in a similar manner to a direct-fired air heater. Off-peak electricity can also be used with individual room storage heaters. In the simplest type, heat emission is unrestricted and close control is not possible. Better control is given by units which have dampers for regulating the convection air flow through the heater. Further improvement is possible by using a multi-speed fan to blow air through the airways of the heater.

Another method of heating buildings is to use underfloor heating cables. These usually consist of an electrical wire heating element wound on a rayon core and fitted in a flexible

Below: a ducted hot-air central heating system.

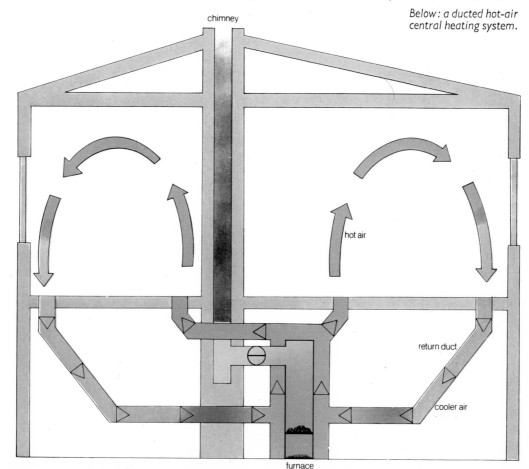

chimney

hot air

return duct

cooler air

furnace

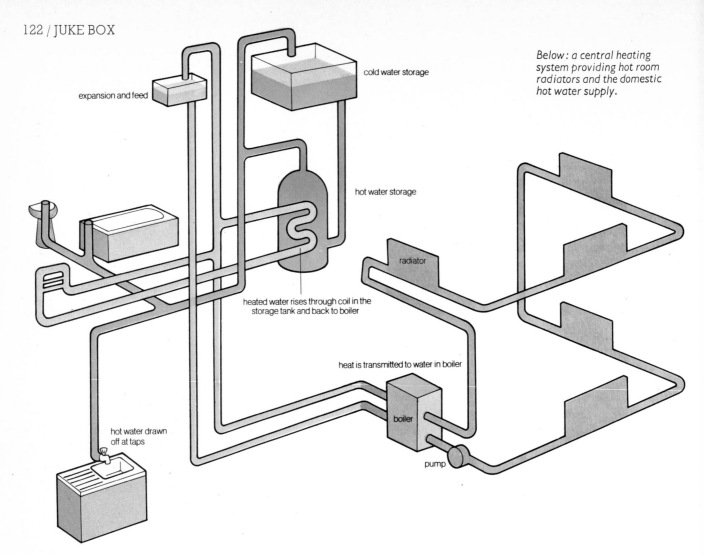

expansion and feed

cold water storage

hot water storage

Below: a central heating system providing hot room radiators and the domestic hot water supply.

radiator

heated water rises through coil in the storage tank and back to boiler

heat is transmitted to water in boiler

boiler

hot water drawn off at taps

pump

PVC tube. Such cables are easy to install under wooden floors and can even be embedded in concrete floors.

Room heaters In electric room heaters an electric current is passed through a coiled wire element to generate heat. Radiant heaters have elements which are designed to glow red during operation and the heat radiates directly into the room from the element. In bar heaters a reflector is positioned behind the glowing element to project heat into the room by reflection. The element of a convector heater operates at a lower temperature and is mounted in a housing which has an air inlet beneath the element and an air outlet above it. Air next to the element is heated and expands so that it becomes less dense than the surrounding cooler air. Being less dense, the hot air rises up through the heater body and out through the air outlet; it is replaced by cool air drawn in through the air inlet. A current of air is thus set up through the heater which distributes the heat throughout the room. Fan heaters are similar to convector heaters except that an electric fan is used to drive the air past the heating elements. Another form of electric room heater is the oil-filled type. An immersion element mounted inside the heater at its base heats up the oil in the radiator and, like a radiator in a central heating system, the heat is transferred to the adjacent air by conduction and then throughout the room by convection.

Free standing paraffin [kerosene] heaters are useful in applications where economy is important or an electricity supply is not available, for example in the heating of greenhouses. These usually have an adjustable cylindrical wick situated at the base of a chimney tube so that as the hot combustion products pass up the tube, air for combustion is drawn in around the wick. The wick dips into a tank of paraffin [kerosene] at the base of the heater. With proper adjustment of the wick the heater will operate efficiently with a smokeless blue flame.

Hot water supplies Domestic hot water supplies are usually derived from a central heating boiler or from an electrical immersion heater. In the former case, heat is transferred from the boiler to the water tank by means of a water-to-water heater exchanger so that the hot water supply is kept separate from the water circulating through the radiators. An immersion heater consists of a metal tube enclosing an electrical heating element. The heater is generally fitted at the base of the hot water tank and the heat distributed throughout the tank by convection. Some hot water heaters, usually operated by gas or electricity, heat the water only as it is required. Such heaters simply require connection to the cold water mains.

Heaters of various types are widely used in industry. For example, furnaces are used in the steel industry and steam-generating boilers are an important part of power stations.

JUKE BOX

The word *juke* originally came from the dialect of the Black people living on the south-east coast of the United States. It meant 'wicked' or 'disorderly', and probably was derived from a similar African word. A *juke joint* is the sort of place where one expect to find a juke box.

A juke box is an automated record player, operated by inserting coins. Its antecedent was the coin operated musical box. Early juke boxes were called *nickelodeons*, because in the USA it cost a nickel (five cents) to operate them. They only played one side of a 78 rpm record, and the *stylus* (incorrectly, *needle*) had to be replaced often because it was made of steel.

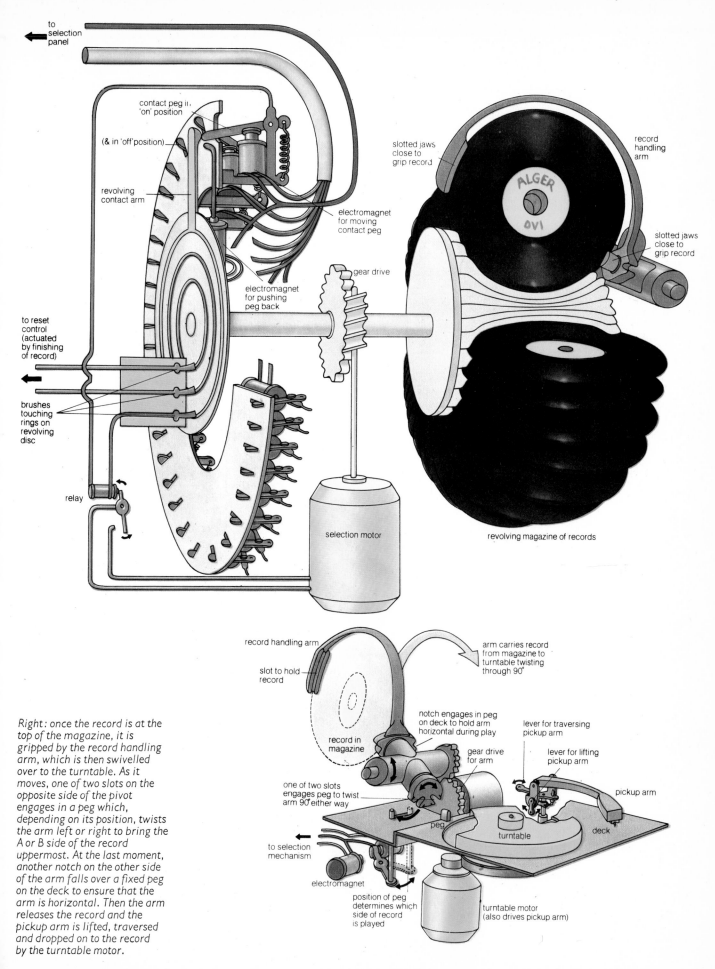

to selection panel

contact peg in 'on' position

(& in 'off' position)

revolving contact arm

electromagnet for moving contact peg

electromagnet for pushing peg back

to reset control (actuated by finishing of record)

brushes touching rings on revolving disc

relay

gear drive

slotted jaws close to grip record

record handling arm

slotted jaws close to grip record

ALGER
OVI

selection motor

revolving magazine of records

Right: once the record is at the top of the magazine, it is gripped by the record handling arm, which is then swivelled over to the turntable. As it moves, one of two slots on the opposite side of the pivot engages in a peg which, depending on its position, twists the arm left or right to bring the A or B side of the record uppermost. At the last moment, another notch on the other side of the arm falls over a fixed peg on the deck to ensure that the arm is horizontal. Then the arm releases the record and the pickup arm is lifted, traversed and dropped on to the record by the turntable motor.

record handling arm

slot to hold record

record in magazine

notch engages in peg on deck to hold arm horizontal during play

gear drive for arm

lever for traversing pickup arm

lever for lifting pickup arm

pickup arm

one of two slots engages peg to twist arm 90° either way

arm carries record from magazine to turntable twisting through 90°

peg

turntable

deck

to selection mechanism

electromagnet

position of peg determines which side of record is played

turntable motor (also drives pickup arm)

During the early 1950s, a so-called 'battle of the speeds' took place in the record industry, and for a while records were widely available in three speeds and three sizes. Some juke boxes were built which could play a mixed selection of records, but nowadays nearly all juke boxes play seven-inch (17 cm) records at 45 rpm, although some machines are designed to play records at 33⅓ rpm as well.

Juke boxes can be divided basically into two types of record selecting mechanisms. The first type, typified by the Seeburg 'Selectomatic', has all the records standing vertically in a row, and the selector travels up and down the row. This design is also available in a conventional wooden cabinet for use in the home, playing full-sized LPs, or for continuous playing of background music in public places. The second type, used by Wurlitzer and others, has all the records in a magazine which revolves around the selecting device.

Credit cycle When a coin is deposited, the mechanism decides what denomination it is by weighing it and measuring it. The coin also passes over magnets which reject any alien metals. After the coin has been shunted into its respective channel according to its worth, it strikes a credit switch, normally two thin metal contacts touching together as the coin passes, closing a circuit to the credit unit.

The job of the credit unit is to establish the right number of plays for the coin deposited. This is done in the majority of cases by a toothed wheel which steps around an appropriate number of teeth. As each selection is made the credit unit cancels a credit by moving the wheel one tooth nearer to its rest position. When the last selection is made, a lever attached to the wheel breaks contact with the selection circuit.

Selection mechanism There are two ways of programming the selection mechanism. The *pin selector* has a bank of pins, one for each side of each record; when one of these pins is lifted or punched up by a small solenoid switch, the scanning device can 'find' the selected record. The other types uses ferrite discs called *toroids* in a *magnetic core memory unit*, similar to that in a computer.

Each disc is shaped like a small washer and can exist in either of two magnetic states which can be called the 'no' or 'off' state and 'yes' or 'on' state. Initially, all the discs are in the 'no' state. When selecting a record one and only one disc (corresponding to that record) will be 'addressed' and in so doing its state will change to the 'yes' condition. For the machine to discover which disc has been changed, each is scanned in turn—passing a current along a wire which passes through the centre of each disc individually. This current will not affect any of the discs in the 'no' state, but when passing through one in the 'yes' state causes it to change back into a 'no'. This change in magnetic state induces a small pulse current in a sensing wire which passes through the centres of every disc. The pulse is detected by a silicon controlled rectifier (a semiconductor device which acts as a switch) which locks the mechanism to the associated record.

From this point on the action is mainly mechanical. The record is pushed out of the rack into the selector by an arm and played vertically, or lifted out of the magazine, placed on the turntable and played horizontally. In the case of the mechanism which plays the record vertically, there is a stylus on each side of the tone arm for playing either side of the record and stylus pressure is maintained by a spring. At the end of the record the unit rejects automatically, as in a domestic record changer. During the mechanical operation the pin or the toroid is restored to its rest position. If a record is selected twice before it is played, it will play only once; the records are played in the magazine sequence rather than the sequence in which they are selected.

Remote selection The *remote selector*, or *wallbox*, is used in places so small that the main juke box must be located in a remote place, or so large that there must be more than one selection location for convenience. The wall box works much like a telephone dialling system. After credit is established in the normal way, a motor revolves, passing a wiper over a printed circuit board. The wiper sends a train of impulses to the main box, where they are received on a *stepper*, which consists of *uniselectors*, one each for letters and numbers.

Left: in this design the arm which removes the record from the revolving magazine lays it down on the turntable, selected side up. The turntable is belt driven to isolate it from vibration.
Opposite page: key-cutting machines. A rotary file removes metal from a blank key, using the key being copied as a template. The latest types (top) are automatic and have covers.

KEY CUTTING MACHINE

A key cutting machine is essentially a rotary file which follows a template provided by the key being duplicated. The machine is fitted with two vices, one of which holds the key to be copied and the other the blank. The carriage has a profile follower at one end which follows the configuration of the key, and a rotary filing wheel with a sharp cross section at the other end which removes metal from the key blank until it matches the profile of the key being copied. An adjustment is provided for raising or lowering the profile follower.

Electric and hand-cranked models are available. The vice jaws locate on the flute (longitudinal groove) on the side of the key, and the key must be pushed in all the way up to the 'ear' of the key. Double-sided keys are copied by turning over both the key and the blank in the vice jaws, copying one side at a time. Care must be taken that the jaws locate on the corresponding flutes of the key and the blank, otherwise the notches which make up the profile of the key will be cut too deeply or not deeply enough.

The latest model of electric key-cutting machine is fitted with a metal cover which encloses the device. The operator pushes the key and the blank into appropriate slots and flips a switch; the machine makes its traverse automatically and the cover prevents the metal filings from flying into the eyes of bystanders.

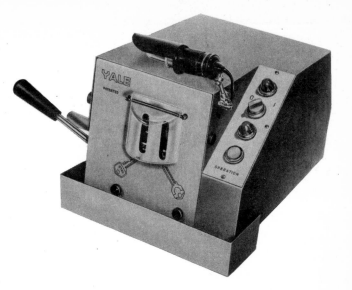

LASER (and maser)

Lasers and masers are devices for producing intensified electromagnetic radiation by a process involving energy states within the atoms of materials.

The word *maser* is derived from the initial letters of the phrase 'Microwave Amplication by the Stimulated Emission of Radiation', and it is the process of stimulated emission, first predicted theoretically by Albert Einstein in 1917, that accounts for certain unique characteristics of the radiation produced.

Gamma rays, visible light and radio waves are all forms of electromagnetic radiation differing only in respect of *wavelength*, which is the distance between successive maxima on the wave profile. Radio waves range from 10,000 metres to 1 metre wavelength and the range occupied by visible light and infra-red radiation is of the order 0.1 cm to 0.0001 cm. The range from 1 metre to 0.1 cm in the electromagnetic *spectrum* (wavelength range) is called the microwave region.

The development of the master in the late 1950s made possible for the first time the generation and amplification of single wavelength radiations in the microwave region and the range was extended in 1960 with the invention of the optical maser to produce single wavelength radiations in the

A 4000 watt carbon dioxide continuous wave laser. CO_2 only forms about 5% of the total gas —there is also about 5% nitrogen and the remainder is usually helium. CO_2 atoms are excited into their higher energy state (pumped) via energy transfer from the nitrogen. The helium is present to stop the CO_2 from absorbing the radiation created by stimulated emission.

visible region of the electromagnetic spectrum. The optical maser is generally referred to as a laser from the expression 'Light Amplification by the Stimulated Emission of Radiation'.

Stimulated emission Before considering the meaning of stimulated emission we must consider the process by which a conventional light source, for example the sun or a filament lamp, produces light energy.

An atom of any material can be considered to consist of a central mass or *nucleus* around which tiny particles, *electrons*, travel in orbits. The electrons can only occupy certain orbits, and corresponding to each of these is a definite energy of the atom. When the electrons are as close to the nucleus as possible, the atom possesses the least possible amount of energy and is said to be in the *ground state*. If the electrons occupy orbits of greater radius, the atom is said to be in an *excited state*.

An atom will normally be in the ground state but can be excited by supplying energy to the material in the form of heat. Increasing the temperature increases the energy of motion (kinetic energy) of the electrons and when this energy becomes equal to the energy corresponding to an excited state of the atom, the atom moves to this excited state. A number of excited energy states are possible for any atom or configuration of atoms.

Once in an excited state, the atom has a tendency to return to a lower energy state by emitting some part of its energy content, which is given out as light energy. According to the quantum theory, the amounts of energy involved in these transitions are considered as finite bundles or packets of energy called *quanta*, the quantum of light energy being termed a photon. The wavelength of the emitted *photon* of light depends on its energy which, in turn, depends on the nature of the energy transition involved. This is the process by which a heated filament in a lamp, for example, emits light.

The light thus produced contains quanta of a wide range of wavelengths, since the number of energy states and hence possible transitions is high. Also the time interval between the gaining of thermal energy and the emission of this energy as light differs for the various energy states. Thus the photon will be emitted with the associated wave motions in a generally random manner—out of step or *out of place* with each other. Such a process is termed *random* emission and the emitted light is said to be *incoherent*.

There is another process by which excitation of an atom can occur. When a quantum of radiation strikes an atom there is a possibility that the atom will absorb energy and be excited from the ground state to a higher energy state. This process is termed *absorption* and once excited the atom may at some later stage emit a photon and hence return to the ground state—*spontaneous* emission.

A third emission process can occur. If an atom in an excited state is bombarded by a photon or a wavelength or energy content corresponding to a difference in energy levels appropriate to the bombarded atom, the atom may emit a photon and reduce its energy to the appropriate lower energy level. This is the reverse of the absorption process and the photon thus emitted will be of the same wavelength and of the same phase as the photon originally striking the atom. The original photon continues on its path and the emitted photon adds to it. The process is repeated at each collision with an appropriately excited atom and the emitted photons all add together giving an intensified or amplified beam of photons, all of which have their *wave motions* in phase with each other. The beam of light thus produced is said to be *coherent* and the emission process is termed *stimulated* emission. When this situation prevails light amplification is produced by the stimulated emission of radiation.

For an amplification of radiation to be achieved in practice, the system of atoms must contain an excess of atoms in high energy states, since those in low energy states will absorb

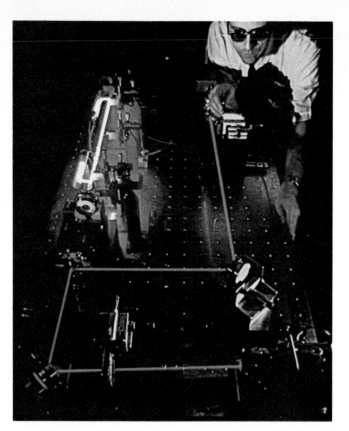

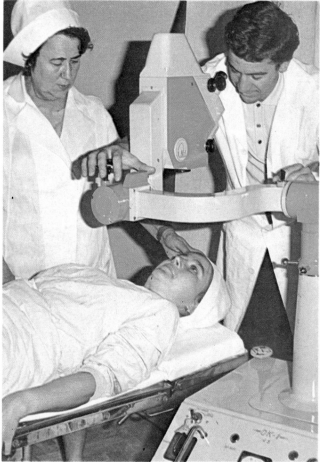

Left: laser beam in detector of optical photocomputer memory system.
Below: laser apparatus in a Soviet hospital being used for the treatment of detached retinas. The laser can be precisely aimed at the detached section, which it 'welds' back in place with one burst.

energy from the beam and thus reduce the beam intensity. Since in normal circumstances low energy atoms will outnumber those in an excited state, the practical problem that needs to be solved is the means of reversing this situation so that amplification may be obtained.

The ammonium maser The first device in which the process of stimulated emission was used to produce amplification of radiation was the ammonium maser.

The ammonia molecule consists of four atoms, one of nitrogen and three of hydrogen, combined together. The molecule possesses a large number of excited energy states but two energy states are of particular interest. These two states, one of higher energy than the other, are such that when the molecule goes from the higher state to the lower, radiation of wavelength 1.2 cm is emitted. If the ammonia can be made to contain a larger concentration of molecules in the more energetic state than in the lower state, it may be possible to amplify radiation of 1.2 cm wavelength by stimulated emission.

Ammonia is in fact a substance in which it is a relatively simple process to produce excited molecules, which must outnumber the low energy molecules if amplification is to be achieved. To produce ammonia with the required enrichment of high energy molecules, a filtering system is used. The low and high energy molecules behave differently when subjected to an electric field and this can be utilized to obtain a very effective separation of excited molecules from the gas.

The system is arranged so that excited molecules pass into a box or cavity which constitutes the amplifier. If a weak signal wave of 1.2 cm wavelength is passed into the cavity at one end, the result of amplification by stimulated emission from the excited ammonia molecules emerges at the other end.

A signal wave making one passage through the cavity de-excites very few of the excited molecules and the amount of amplification produced is therefore small. This is improved by giving the cavity highly reflecting walls so that the radiation fed into it is able to traverse the cavity many times.

Each traverse of the wave results in some stimulated emission from the excited molecules and the wave grows in intensity. The spontaneous emission during this process is negligible.

The ruby laser The next development of stimulated emission amplifiers was a device capable of producing amplification of shorter wavelength radiation, that is, visible light.

The first successful laser used a crystal of ruby. Ruby is a crystal of aluminium oxide in which some of the aluminium atoms are replaced by chromium atoms. It is the presence of chromium that gives the ruby its characteristic deep red colour and it is the suitably excited chromium atoms that account for the amplifying effects in the ruby laser. The chromium atoms are raised to excited energy states by absorption of light energy from an external flash lamp source which is said to 'pump' the chromium atoms to higher energy levels. The ruby crystal rod has its ends polished flat and coated with a reflecting material.

Many of the excited chromium atoms give out their excess energy by spontaneous emission and the light emitted escapes through the sides of the rod. A few of the atoms will give out light in the direction of the mirrored ends of the rod and this light can then be reflected back through the crystal. When this radiation strikes an excited chromium atom,

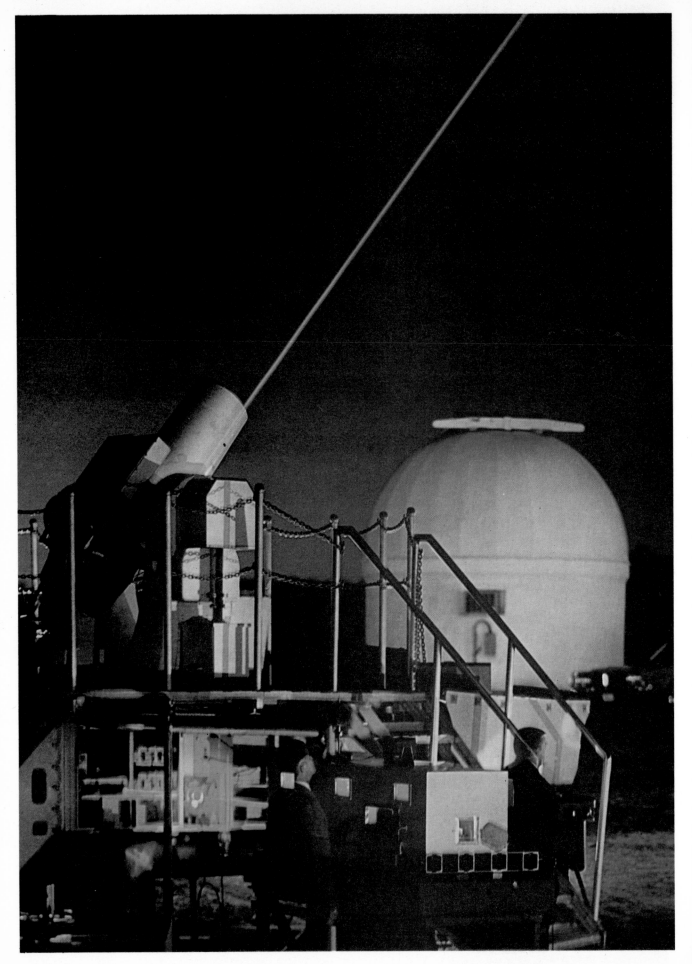

Opposite page: this laser beam is being aimed at the Moon and the reflected beam detected with a telescope. Because a laser emits a nearly perfectly parallel beam of light, the beam spread (even after travelling 240,000 miles, 384,000 km) is small enough and the reflected light therefore bright enough to be observed. This enables the Earth-Moon distance to be determined with great accuracy by recording the time taken by the light to return to the Earth.
Left: cutting through sheet metal with a carbon dioxide laser.

stimulated emission can occur, amplifying the radiation, and thus the light travels back and forth between the reflecting ends of the rod, becoming more and more intense with each traverse.

The process of exciting the chromium atoms imparts heat to the ruby crystal, and for continuous operation adequate cooling of the device must be arranged.

Since the development of the ruby laser, many other crystal systems have been used in laser devices and by suitable choice of crystal, laser beams of a range of wavelengths can be obtained.

The gas laser Another class of laser devices employs gases as the amplifying medium. The first such device was one containing a mixture of helium and neon. As with the crystal laser the problem is to energize the atoms into excited states from which emission can be stimulated.

Gas atoms can be excited by passing an electric current through the gas, a technique that is responsible for the production of visible light in neon signs and sodium street lighting. By subjecting a mixture of helium and neon to an electric current it is found possible to excite the gas atoms into energy states from which emission can be stimulated giving radiation of a wavelength of 1.15 micrometres

(0.000115 cm), which is in the infra-red range of electromagnetic radiation. The stimulated radiation is again reflected back and forth through the gas chamber, increasing its intensity as more and more atoms are stimulated to emit radiation.

Other combinations of gases have been used since this initial development, and among the more interesting devices is the one that uses a mixture of argon and krypton. This device is of interest since it produces amplified radiation from more than one region of the visible range of wavelengths with the result that the emitted beam contains several colours and the net result is to give the appearance of white light.

The advantage of the gas laser as compared with a crystal laser is that it can be operated continuously for long periods.

Applications of lasers The remarkable properties of lasers—the high intensity and single wavelength nature of the emitted light—have already led to a large number of applications in a number of fields of activity, and many other applications are envisaged.

An optical lens can be used to focus a laser beam on to a very small area, and this technique has been used in several industrial applications. Holes have been produced in very

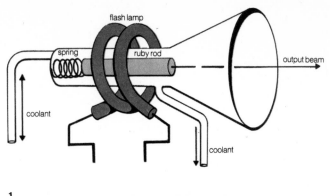

Right: chromium atoms in the ground state (1) are excited by light from the flash tube (2); some of them emit photons which stimulate further photon release (3). These are reflected back and forth between the silvered ends (4, 5) and finally emitted as a powerful beam. Below: this laser beam is used in the motor industry for welding.

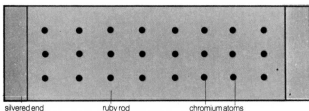

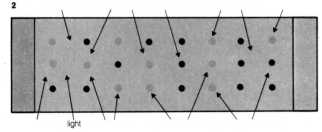

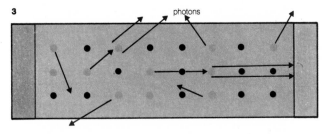

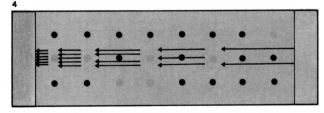

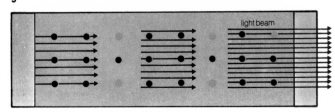

hard substances including steel and diamonds, and as such a tool the laser has the advantage that no physical contact with the material is required, very small holes may be produced without affecting the surrounding material and the process is very rapid. These advantages have also resulted in the use of laser beams in welding operations.

In the field of medicine, lasers have been used successfully in the treatment of detached retinas and show promise of being useful in the treatment of cancer. The condition of detached retina occurs in the eye when the fluid within the eyeball seeps through a hole in the retina. The retina can eventually become detached from the back of the eyeball, causing blindness. Flaws in the retina have in the past been treated by surgery but can now be welded by a flash of laser light lasting for about a thousandth of a second. The light can be precisely aimed and it rarely requires more than a single pulse to attain action in the correct location.

It is in the field of communication that the laser has an enormous potential. In its coherent, single wavelength nature radiation from a laser resembles that from a radio transmitter. If the same techniques are evolved for impressing signals on laser beams as are used for radio waves (such as amplitude modulation) then lasers can be used for communication. The amount of information that can be carried by a wave of electromagnetic radiation increases as the frequency increases. The frequency of a laser beam is so much higher than radio frequencies that it is possible in principle to carry many million radio transmissions on a single laser beam. Also the greater directionality of a laser beam as compared with conventional radio transmission means that less power would be required to communicate between two stations.

The notion of all the world's broadcasting requirements being satisfied by using a single laser beam is an exciting one, but several problems remain to be solved before this becomes a practical reality. The most serious of these problems is the fact that laser light, like any other form of light, is stopped by fog and cloud. In outer space this obviously creates no problems but for terrestial communications it will be necessary to send the beam along tubes or pipes to overcome the atmospheric difficulties. Use of glass fibres to carry the laser beam is envisaged but at present there is great difficulty in manufacturing glass fibres of sufficient quality to enable them to be used over long distances.

Another technique which utilizes the coherence of laser sources is holography, by which means it is possible to produce images and objects and scenes in seemingly perfect three-dimensional realism. It is also possible by holography to store on a single piece of film a vast amount of information which can be subsequently retrieved by illuminating with laser light. This technique could well render microfilm storage of printed matter obsolete. Computer technology is also likely to make great strides following the application of lasers to the problems of data storage and retrieval. Certain materials undergo a change of colour when exposed to a laser beam, and if such a *photochromic* material is subjected to a laser beam of varying intensity, a series of coloured dots may be 'written' on the crystal and stored there. By the use of a weaker beam the information can be subsequently read out of such a memory. The interest in this form of data storage derives from the fact that the laser beam can be focused to extremely small diameters, enabling an enormous amount of information to be stored in a very small area.

Lasers in thermonuclear fusion One enormous potential of the laser is in thermonuclear fusion—considered by many scientists to be the energy source of the future. Much research has been done in this field and it seems to be only a matter of time before a working prototype is developed.

At very high temperatures and pressures hydrogen gas atoms are stripped of their electrons and their nuclei are packed closely together. If the temperatures and pressures can be pushed high enough, two hydrogen nuclei can be made to fuse together to form a helium nucleus—in the process there is a reduction in mass which means a release of energy.

By bombarding a frozen pellet of hydrogen on all sides with an intense 'pulse' of laser light, its temperature can be rapidly raised to over 10,000,000 °C. The pellet immediately vaporizes and expands with terrific force. At the same time, this causes an equal and opposite implosion which compresses the hydrogen to the pressures necessary for fusion to take place.

The major drawback to this particular fusion technique has been developing sufficiently powerful lasers. Military research has, however, played an important part in this through their interest in developing a 'death ray'. One particular type—the gas-dynamic laser—has brought down aircraft targets up to two miles away, but military applications are limited at such enormous powers because of the atmosphere. One problem is that such an intense beam heats up the air, which makes the beam defocus.

LENS

The principle behind the working of a lens is the *refraction* or bending, of light as it passes from air into glass. If a light ray strikes the glass surface face on (along the *axis* of the lens), it is undeviated, but the angle by which the ray is bent progressively increases as the ray strikes the glass at more and more of a grazing angle, the amount of the deviation being measured by the *refractive index* of the material. Thus, a group of light rays radiating from an object, passing from air into the convex face of a magnifying lens, is reconcentrated into the glass. They are further concentrated as they exit from the lens, and can form an image on a nearby screen. The more the lens surfaces are curved, the greater their concentrating effect, so that the image is formed closer to the lens.

The *focal length*, f, of a lens is defined as the lens-to-screen distance for the image of a very distant object. The more the concentrating ability of the lens, the greater its *power* is said to be, and this is measured by the reciprocal of the lens's focal length in metres ($1/f$), expressed as so many *dioptres*.

As a distant object is brought nearer to the lens its image is formed at a greater distance from the lens than its focal length and it is necessary to increase the lens-to-screen distance by backing off the screen. The eye contains two lenses, the *cornea* and the *crystalline lens*, which together form an image of the outside world on a light-sensitive screen, the *retina*. The eye forms images of objects at various distances, not by changing the lens-to-screen distance, which is the fixed depth of the eyeball, but by changing the focal length of the crystalline lens located behind the pupil. It does this by altering the curvature of the lens, flexing it with muscles attached to its edge. Because

Astigmatism is a defect of lenses which are not symmetrical and results in the focal length being different in one direction, so that an image in a straight line will change direction slightly from one side to the other. The human eye can develop astigmatism, which is corrected by spectacle lenses.

astigmatic image

optical axis

lens

point source

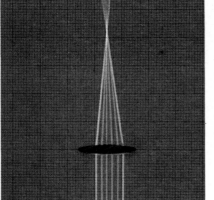 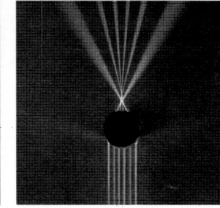 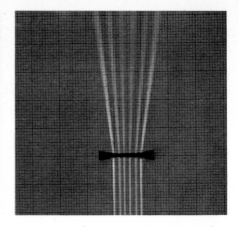

The behaviour of various types of lenses is shown in the sequence of photographs above. At far left is seen the effect of passing parallel light through a convex (converging) lens. The most extreme rays of light are refracted the most, and the more central ones the least, but all rays cross at the focal point. A thick convex lens, however, shows the effect of spherical aberration (next picture): the extreme rays are focused closer than the central rays. The focal point is also closer to the lens, though the aperture is the same. A concave (diverging) lens spreads the light out, as shown next. A convex lens can be used as a magnifying glass, but a concave lens has the effect of making objects appear smaller. The ray paths show why this is so: to an observer looking through the lens, the light which was originally parallel, as from a distant object, now appears to have come from a smaller point very much closer, where the rays would meet if traced back. This image cannot be shown on a screen—it is a purely visual effect, and is known as a 'virtual' image. The image from a convex lens, however, is 'real' and can be seen on a screen. The next two photographs above show 'pincushion' and 'barrel' distortion respectively, due to the magnification of the lens varying with the distance from the centre. The blurring is caused by spherical aberration. The last picture shows the light path through a compound lens, designed to reduce aberrations.

the eye can only do this up to a point fixed by the flexibility of the lens material it is impossible to focus one's eye on objects closer than a certain minimum distance, about 10 cm (4 inches) in young people. The lens material becomes less flexible with age, so that the *minimum working distance* increases to 50 cm (20 inches) or more, a condition known as long-sightedness which forces the aged to read a newspaper at arm's length. In the extremely long-sighted it is necessary to pre-concentrate the light rays with a spectacle lens before they pass into the eye.

In the somewhat less common condition of short-sightedness the cornea overconcentrates the light rays, so a spectacle lens must be used to diverge the rays before they pass into the eye. Such lenses, *diminishing* lenses, have surfaces which are concave.

Focal ratio The light grasp of a lens increases with its area. The light, however, is spread into an image whose size increases with the focal length of the lens, so an image appears bright if formed by a lens of large diameter and small focal length. The quotient of diameter and focal length is called the *focal ratio* of the lens. A lens whose diameter is half its focal length has a focal ratio $f2$ (pronounced 'f-two'). The eye, like a camera, controls the brightness of the image formed on the retina by altering the diameter of the pupil, giving a range of focal ratios from $f10$ to $f3$.

If two lenses of the same diameter, but of different focal lengths—say $f2$ and $f8$—are taken, it will be found that the $f2$ lens gives a small, bright image close to the lens while the $f8$ gives a more magnified image further away.

The *resolving power* of a lens, its ability to give a finely detailed image, improves with increasing diameter. This can be tested by looking at a scene through a pinhole, which reduces the eye's effective diameter from the normal 4 mm or so to much less than a millimetre. The scene then appears blurred, though the depth of field—range of objects in focus—is much greater.

Lens design Simple lenses are made of just one kind of glass, ground so that the surfaces are parts of spheres. It is a matter of experience that the images seen through the edge of such a lens—a simple magnifying glass, for example—are less distinct than those seen through the central area. Moreover, the image often has a bent appearance over its area: the image of a newspaper, for instance, seems to curve away at the edges of the glass. Often it is noticeable that the edges of a white object seen on a black background appear coloured, blue to one side, red to the other.

All these deficiencies (known as *aberrations*) illustrate the inability of a simple, single component lens with spherical surfaces, to focus light of all colours, passing through different parts of the lens at different angles. The job of a lens designer, in specifying a lens for a particular function, is to minimize aberrations within restrictions of cost, size and position of the lens (to make it fit a given optical system), loss of light, and so on.

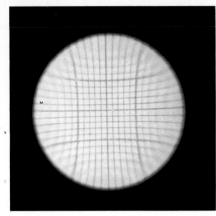 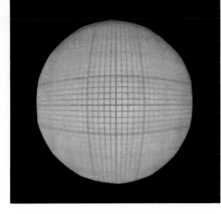 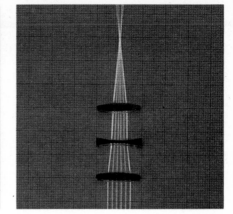

Spherical aberration arises because the rays of light which pass through the edges of a lens with spherical surfaces focus at different points from the rays passing through the central areas. The aberration can be minimized (but not eliminated) by a choice of the curvature of the two surfaces of the lens. Non-spherical or *aspheric* curves can also be used but are avoided in cost-conscious designs as they are more difficult to manufacture.

Images of points of light, such as stars, often have a cometary appearance, with a fainter outward-pointing 'tail' of light, as they get further from the axis, and this aberration is hence called *coma*. It can in fact be eliminated in a simple spherical lens by a proper choice of the curvature of the lens surfaces, but not the same choice which minimizes the spherical aberration. A lens designer must trade off coma against spherical aberration in a compromise choice which minimizes the total effect of both aberrations for the particular system he is working with. To some extent the designer becomes an artist at this point, because there is no unique choice of a *figure of merit* with which to measure the success of the trade-off, and the designer must choose the formulation which matches best the design requirements for the sharpness of the image to be produced.

Chromatic aberration arises from the change of refractive index of glass with colour. It is overcome by combining two or more lenses of different kinds of glass, the spectrum of colour dispersed by the first lens being at least in part re-combined by subsequent lenses. The result is called an achromatic lens. The designer's task is easier if the optical instrument is to be used only in a restricted band of colours. Thus, early photographic refracting telescopes needed to be corrected for chromatic aberration only in the blue region of the spectrum, since photographic material was not sensitive to other colours. Modern focusable camera lenses are corrected for chromatic aberration over the whole of the visible range of the spectrum since they are required to take colour photographs; the focus needs correcting to a fiducial mark usually indicated by a red letter R if infra-red photographs, using longer wavelengths of light, need to be taken.

A designer faced with a multi-lens system has more freedom to choose the degree of curvature of the lens surfaces to minimize other aberrations but may begin to be worried by the loss of light in the glass of each lens and by the reflection of light at each surface. The designer is helped by modern methods of *blooming* lenses, coating them with surfaces which reduce the reflected light losses (at least for certain restricted ranges of colours). Blooming is visible on the lens of, say, a good pair of binoculars, as a coloured appearance of the lens when viewed obliquely.

Photographic lenses consist of several individual lens *elements*, chosen to produce as sharp an image as possible over a given field of view. A vast range of focal lengths from wide angle to telephoto, is available, as well as zoom lenses, in which the focal length is variable.

LETTER ADDRESSING MACHINES

One of the problems regularly encountered in large offices is that of addressing large numbers of envelopes in a short time. The first attempts at mechanizing this task came at the end of the nineteenth century with the growth of the office machine industry. Existing duplicating machines were unsuitable because they were designed to produce many copies of the same message, not a series of different ones, but on the other hand the same addresses were used over and over again by large firms sending regular statements to their customers.

The solution adopted was to prepare a small permanent plate which could print at one pass, thus eliminating the need to write or type the address every time it was needed. The plates were small enough to be handled mechanically, using a simple gravity feed to move each plate to the printing position quickly and reliably. Up to several hundred envelopes per hour could be prepared by this method, compared to the rate achieved by a typist or clerk which was about one per minute.

The first plates to gain wide acceptance were made of metal, and were embossed with the name and address of the recipient. The embossing was done on a typewriter or, more satisfactorily, on an embossing machine. The embossed plates printed the envelopes by pressing an ink-impregnated ribbon against them.

Metal plates are widely used today and have changed little over the years. The advantages are that they are very robust, being capable of producing up to 10,000 impressions, with a print quality that can be almost indistinguishable from typing, although this is not always achieved. They are often used for other repetition work besides just letter addressing. Recently plastic cards have been developed to perform the same functions.

The major disadvantages of plates or plastic cards are that they are heavy to type and the results can be inconsistent if a normal typewriter is used, apart from shortening the life of the typewriter. The embossing machines used range from simple letter-by-letter machines to heavy duty typewriter-like machines which can re-emboss the plate if a mistake is made.

The inconvenience of preparing metal plates, particularly where only a limited use is made of them, created a demand for a plate that was simpler to prepare. Stencil-type plates are much simpler to prepare and are therefore more suitable for a rapidly changing address file. The stencils may be similar to duplicator stencils or hectographic masters, and the simplest machines consist of a stencil and a hand roller which carries the ink.

With large files of addresses it is frequently necessary to select only a certain number of the entire list, and the desired shorter list may be distributed in a fairly random fashion throughout the file. Some method of selecting the required plates is therefore necessary.

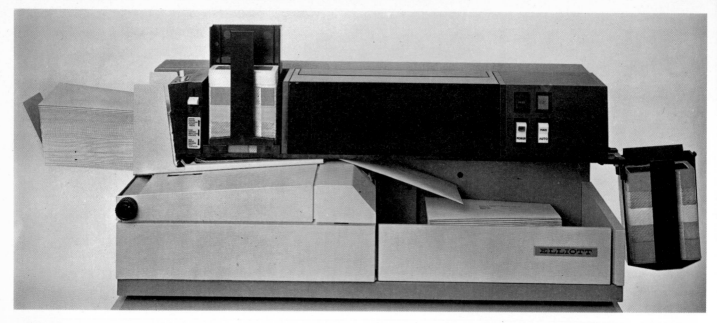

Above: a modern automatic envelope addresser. The envelopes are fed in at the left and pass beneath the stack of address cards where they are printed. After printing the envelopes are stacked and the cards are fed into the hopper on the right. The lower picture shows part of the drive mechanism, the diecast aluminium chassis, and the power supply of the machine.

Selection The simplest forms of selection use tags on the edges of the plates which are sensed by fingers on the addressing machine. Most machines possess a 'skip' facility so that as each plate enters the printing position it can be either printed or skipped according to the machine setting.

More complex selection systems use some form of punched card control, with electronic selection of the wanted plates. These allow address lists to be cross-indexed over forty ways, so that mailing may be subdivided by area, size of company, number of employees, type of products, number of branches and so on; giving a number of permutations that may be several hundred times the length of the mailing list.

The most advanced selection systems use a computer to store the mailing list and select the desired addresses. The computer may print the addresses on to labels for separate attachment or directly on to envelopes. Other machines perform the printing operation independently, using a computer-generated listing.

Selection systems, especially the computor-generated type, are also a convenience for those companies which sell lists of subscribers or members to other companies for the purpose of direct-mail advertising.

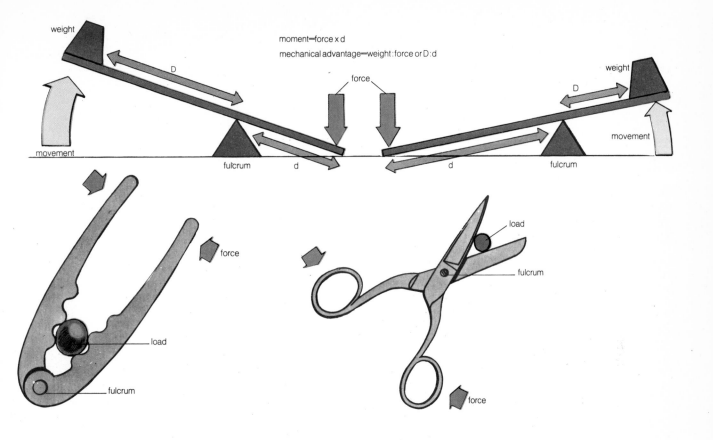

LEVER

The invention of the lever, one of the simplest machines, was of great importance to the development of primitive society, comparable in importance to the invention of the wheel. The lever allowed a man to amplify his limited strength to a level capable of lifting very large and heavy objects. Ancient structures like Stonehenge or the pyramids could never have been built without the use of levers to manipulate blocks of stone weighing many tons.

The lever amplifies force, movement and velocity.

Force The simple lever consists of a pivoted beam, the pivot point being the *fulcrum*. The applied force or *effort* is brought to bear upon one end of the beam, preferably as far from the fulcrum as possible. An amplification of the effort is achieved when the load to be acted upon is nearer to the fulcrum than the effort is.

The amount of amplification of the effort, known as the *mechanical advantage*, is the ratio of the load to the effort. It is also the ratio between the distance of the effort from the fulcrum and the distance of the load from the fulcrum.

Thus an effort of I kg acting at a distance of I metre from the fulcrum would lift a load of 2 kg placed on the other end of the lever 0.5 m from the fulcrum, or a load of 4 kg placed 0.25 m from the fulcrum. In the first example the mechanical advantage would be 4 and in the second example it would be 8.

A small effort acting at a large distance from the fulcrum exerts the same leverage or *moment* as a large effort acting closer to the fulcrum. As the moment is defined as the product of a force and its distance from the fulcrum, it follows that for a simple lever arrangement the moment exerted by the effort must equal that exerted by the load in order to balance it, and exceed it in order to lift the load.

The load and effort may be on opposite sides of the fulcrum, or on the same side. A pair of scissors, for example, which can be thought of as two levers acting in opposite directions through the same pivot, have the load at one side of the pivot and the effort is applied at the other. The mechanical advantage, and thus the cutting force, increases the nearer the material to be cut is to the pivot.

On the other hand, a common design of nutcracker is a good example of the type of lever in which load and effort are on the same side of the fulcrum, and in order to crack a nut with minimum effort (with maximum mechanical advantage) it should be placed as near the fulcrum as possible.

Movement Levers have countless applications where they are used to amplify an effort or force, but another important feature of the lever is that it amplifies movement. In the case of a simple lever with the fulcrum at the centre, such as a see-saw, if one end moves a certain distance the other end will move the same distance. If the fulcrum is placed nearer to one end than the other, however, the two ends will move through different distances, the ratio between the distances being the ratio between the lengths from each end to the fulcrum.

For example if the fulcrum is placed one third of the way along the lever, the end of the longer section will move twice as far as the end of the shorter section. Thus by using a lever arrangement a large movement can be obtained from a small one and vice-versa.

Velocity In addition, as the two ends are moving through different distances during the same time interval, they are travelling at different velocities, the ratio between these velocities being the same as between the distances travelled.

In many machines levers are used to provide amplification of force, movement or velocity, and often a combination of these, as in an ordinary typewriter where the relatively short and slow motion of the keybuttons is converted into the rapid movement of the typebars.

LIGHT BULB

If sufficient electric current can be passed through a conducting filament, the molecules of the filament become excited, the filament gets hot and eventually glows. This is the principle of the light bulb (more correctly known as an incandescent filament lamp).

Light bulbs are so much a part of our everyday lives that we tend to overlook the advanced technology that makes such a simple device practical.

After many experiments with incandescent metallic filaments in the mid-1800s, the first light bulbs to show signs of becoming practical light sources were devised independently by Swan in Britain and Edison in America in 1878. Swan's first lamp comprised a carbon filament made from vegetable fibre sealed in a glass envelope with platinum wire leads. The envelope was evacuated to quite a low pressure but oxygen trapped in the filament material was liberated during operation and, together with the non-uniform cross-section of the filament, led to a short life.

By 1905 however, filaments were being made from a cellulose solution extruded through dies and then carbonized. Lamps using these filaments were commercially available and achieved a light output of about 2 to 3 *lumens* for every watt of electrical energy supplied to the lamp (usually written as lm/W). A lumen is the amount of light falling per second on unit area placed at unit distance from a small light source of one *candela* output. Attention was soon turned to the use of metal filaments in an attempt to improve efficiency and osmium and tantalum were used for a time but each had disadvantages which led to early failures. In 1909 the practical problems of drawing tungsten into a fine wire were solved, and because it has a melting point of 3382 °C (6145 °F) it could be operated at temperatures considerably higher than previously, with resulting benefits in light output.

Gas filling The life of the filament lamp depends upon the rate at which the metal filament evaporates, which in turn depends upon the temperature at which it operates. In 1913 it was discovered that filling the evacuated bulb with an inert gas, such as nitrogen, retarded the rate of evaporation, thus prolonging the useful life. Gas-filling, however, had the disadvantage that convection currents in the gas tended to reduce the filament temperature and hence the light output. Restrictions caused by World War I meant that it was not until 1918 that the technique of forming the filament into a fine coil was introduced.

Coiling the filament and later coiling the coil (1934) maintained the filament temperature and provides the basis of the lamps used today. The standard pear shape is known as a General Lighting Service (GLS) lamp and is available in a range of wattages from 25 W to 150 W with efficiencies ranging from 8 lm/W to nearly 19 lm/W depending upon wattage.

Types of filament lamp While GLS lamps account for the bulk of mass produced filament lamps the range of types runs into many thousands. They include lamps smaller than a grain of rice for medical instruments, lamps larger than a football for lighthouses, reflector lamps with enough energy at their focus to light a cigar, and reflector lamps designed to let most of the heat escape from the back of the lamp, as well as decorative lamps of dozens of different shapes of several colours and ratings. It is impossible to convey the varieties available but one particular type is of special interest as it represents the most significant advance in filament lamp technology since 1934.

Tungsten halogen lamps Originally called 'quartz-iodine lamps' because of the materials used in early examples, the tungsten halogen lamp is conspicuous for its small physical size compared with GLS lamps of similar wattage ratings.

The principle involves introducing a halogen such as iodine (although bromine and fluorine are also used) into the gas filling which combines with tungsten evaporated

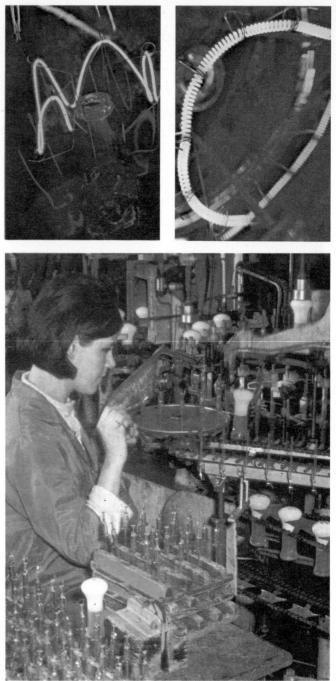

from the filament to form tungsten iodide. The evaporated tungsten is thus prevented from condensing on the lamp envelope and blackening it. The tungsten iodide vapour is circulated within the lamp by convection until it approaches the filament where at a temperature of over 2000 °C (3656 °F) the vapour breaks down and some of the tungsten is redeposited on the filament while the iodine is available for further combination. Because this cycle will only function at temperatures above 250 °C (508 °F) the lamp must be kept small to reduce its heat loss.

The main advantages are that the light output is maintained throughout the life of the lamp and that the filament can be run at a higher temperature, giving 20 to 22 lm/W. Also, because evaporation is reduced a life of 2000 hours is readily achieved. The small physical size lends itself to precise optical control and sizes currently in general use range from 50 to 55 W used for shop spotlights and car auxiliary lighting to 10 kW lamps for floodlighting. It has been suggested that even lower wattage lamps enclosed in a diffusing bulb of

normal size could be the domestic lamp of the future, but present costs of materials make this unlikely.

Manufacture The early lamps were of course handmade, as are a large number of the specialized types today. The GLS lamps of normal domestic size are, however, made on automatic machines at a rate of between 2000 and 4000 per hour.

A specialized high purity glass is fed in a continuous stream to the bulb making machine which first thickens the glass at regular intervals and then blows this thickened portion into a mould to form the bulb shape. The bulbs then move on to where they are cooled, cut off from the ribbon and drop on to a conveyor belt. The process is so quick that a single plant equipped with five machines is serving the whole British lamp-making industry together with a large part of the European requirement.

The tungsten wire for the filament is drawn from a coherent ductile rod made by subjecting tungsten powder to great pressure. The diameter of the filament wire is about 0.014 mm for a 15 W lamp or 0.042 mm for the 100 W.

The two small photographs at top left show (left) a straight (uncoiled) filament and (right) a coiled one. Convection of the gas around a filament reduces its temperature and light output efficiency; coiling it raises the temperature. Left below and above: modern light bulb manufacture is largely mechanized. Left shows insertion and sealing of the filament. Behind the operator, lamps are filled with an argon-nitrogen (90%-10%) mixture, the process shown in detail above.

Because of the difficulty of measurement, the correctness of the filament is checked by weighing measured lengths and the maximum variation is about 2% or, for the 15 W lamp, a tolerance of only 0.00014 mm, or a quarter of the mean wavelength of visible light. The filament is then wound about a mandrel to form the coil and then about another mandrel if it is to be a coiled coil. After annealing, the mandrels are dissolved out in acid, the filaments are re-annealed and subjected to microscopic examination.

Automatic machines mount the filaments to the stem assembly clamping the ends to the lead-out wires. The mounted stem is fed up into the neck of the glass bulb and fused to ensure a perfect seal. The lamp is then evacuated and gas flushed until the residual oxygen is no more than 5 to 10 parts per million when the exhaust tube is sealed. The cap is then cemented to the lamp and the lead-out wires soldered to the contacts.

LIGHT-SENSITIVE GLASS

Light-sensitive glass is a glass that will darken or become tinted when exposed to light. In *photochromic* glass the change is reversible, whereas in *photosensitive* glass it is not. Also, photosensitive glass must be heated to develop the coloration.

Photochromic glass Photochromic glass darkens on exposure to radiation of a particular wavelength, for example ultra-violet light, and becomes clear again when removed from the source of the radiation. It is used to make self-adjusting lenses for spectacles. The lenses will darken in sunlight, which contains ultra-violet light, so that the spectacles can act as sunglasses, but become clear again when the wearer goes indoors or when the sun clouds over. Photochromic lenses can be used for spectacles obtained on prescription.

Above: an image fixed in photosensitive glass by laying a leaf on the glass, exposing the glass to sunlight and then reheating.
Above right: spectacle lenses made of photochromic glass. The upper picture shows the glasses before exposure to light and the lower picture shows them after exposure.

Photochromic glass contains extremely small crystals, about 50 angstroms (50 hundred millionths of a centimetre) in diameter, of silver halides (halides are salts derived from halogen) dispersed throughout the glass composition together with a small amount of copper oxide sensitizer. The proportion of the silver halides (silver chloride, bromide and iodide) one to another determines the wavelength of the radiation to which the glass will be sensitive. Glass containing predominantly silver chloride is sensitive to ultra-violet light, whereas addition of silver bromide and iodide brings the sensitivity into the visible part of the spectrum. When photochromic glass is exposed to ultra-violet light, the energy of the radiation splits the silver chloride molecules into silver atoms, which cause the darkening, and chlorine atoms. This process is very much the same as exposing a photographic film with visible light, but there are three important differences. Firstly, the silver crystals formed in photochromic glass are more than ten million times smaller than those of a photographic emulsion, secondly glass is relatively inert and so the reactive chlorine atoms are not removed by side reactions, and thirdly glass is impermeable so the chlorine atoms cannot migrate away from the reaction zone. When the source of ultra-violet radiation is removed, the silver atoms and chlorine atoms simply recombine to form silver chloride, and the glass clears. The basic composition of photochromic glass is usually that of an alkali metal *borosilicate* or *metaphosphate* glass.

Apart from its use in spectacle lenses, photochromic glass can be used to make aircraft windows in order to protect pilots against glare from the sun. Other proposed uses include optical systems for information storage and display, coatings for radiation control, bottles for light-sensitive drugs and food, and glazing buildings, though this would be expensive.

Photosensitive glass The phenomenon of photosensitivity

in glass was first discovered in ruby glass, which is ruby-coloured glass made by incorporating copper or gold compounds in the glassmaking composition. The glass is clear when first manufactured, and only develops its characteristic colour when it is reheated to a temperature between its *annealing* and *softening* temperatures. It was discovered that if the glass was irradiated with ultra-violet light before reheating, the final colour of the glass was much darker than it would otherwise be. Modern photosensitive glass compositions contain copper, silver or gold compounds and only develop a colour on reheating if they have first been irradiated; otherwise they remain completely clear. Thus, if an image from a photographic negative is projected on to a sheet of photosensitive glass using ultra-violet light and the glass is then reheated, only those areas where light has fallen will darken and the image will be fixed in the glass.

Gold-based photosensitive glass compositions contain a *sensitizer*, usually a compound of the metal *cerium*. When the glass is irradiated with ultra-violet light before reheating, the trivalent cerium ions (three positive charges) in the composition lose a further electron to form tetravalent cerium ions (four positive charges). The electron remains trapped when the glass is cold, but on reheating the glass it reacts with gold ions in the composition to give gold atoms, which cause the glass to become coloured.

Copper and silver based photosensitive glasses do not normally require sensitizers, but the mechanism of coloration is very much the same as for gold-based glasses. It is the formation of metal atoms in the glass which causes the final coloration.

Photosensitive glasses are ideal for making optical scales (or *reticles*) because the particles of metal which form the image are extremely small. By carefully controlling the cooling of the glass after reheating, the size of the metal particles can be kept below the wavelength of visible light.

Silver-based photosensitive glass differs from photochromic glass in that the former contains silver ions whereas the latter is composed of silver halide crystals. The development of the colour in photosensitive glass is irreversible because, once formed, the metal atoms cannot react again to form ions.

METAL DETECTORS

Metals have one important property possessed by no other elements that enables them to be readily distinguished with suitably sensitive apparatus—this property is their high electrical conductivity. By generating an alternating magnetic field in the vicinity of a metal object, electric currents are induced in the object which in turn set up a magnetic field around the object which distorts the original field. By detecting this distortion, the metallic object can be located. Types of detector that employ this principle of *induced magnetism* include: *balanced search coil* units, *field search* units and *pulse magnetization* units.

One particular class of metals—the *ferromagnetic* materials —can also be detected using a different technique. Ferromagnetic materials have a high permeability, that is, they offer less 'resistance' to the flow of magnetic flux through them than any other material. A magnetic field, such as the Earth's magnetic field, generates lines of flux which will take the path of least resistance and therefore concentrate in the vicinity of any ferromagnetic material, causing a distortion of the general magnetic field which can be detected. Devices operating on this principle are known as *magnetic search units*.

Balanced coil   Balanced coil search units have two identical search coils, each with a primary and secondary winding. The primary windings are driven in series by an alternating current and so generate an alternating magnetic field. With the two coils placed over a non-metallic medium, the voltages induced in the secondary windings by this alternating field will be identical. In this situation, when the two secondaries are connected in opposition, there is no signal. A metallic object, such as a coin, will, however, produce an induced magnetic field which interacts more strongly with the secondary to which it is closest. A nett signal (the difference between the signals in the two coils) is then produced which, when amplified and displayed, indicates the presence of a metallic object.

The problem with this type of unit and the others that operate on the principle of magnetic induction is that, when used over earth, variations in the earth's conductivity can affect the readings. Such detectors tend to be useful only

Left: an American archaeologist using a metal detector in a systematic research of the original town of Petra in Jordan.

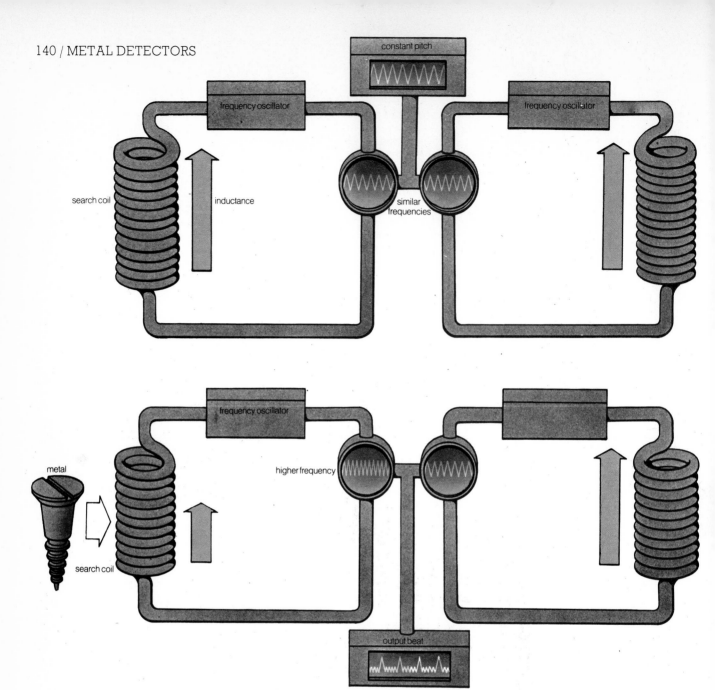

constant pitch

frequency oscillator

frequency oscillator

search coil

inductance

similar frequencies

frequency oscillator

metal

higher frequency

search coil

output beat

Above: diagram demonstrating the principle of a heterodyne type of metal detector. Two coils form the inductances in separate but similar oscillator circuits. A metal object brought close to the search coil influences the inductance in that coil, altering the oscillator frequency and generating a beat.

over small distances of the order of six inches (15 cm).

Heterodyne unit The heterodyne search unit also uses the principle of induced magnetism but consists of two coils with only one winding in each. They are separately connected to two oscillator circuits in which they form the inductive components. The two oscillators are initially adjusted to have the same frequency so that the two coils produce an alternating magnetic field at that one frequency. A metal object placed close to one of the coils changes the inductance of that coil and therefore also the frequency of oscillation of that circuit. By mixing the signals from the two circuits, a beat frequency is generated (in the same way that two similar musical notes produce beats). This beat frequency can be reproduced through earphones thus enabling the operator to locate the object.

This type of detector can be very sensitive, but again, variations in the earth's conductivity can affect the results and it is difficult to obtain sufficiently stable oscillators.

Field search unit The field search unit operates similarly to the balanced coil unit. It employs a loop of wire driven by a static high power oscillator which can generate an operating field over an area the size of a tennis court. Such a device, since it reacts to variations in the earth's conductivity as it does to the presence of metal, has an interesting archaeological use. If the detector readings of an archaeological site are plotted on a map, the outline of the buildings is often apparent from the variations in conductivity.

Pulse magnetization units Pulse magnetization units rely on the fact that a magnetic field takes finite time to propagate through air or earth—in fact the speed of propagation is the same as for electromagnetic radiation, that is, the speed of light. Such units operate as magnetic 'radar' units. A short high power pulse is generated in a search coil, and after the pulse is cut off the unit goes from a transmit mode to a receive mode. If there is a metallic object within the field, this object generates its own magnetic field by the process of induced magnetism, and this is detected a finite time after the transmitted pulse. Such units can be very effective, but again there are limitations due to the earth's conductivity.

Magnetic search units Iron, steel and other ferromagnetic materials are very much easier to detect using a magnetic search unit. This will find a one inch (2.54 cm) nail at a distance of about 24 inches (60 cm) or a car at 60 ft (18 m).

All objects on the Earth are in the Earth's magnetic field. Where there are not metallic objects this field can be considered as constant in strength and uniform—that is, in the same direction. Any metallic object will, however, distort the Earth's field because the lines of flux will tend to take the path of least 'resistance' (*magnetic reluctance*) and be concentrated in the vicinity of the object. To detect this distortion in the Earth's field a magnetometer system is used. This is capable of detecting the difference in magnetic field strength at two points and is sensitive to field differences of the order of 1/100.000 part of the Earth's field.

In practice, two magnetometer probes are used. These are fixed in a tube about 12 inches (18 cm) apart and carefully aligned along the same axis. The tube is suspended by its own weight so that it measures the vertical component of the Earth's field. Being largely independent of the qualities of the earth over which it is used, such search units are extremely powerful even where non-magnetic coins and so on are sought because magnetic objects usually exist there as well.

METER, electricity

During the latter part of the nineteenth century most electricity supplies were direct current (DC), and so the majority of electricity meters were DC instruments. Ferranti pioneered a mercury motor ampere-hour meter, and a similar type developed by Hookham was a true watt-hour meter. Thompson's commutator motor meter, Wright's electrolytic meter and the Aron clock meter also enjoyed fairly widespread usage.

Ampere-hour meters depend only on the amount of current being consumed, and registration in watt-hours or kilowatt-hours (kWh or 'units') assumes a constant nominal value of supply voltage. Such meters are no longer used on public power supply systems, which are almost always alternating current (AC).

Single phase AC meters The basic AC watt-hour meter

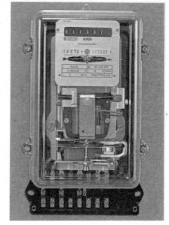

Left: a magnetic search unit works on the principle that certain materials distort the Earth's magnetic field, which can be measured.
Above: a mercury motor meter made in 1897, shown with a modern meter alongside. The two smaller pictures also show modern meters.
On the left is a single-phase domestic meter with its cover removed; the magnetic brake and its adjusting screw are on the left. The other meter is a kWh meter for 3 phase 4 wire operation.

consists of a horizontal aluminium disc with an electro-magnetic coil mounted above and below it. The driving torque is derived from the interaction of eddy currents induced in the disc and the flux from the two coils, which are arranged so that their fluxes are displaced relative to each other both in space and time. The speed of rotation of the disc is proportional to the amount of power passing through the meter, and a gear at the top of the disc spindle drives the counter dials.

In the simple single phase AC meter the flux in the upper electromagnet is proportional to the supply voltage, and that in the lower to the load current. The turning force or torque acting on the rotor disc is proportional to the product of the voltage and current and the *power factor* of the electrical load being metered. Power factor refers to the amount by which the load current waveform lags behind or leads the supply voltage waveform. It depends on the nature of the load, and can vary between I (zero lag or lead) and 0 (90° lag or lead). Inductive loads have a *lagging* power factor, and capacitive loads a *leading* power factor.

To translate the torque acting on the rotor disc into proportional speed of rotation a permanent magnet brake is fitted which sets up eddy currents in the disc as it passes through an airgap in the magnet. These eddy currents are proportional to the speed of rotation of the rotor disc, and by Lenz's law link with the permanent magnet flux to oppose the rotation. The full load speed of the disc is adjusted by altering the position of the brake magnet. The meters also contain various devices which compensate for variations in operating conditions so as to maintain overall accuracy within about 2%.

Polyphase kWh meters Industrial power networks, either 3 phase 3 wire or 3 phase 4 wire, are metered by *polyphase* meters having two driving elements (3 wire) or three driving elements (4 wire), each acting on a separate rotor disc mounted on a common spindle. Alternatively the drives can be concentrated on a single disc.

kVAh meters Kilovoltampere-hour (kVAh) meters register independently of the nature of the load, that is, independently of the power factor. One method employs basically two polyphase kWh meters, one of which is cross-connected so as to cause it to meter the reactive or wattless component (kVArh) of the load. The two meters are mechanically coupled through a multiple differential mechanism which

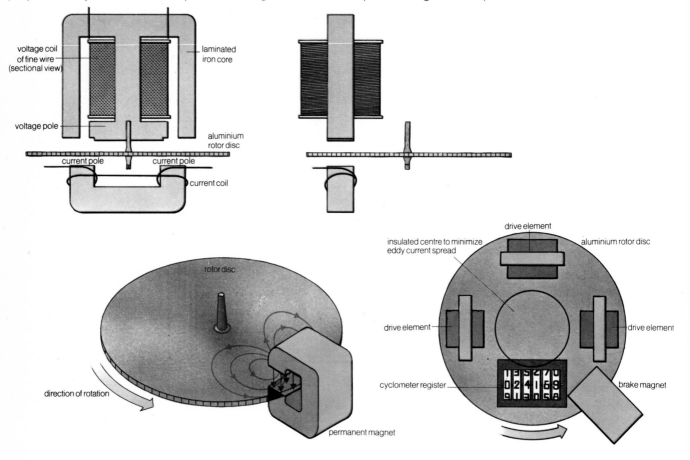

The upper pair of diagrams show cross-sections of a single phase meter, seen from the front and from the side. The lower two diagrams show the eddy currents induced by the permanent magnet brake, and a plan view of a polyphase meter for 3 phase 4 wire use.

summates or computes the true kVAh. Another approach involves the conversion of the AC values of voltage and current to equivalent DC values. These are summated and reconverted to equivalent AC values which are independent of load power factor and can be metered as kVAh.

kVAh metering is important where large industrial consumers are concerned, because the reactive component which results from a power factor less than unity (I) will not be registered on an ordinary kWh meter, and so in effect energy would be consumed without being registered on the meter.

Maximum demand indicators Another complication which arises in the supply of electricity to large consumers is that of maximum demand. Actual energy consumed over a period

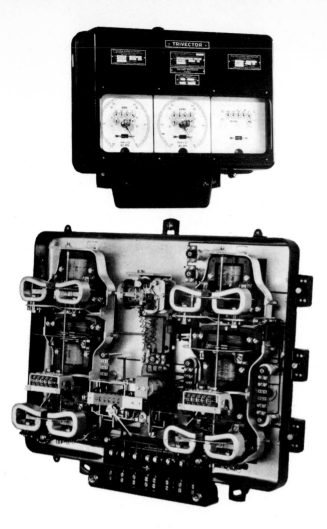

Two types of 'Trivector' meter which indicate kWh, kV Ah and reactive kV Ah, and incorporate maximum demand indicators. The upper meter has dial type counters, and the lower one (with its cover removed) has digital counters. When large loads are to be metered it is often impractical to connect the meters directly, so they are driven by transformers whose outputs are directly proportional to the load current and voltage.

of time is indicated by the kWh metering, but there may be peak loads at various points during the day and the supply network must be capable of handling these peaks. This may involve the supply authority in extensive capital expenditure on plant and cabling to cope with the maximum demand, and so the consumer is billed for the actual power delivered plus a charge based on maximum demand during the billing period. This is to help offset the capital expenditure and to encourage the consumer to spread the demand more evenly.

The Merz pattern of maximum demand indicator (MDI), still widely used, embodies a pusher or drive arm which is continuously driven from the meter, except for a few seconds (usually at 30 minute intervals) when the arm is decoupled and returned to zero by a spring. The drive arm controls the angular movement of a slave pointer, manually reset at regular intervals (usually monthly), so that this pointer will indicate on a circular scale the maximum value of kW or kVA averaged over any single 30 minute interval during the billing period. The coupling and decoupling of the drive from the meter is time-switch or relay operated.

When the MDI is separately housed it is actuated by electrical pulses from the master meter transmitted over 'pilot' wires from a device which produces a pulse for each revolution of the meter rotor. Summation of two or more circuit MDIs is possible, and the display can take the form of a digital printout.

Prepayment meters   The prepayment meter, an almost exclusively British development, followed the earlier widespread use of coin operated gas meters for domestic use.

The inserted coin is checked for size and weight, and if accepted it passes into the coin till, releasing the mechanism which is turned manually through a certain distance which is predetermined by a price setting device. This movement is

stored as credit in a differential gear and causes the load current switch to close, if open, or to remain closed. The rotation of the meter rotor actuates the differential mechanism in the opposite sense, that is, by reducing the stored credit eventually to zero at which point a spring-loaded mechanism will trip the switch and cut off the load current.

The more complex two part tariff prepayment meter incorporates a synchronous motor driving a timing mechanism which contributes to the total reduction of credit according to a preset fixed charge rate. When credit is exhausted the fixed charge device will build up a debit which must be cleared, by inserting sufficient coins, before the switch will close.

Testing   Statutory regulations commonly require ordinary meters to be tested by one of three methods. In the first method, the registration of the meter is compared with that of a certified test meter, both meters being run on the same load for the same period of time. The second method involves the comparison of the rotor revolutions of the meter under test with those of a certified meter connected to the same load, with an additional comparison of the dial readings.

In the third method, the test load is measured by a precision indicating wattmeter and the rotor revolutions are timed with a stopwatch. A dial test is also carried out. Meters which are within the specified limits of accuracy are officially sealed and certified, usually for up to 15 or 20 years service.

METER, gas

Gas meters are instruments for measuring the volumetric flow rates of gases in the gas supply industry. Various types are available depending on the application. Where large flow rates need to be measured, for example, at the point of supply (the gas works) and by large industrial consumers, rotary (including turbine), orifice and heat capacity meters are employed. These are called *flowmeters*.

Small industrial and domestic consumers nowadays use *positive displacement meters*. These are robust and accurate meters which can be suitably adapted for use with prepayment and coin operated mechanisms.

History   The first gas meters were invented by Clegg in about 1815 and were known as *wet meters* because they depended on a certain quantity of water or other liquid to ensure their working. Wet meters were used in the gas industry for over a hundred years but have now been superseded by dry meters.

Basically they consist of a measuring drum, casing and a counter mechanism. The measuring drum is mounted axially on bearings so that it can freely rotate and is divided into three or four radial compartments. This is positioned within the watertight casing and filled approximately half full with water. There is one inlet and one outlet aperture associated

Below: this positive displacement meter is suitable for measuring small flow rates.
Bottom: turbine gas meter showing internal components. The cone (in green) at left directs the gas over the turbine blades (white), causing them to turn. The turbine shaft can be linked to a rate counter or an integrator showing total volume passed.

with each compartment so positioned that, at most, only one aperture of each compartment is above water. The compartment whose inlet aperture is above water receives gas from the common gas input supply to the meter, thus expelling any water in this chamber. This drives the drum round until the input aperture is immersed by which time the adjacent compartment is receiving gas—thus maintaining the rotary motion. Meanwhile, the gas in the first compartment is driven out through the outlet aperture by the water entering the submerged inlet. Knowing the volume of each chamber and counting the revolutions of the drum with the counter mechanism enables the quantity of exhausted gas leaving the meter to be measured.

The water acts as a perfect seal preventing unregistered gas flowing through the meter, although this is only true while the water level is maintained. Water evaporation is a major problem with this type of meter because this leads to a reduction in the water level. Also, condensation further along in the gas pipes can lead to partial blocking of the gas and a phenomenon known in the old days as 'jumpy lights'. In winter there was always the problems of freezing. They were, however, mechanically simple devices that maintained their accuracy over many years and are now used only as a standard to check other meters against.

Positive displacement meters The modern positive displacement meter is based on a design originally developed between 1830 and 1850. Essentially they consist of two chambers which are alternately filled and emptied. Their reciprocal actions are connected via a linkage mechanism to the inlet and outlet valves associated with each chamber to control the gas flow, and also to a counter mechanism. Knowing the displacement volume of each chamber enables the flow rate to be measured.

In modern types, the chambers are separated by a leather

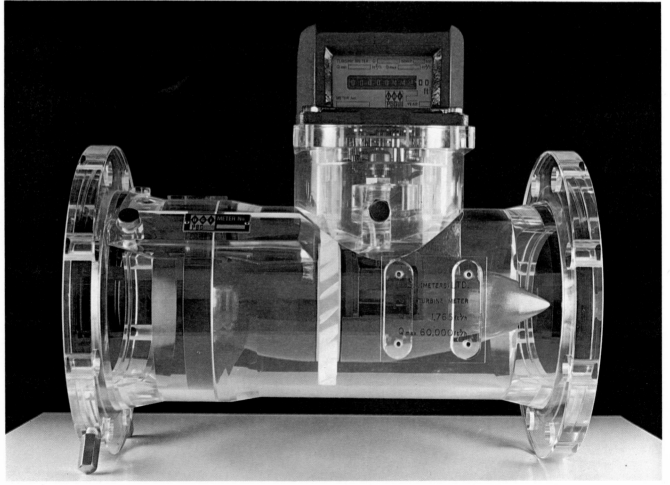

(or synthetic material such as a high nitrile rubber) diaphragm which can be alternately stretched and relaxed. Also, to provide a more continuous flow of gas, two such double chambers are combined into one unit. These meters have good accuracy at low flow rates (hence their value in domestic installations), they are easy to install and service, and maintain their accuracy over many years.

Rotary meters The main type of rotary meter—sometimes called a *lobed impeller meter*—consists of two or three precisely shaped and interlocking impellers (an impeller is *driven* by a fluid flow whereas a propeller drives). They are

positioned inside a carefully shaped chamber such that at no time can the gas flow straight through the meter unregistered. The pressure of the gas entering the meter forces the impellers to rotate thus turning a counter mechanism. Each rotation of the impellers corresponds to a fixed volume of gas flowing through the meter.

There is very little pressure loss across this type of meter especially with high flow rates where frictional forces are less. They can be constructed from various materials but tend to be easily damaged by impurities in the gas (for example, water vapour). They are relatively costly to manufacture because of the precision engineering required.

The latest development is the turbine meter used for industrial applications. Basically it consists of a turbine blade which is operated by gas pressure. A mechanical or electronic sensing device monitors variations in temperature and pressure and a correction factor is then put into the reading on the meter. Such a meter, the size of two domestic meters, has replaced much larger installations the size of a small room.

Coin operated mechanisms Prepayment mechanisms were first applied to wet meters in 1887. Here, an official of the gas company sets the meter to pass a fixed quantity of gas for which payment is made in advance. The quantity paid for is indicated on a dial, and when this has been used the counter mechanism shuts off the gas supply to the meter by closing a valve inside the meter which is operated by pressure in the mains. Coin operated mechanisms operate in a similar way, but the quantity of gas to be passed is set by the consumer. By entering a coin into the mechanism and turning a key the quantity of gas to be passed is registered.

Coin operated meters revolutionized the gas industry in Britain. They put the benefits of gas heating and lighting within the reach of the poorer classes, because the formerly prohibitive installation costs could be incorporated into the unit cost of the gas supplied.

Below left: a modern laboratory water meter—their high accuracy ensures their use as standards for testing and calibrating other meters.

METER, parking

The control of the movement and parking of vehicles in urban areas has been practised for at least 2000 years. In ancient Rome the authorities provided off-street parking areas for chariots, and traffic congestion became so bad that Julius Caesar banned vehicles from business areas of the city during certain hours of the day. This ban did not apply to vehicles on religious or State business.

The parking meter was invented by Carl Magee, and first used in 1935 in Oklahoma City. Meters were first installed in New York in 1951 and in London in 1957, and are now commonplace throughout the world.

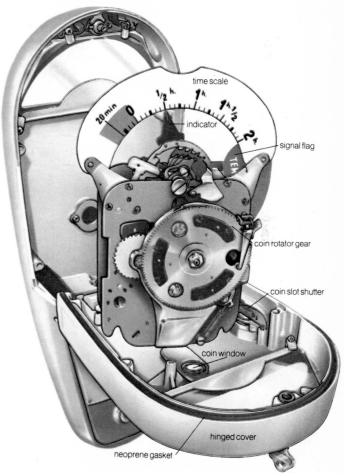

This type of parking meter has to be wound once a week. The last two coins to be inserted can be seen through windows so the attendent can check that they are correct. The case is waterproof, all joints being sealed with neoprene gaskets, and the coin slot has a spring-loaded shutter to keep out dirt.

Below right: a 1901 French taxi meter.
Bottom: a variable-tariff meter which can run at four rates. The speedo cable or a clock, whichever runs faster, operates a solenoid which advances the reading. The slower drive freewheels.

Mechanisms The basis of the parking meter is a clockwork mechanism, which is wound by an attendant about once a week in the case of the 'automatic' type, or by the user (by turning a knob after inserting a coin) in the case of the manual type. In either case insertion of the coin begins the timing cycle, elapsed time being indicated by a pointer and scale arrangement. When the bought time period has expired a penalty 'flag' is displayed in the meter window.

Meters may be either *cumulative* or *non-cumulative*. The cumulative type add the bought time to any unexpired time left on the meter by the previous user, but the non-cumulative meters, on insertion of a coin, cancel any unexpired time and register only the amount of time bought by the coin.

Construction The meters are usually mounted on steel posts at a height of from 3 to 5 feet (0.9 to 1.5 m) above the ground, or else on wall brackets where convenient. As the meters are exposed to a wide range of weather conditions, the bodies are made of non-corrodible alloy, usually diecast aluminium alloy, and the joints are sealed with waterproof gaskets.

The windows are of toughened glass or unbreakable acrylic material, and the clock mechanisms of brass, aluminium alloy, or stainless steel. A pair of adjacent parking spaces may be served by a pair of meters mounted on a single post, or a duplex meter which has two meters housed together in a single body, either back-to-back or side by side. Strengthened meter bodies are used in areas where there is a high risk of vandalism or attempted theft.

Coin collection There are several types of coin receptacles in use, the simplest of which is merely a small metal box which is emptied periodically by an attendant. Where more security is required, the coins pass into a coinbox which

locks automatically as it is withdrawn from the meter. Two keys are needed, one which the attendant uses to open the meter to remove the full coinbox and replace it with an empty one, and the other which is kept by the local authority to open the coinbox. The clock mechanism is usually separated from the coin receptacle, and a separate key is used for access to the mechanism.

An alternative collecting method involves the use of a collecting trolley, the coins being jettisoned from the meter down a flexible metal tube and into a locked canister on the trolley. A variation of this method which provides greater security uses a locked coinbox, which is withdrawn from the meter and pushed into a receptacle on the canister. A key device in the receptacle unlocks the box and the coins are released into the canister. The box is automatically relocked as it is taken from the receptacle to be replaced in the meter.

METER, taxi

The fare charged for a taxi journey has to depend not only on the distance travelled but also on the time spent waiting in traffic jams or elsewhere. In addition, different tariff rates may apply on different occasions, such as after a particular time at night or for certain journeys; there may be a special rate which applies if the taxi has to wait for the passenger

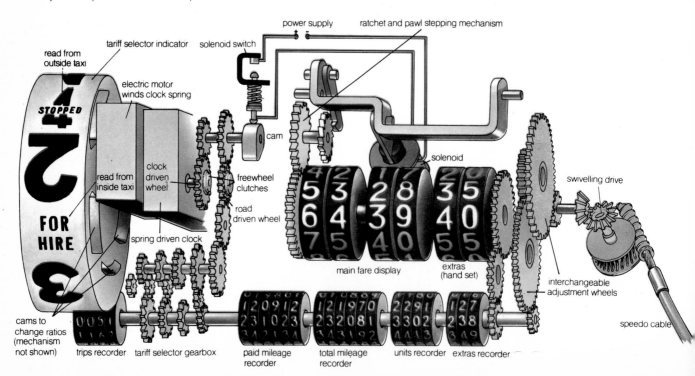

for any reason. The *taximeter* which records the fare must therefore take all these factors into account, but at the same time must be easy to operate though difficult to tamper with.

A typical design of taximeter takes its rate of operation from the speedometer cable of the taxi. This is almost always a *Bowden cable*, which has a central core rotating within an outer sheath. Both cables are flexible.

So that the speedometer still operates, a 'T' piece is screwed on to the end of the normal cable, bearing an extension on one arm of the T which goes to the speedometer as usual, and a reduction gear on the upright of the T. A cable from this goes to the taximeter input shaft.

This drives a set of adjustment gears, which are chosen for the vehicle in question—different car manufacturers have different rates of speedometer cable revolution. These turn a shaft on which are various mechanical counting devices, such as *odometers*. These are not reset after each journey, but record the total distance travelled.

From this shaft, another gearwheel transfers the distance travelled to a cog on a shaft whose job it is to produce the final rate to be charged. Also on this shaft is a cog driven by a clock movement. The final rate is determined by whichever of these cogs travels faster. Each has a free wheel mechanism, like that of a bicycle, so that only one turns at any time.

METRONOME

The metronome is an instrument used for visually and audibly indicating the *tempo* or speed of music. In its simplest, clockwork, form it consists of a pivoted pendulum; below the pivot is a fixed weight, and above it a sliding weight. The beat is altered by sliding the weight up and down.

The history of the metronome goes back at least to the seventeenth century, when Etienne Loulié described in his *Elements ou Principes de Musique . . .* (Paris, 1669) an instrument he called the *chronomètre*. This consisted of a metal bullet suspended on a cord so that it could swing from side to side. Provision was made to vary the length of swing to give 72 speeds. Other people, such as Joseph Sauveur, originator of the word acoustics, proposed similar instruments. Robert Bremner, the Scottish music publisher and instrument maker, wrote in 1756 that it was necessary to find a method whereby musical 'time in all churches may be equal', saying that a pendulum eight feet eight inches (2.67 m) long would by its double swing or *vibration* fix the length of the semibreve [whole note] (the longest note in common usage). He also suggested that such a pendulum be hung at the end of each school hall where church music was sung.

John Harrison, known for his invention of the perfect marine chronometer, noted the connection between time keeping and the tempi of music in 1775. In 1800, a German named Stöckel made a metronome which had a bell struck by a clock-type hammer.

Finally, in 1814, a German-born master organ builder living in Amsterdam, Dietrich Nikolaus Winkel, perfected the first reliable metronome having the double pendulum feature described above. This feature allowed a compact device to be built with then unrivalled accuracy. As a leading maker of clockwork barrel organs, Winkel was both musician and mechanic. His first instrument was a hinged box about 1 ft (30 cm) square containing a small infinitely adjustable double-weighted pendulum kept in motion by a clock-type escapement, powered by a small weight on the end of a cord wrapped around a drum.

Winkel demonstrated his instrument to Johann Nepomuk Maelzel, an unprincipled fellow who had a habit of capitalizing on the work of others. Maelzel for a time enjoyed the friend-

ship of the great composer Beethoven, who was concerned that his music be played correctly. He was the first composer to include metronome markings in his scores, and one of the movements of his *Eighth Symphony* is said to have a beat inspired by the metronome.

Maelzel patented the metronome in his own name and set up a factory to produce it as the Maelzel Metronome, the name by which it is still known today, in spite of an investigation in Paris which upheld Winkel as the real inventor, and legal action over Maelzel's various misdemeanours, including stealing Beethoven's music and publishing it as his own.

Since then many kinds of metronome have been built. One popular nineteenth-century model was called Pinfold's Patent Metronome and was a pocket model with a pendulum which was reeled up like a tape measure; others can strike a bell at the first beat in a measure. Solid-state electronics have made possible small, extremely accurate metronomes; one model has a frequency generator which will emit a tunable A at the press of a button. Another model, which looks like a portable radio, can be set to beat any time signature, however complicated. An advantage of electronic metronomes is that, unlike the familiar pyramid-shaped Maelzel model, they do not have to stand on a level surface in order to work accurately.

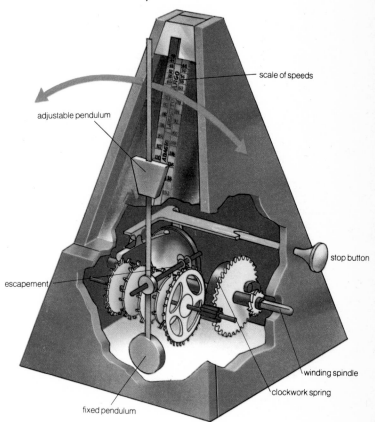

Above: the familiar modern pyramid-shaped metronome. Its speed is adjusted by sliding the top pendulum up or down.

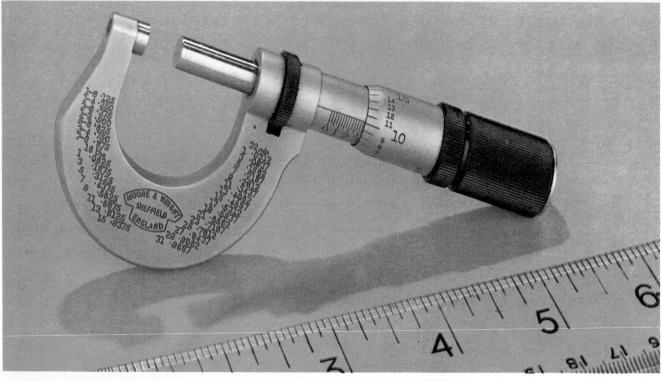

MICROMETER

A micrometer caliper is a measuring device widely used in engineering to measure the diameter of round objects or the thickness of flat pieces. It consists of an accurately ground screw or *spindle* which is rotated in a fixed nut. The end of the spindle advances or retracts, opening or closing the distance to an *anvil*.

The thread grinding to the spindle is done with an almost diamond-hard formed stone grinding wheel and the thread is inspected rigorously. The pitch diameter taper tolerance and the uniformity of thread and pitch must be held to within 0.00005 inch (1.20 micron). With the micrometer assembled, the anvil and spindle faces are ground parallel and then lapped. The tolerance to which lapping must be held is about three *lightbands* with flatness tied to within one lightband. (A lightband is a unit of measure used in interferometry, the measure of very small distances by the use of light. A lightband is the wavelength of *sodium*, a yellow colour: 0.0000011 inch, or about $\frac{1}{4}$ micron.)

A micrometer spindle which measures in inches has a pitch of forty threads to the inch, so that one complete turn of the spindle advances the spindle face exactly one-fortieth or 0.025 inch. The spindle revolves inside a fixed nut covered by a sleeve which is marked with a longitudinal line having forty graduations. The outside shell of the micrometer spindle is called the *thimble* and has a bevelled edge which covers these graduations as it is turned; on the bevelled edge are 25 graduations which correspond to thousandths of an inch.

When the micrometer is closed, only the zero line can be seen on the sleeve next to the bevelled edge of the thimble. A measurement is taken by reading the number of longitudinal graduations uncovered and adding the number of thousandths above the zero mark on the thimble. Thus, for example, measuring the size of a piece of ground bar stock of 0.259 inch, ten longitudinal lines will be uncovered on the sleeve ($10 \times 0.025 = 0.250$ or $\frac{1}{4}$ inch) plus nine lines on the bevelled edge of the thimble.

Metric micrometers are used the same way. The pitch of the spindle thread is 0.5 mm, one revolution of the thimble advancing or retracting the spindle $\frac{1}{2}$ millimetre. The longitudinal line on the sleeve is graduated from zero to 25 milli-

Top of page and opposite page: the machinist's or inspector's micrometer, measuring zero to one inch. The 'mike' in the photo is open to .335 inch. Inside micrometers are also available for measuring the size of holes; they consist of the thimble mechanism and a set of interchangeable rods of precise lengths.
Above: an inspector using a larger size micrometer to measure the size of pistons for a Diesel engine.

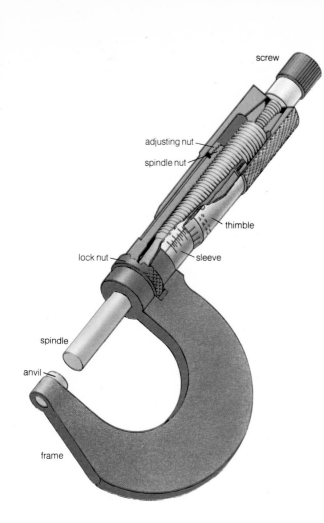

screw

adjusting nut

spindle nut

thimble

lock nut

sleeve

spindle

anvil

frame

MICROPHONE

A microphone is an electro-mechanical device for converting sound energy into electrical energy. It usually has a flexible diaphragm which moves in response to minute changes in air pressure caused by sound waves. There are basically two ways of utilizing these pressure variations; *pressure operation* and *pressure gradient operation*.

Any microphone whose diaphragm is open to the air on one side only is said to be *pressure operated*. The magnitude of the force on the diaphragm depends on its area and the instantaneous pressure in the sound wave (pressure is force per unit area). When both sides of the diaphragm are exposed to the air, the microphone is said to be pressure gradient operated. The diaphragm only moves when there is a *difference* in pressure between the back and the front, the pressure difference being due to a difference in *phase* (one wave has to travel further to reach the far side of the diaphragm and thus arrives later).

Types of microphone   There are various ways in which sound energy can be converted into electrical energy—the most important types being the carbon, moving coil, ribbon, electrostatic and crystal microphone.

In the *carbon microphone*, pressure variations in the sound waves cause the diaphragm to vibrate. The resultant alternating pressure on the carbon granules held behind the diaphragm causes a change in their electrical resistance about its mean value. This imposes an alternating current on the steady current drawn from a battery and, by means of a transformer, the alternating current can be 'tapped off' as output. Although such microphones do not provide 'high fidelity', they are cheap to produce and sturdy and are used extensively in telephones.

The *moving coil microphone* works on the electromagnetic principle. A coil attached to the vibrating diaphragm moves in the magnetic field of a strong permanent magnet and the relative motion induces a current in the coil, which is directly related to the incident sound wave.

The *ribbon microphone* also works on the electromagnetic principle. A corrugated metal ribbon suspended between the poles of a permanent magnet acts as the diaphragm and, like the moving coil, it receives an induced current which is related to the incident sound waves.

The *electrostatic microphone* is really a capacitor with one plate fixed and the other acting as the flexible diaphragm. As the diaphragm vibrates, the capacitance varies about its mean value. With a constant voltage applied between the plates, the amount of stored charge will vary with changes in capacitance, causing an alternating current to flow related to the incident sound wave. With electret microphones the principle is the same, but the electret material contains its own built-in voltage source and no external voltage is required.

The *crystal microphone* relies on the piezoelectric effect—that is, when a piece of Rochelle salt is bent or twisted by the application of an alternating force an alternating current is produced. A 'sandwich' of Rochelle salt held within the microphone produces an alternating voltage when excited by an incident sound wave.

Frequency response   The frequency response of a microphone is assessed by subjecting it to sounds of various pitch (frequencies) while monitoring its electrical output. The two resultant sets of figures, frequency, measured in *hertz* (the number of pressure variations, or cycles, per second) and the relative electrical outputs, measured in decibels, or dB (a unit used to compare sound intensities), are plotted graphically.

A horizontal line on a frequency response graph means that the microphone produces the same electrical power for all frequencies. In practice this is not achieved, but modern design techniques have enabled the production of microphones that come very close to this perfect situation within

metres and each millimetre is subdivided in half. The thimble is graduated in fifty divisions so that each graduation equals 1/50th of 0.5 mm or 0.01 mm.

At an extra cost, some micrometres have an additional *vernier* scale on the thimble so that they can be read, for example, to one ten-thousandths of an inch (0.0001 inch).

Micrometers are made in graduated sizes. The most common micrometre measures zero to one inch (or zero to 25 millimetres); the next size is for measuring one inch to two inches (or 25 to 50 mm), and so forth. Thus the spindle, thimble and other parts are identical on all micrometers, and the size of the body of the instrument is the variant. When learning to use a micrometer, some practice is necessary to get the 'feel' of the device—for example, the thimble is never tightened as though it were a clamp. This would result in excess wear of the threads and inaccurate measurement. The object being measured should slip between the anvil and the spindle face without looseness but without excessive tightness.

Adjustment to the micrometer is seldom necessary, but with constant use play sometimes develops in the spindle due to wear of the spindle nut. By backing off the spindle an adjusting nut becomes accessible and by slightly tightening this nut, play is eliminated. Sleeve adjustment is possible by having spindle and anvil faces in contact and rotating the sleeve very slightly by means of a tiny spanner, supplied with the micrometer, until the line on the sleeve coincides perfectly with the zero line on the thimble.

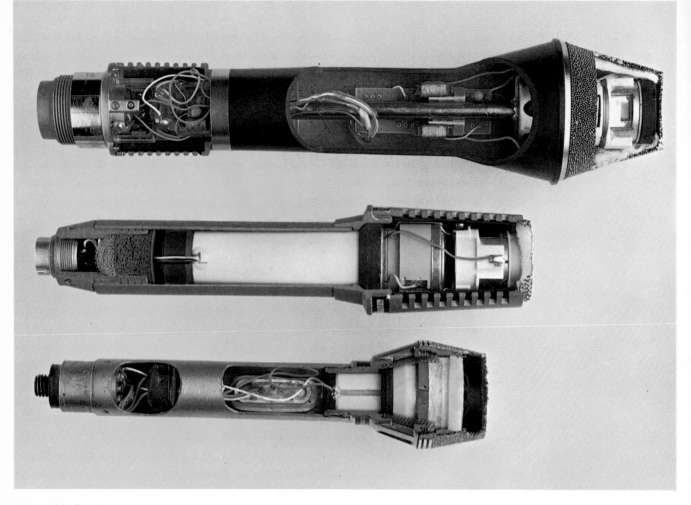

the audible frequency range.

Directional properties A *polar diagram* of a microphone is a graph of the relative voltage outputs for sounds arriving at different angles. Angles are measured from the microphone axis and separate polar diagrams are drawn for various frequencies and planes. In practice three different types of polar diagram are obtained—a circle, figure of eight and heart shape (*cardioid*).

A circle represents an *omni-directional* microphone, that is, one which is equally sensitive to sounds arriving from all angles. A figure of eight illustrates the characteristics of a *bi-directional* microphone which is sensitive to sounds arriving at the front and the back of its diaphragm but not to those arriving from either side. A heart shape illustrates the characteristics of a microphone which is most sensitive to sounds arriving from the front.

A circle (really a sphere when three dimensions are considered) is usually associated with a *pressure operated* system. All pressure operated microphones have a similar set of polar diagrams. They are omni-directional at low frequencies, tending to a more one-sided response at higher frequencies. This change in response is due to the *obstacle effect*. This effect is caused by the fact that objects tend to reflect sounds whose wavelengths are smaller than the dimensions of the object. Thus an object the size of a microphone only starts to reflect when the frequency exceeds about one kHz (1000 cycles per second).

A figure of eight usually illustrates the characteristics of a *pressure gradient* operated system. As mentioned earlier, a pressure gradient operated microphone has both sides of its diaphragm exposed to the air, the diaphragm only moving when there is a difference in air pressure between the two sides. This only occurs when there is a difference in 'phase'

between the sound arriving at the front and the sound arriving at the back of the diaphragm, this in turn being dependent on the angle of incidence of the sound wave. If the sound originates from a point directly in front or directly behind the microphone a maximum pressure difference is obtained. If the sound, however, approaches the microphone from either side, there is no difference in phase as the sound waves have travelled equal distances in order to reach the two sides of the diaphragm. This means that there is no pressure difference and no movement of the diaphragm. Sounds approaching from other angles produce pressure differences between the maximum and minimum values.

The cardioid polar diagram indicates that a microphone is uni-directional. This type of response is particularly useful for discussions, where extraneous noises, like those produced by an audience, are to be minimized. It is obtained by combining the characteristics of pressure operation and pressure gradient operation. At the front of a cardioid microphone the two elements operate in phase so that their electrical outputs add thus producing an electrical signal, whereas at the back of the microphone the elements operate out of phase, resulting in the complete cancellation of the electrical outputs.

It is possible to obtain a cardioid response by using a single element which combines both the characteristics of pressure operations and pressure gradient operation. This technique has been evolved in *electrostatic microphones* which have holes bored in their rear plate. The discovery of the cardioid electrostatic microphone then led to the development of an extremely versatile microphone, which possessed two diaphragms and behaved like two cardioid microphones placed back to back. The resulting microphone could have a uni-directional, a bi-directional or cardioid response.

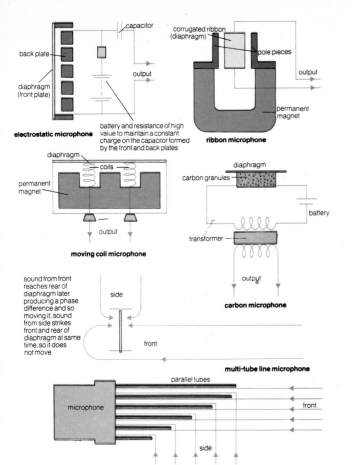

electrostatic microphone

back plate

diaphragm (front plate)

capacitor

output

battery and resistance of high value to maintain a constant charge on the capacitor formed by the front and back plates

ribbon microphone

corrugated ribbon (diaphragm)

pole pieces

output

permanent magnet

moving coil microphone

diaphragm

coils

permanent magnet

output

carbon microphone

diaphragm

carbon granules

battery

transformer

output

multi-tube line microphone

sound from front reaches rear of diaphragm later, producing a phase difference and so moving it; sound from side strikes front and rear of diaphragm at same time, so it does not move.

side

front

microphone

parallel tubes

front

side

Opposite page: cutaway views of a dynamic cardoid microphone and two double-system dynamic cardoid models.
Left: diagrams showing the basic configuration of electrostatic, ribbon moving coil and carbon type microphones.
Below: a selection of modern microphones. At left is a dynamic 'gun' microphone, which has highly directional characteristics for picking up distant or localized sounds. Top right is a dynamic noise-cancelling microphone with hypercardoid polar plot and below that is a dynamic omnidirectional type with spherical polar plot.

Highly directional microphones There are circumstances when sound must be picked up from a distance. To achieve this, a system which will boost the required sound and reject unwanted, ambient, sounds is required. Two systems in use today are the *parabolic reflector* and the *line (rifle) microphone.*

Parabolic reflectors, like those used in radar, concentrate the required information at their focal point. A microphone, normally pressure operated, positioned at the focus then converts the concentrated sound energy into an electrical signal. Because of the obstacle effect, only very large reflectors are capable of reflecting the whole audible frequency range, but as a compromise between efficiency and portability, reflectors are usually about three feet (1 m) in diameter. They are built of aluminium backed with thick sponge rubber to reduce the noise produced by rain when they are used for outdoor work.

The line or rifle device consists of a large number of narrow parallel tubes terminated by a microphone. Sounds approaching the microphone along its axis (that is, parallel to the tubes) effectively have the same distance to travel in order to reach the diaphragm and thus arrive in phase, but sounds approaching from any other angle travel different distances, thus arriving at the diaphragm out of phase. This results in the effective cancellation of the sound wave. It is also possible to achieve highly directional results with a single tube which has a large number of holes (sound entrances) evenly distributed along its length. Like the multi-tube microphone it utilizes the fact that sounds approaching the microphone from any direction other than that from the front arrive at the diaphragm out of phase. Only a single row of holes is needed to cancel sounds from any unwanted direction.

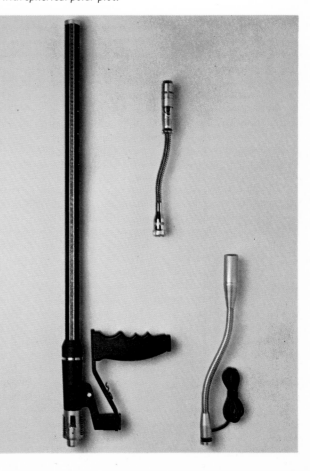

MICROSCOPE

The use of two lenses gives the light microscope its alternative name of *compound* microscope. It is based on the principle that the image formed by one lens (the *objective*, next to the specimen) is further magnified by a second lens (the *eyepiece*), to give much higher powers than could be obtained by a simple magnifying glass alone. The microscope, which has uses in all branches of science and technology, is one of the most valuable instruments ever invented.

The compound microscope was probably invented by Hans and Zacharias Janssen in Holland about 1600. It became well known through the work of Robert Hooke in England, especially through publication of his book *Micrographia* in 1665. During the eighteenth century the instrument was largely a plaything, but by 1830 the objective had been improved from a simple lens to a high-quality *achromatic* and *aplanatic* system (where the image is free from false colours round the image parts and from other defects which obscure detail), largely through the work of J J Lister. Many other people contributed to the perfection of the instrument in the next 50 years, until by 1880 Ernst Abbe had established and applied the theoretical framework which enabled the microscope to give a usable range of magnification from about × 30 to × 2000.

This maximum useful magnification cannot be exceeded because of the nature of light itself—a form of electromagnetic radiation. Because of its wave motion it simply bends round very small objects without itself being affected: this makes very small objects invisible. Therefore, although it would be possible to make a microscope, which would magnify more than × 2000 (for example, just by using a stronger eyepiece), because the *resolution* (ability to separate fine detail) is fixed not by the instrument but by the nature of light itself, further magnification would give no better resolution. Thus to resolve details closer together than 2500 angstrom units, or 0.000025 cm, an electron microscope must be used, as electron waves have a shorter wavelength than light waves.

Illumination In use, the microscope is arranged for either transmitted illumination, where the light goes through the specimen, or for reflected illumination, where the light bounces back from the specimen. Transmitted lighting is more usual nowadays. The light source for most work is some kind of electric lamp, as daylight is not reliable in quality or intensity. In some instruments the light is built into the *stand* itself, which is very convenient in use. If a separate lamp is used, the light has to be carefully directed up through the specimen by means of a mirror. On advanced instruments the lamp is complicated, with a lamp condenser and lamp iris controlling the light from a compact-filament bulb, usually operated at low voltage. A wide variety of lamps is available, to give selected wavelengths of illumination, some at very high intensity.

For the best results, a *substage condenser* is used to focus a corrected beam of light accurately on the specimen on the *stage*. Although for low power work this is not vital, for other requirements the illumination is not sufficiently intense nor sufficiently controlled without this condenser, which also has an iris diaphragm used to control the angle of the light reaching the specimen. After passing through the specimen slide, the light reaches the *objective*. This may be a very complicated and expensive lens system, containing as many as 14 separate lenses, some as small as 1 mm in diameter, and very carefully arranged relative to each other. The higher the magnification of the objective, the more complicated it becomes; in addition there are special types such as *flatfield* objectives for photomicrography, and *apochromatic* (having both spherical and chromatic aberrations corrected to a high degree) objectives. Each type makes further demands on the designer and is more complicated. For convenience, several objectives are usually carried on a revolving

reflecting prisms

displacement lens

objective barrel

objective optics

specimen

condenser

aperture diaphragm

centring insert

light source

eyepiece optics

urret

pecimen axis
ontrols

focus control

base

field diaphragm

collector

lamp

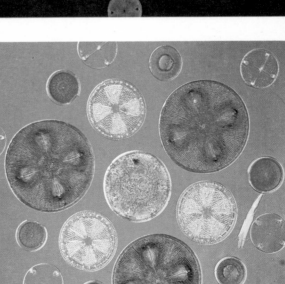

Right: this Nikon microscope has a device like a shadowgraph attached so that what the technician sees can be projected on to a screen. The projection can be overlaid with original working drawings and comparisons made for quality control.

nosepiece, which allows different magnifications to be chosen simply by turning to a different position.

On a modern research microscope the range of powers of the objectives might be ×3, ×6, ×10, ×20, ×40 and ×100. The highest power would be of the kind called *oil-immersion*, because it has to be used with a drop of oil between the front of the objective and the slide being viewed. At the top end of the *body-tube* is the eyepiece, and these also are available in various powers. A usual range nowadays would be ×6, ×10 and ×15. The total magnification is given simply by multiplying the power of the objective by that of the eyepiece, so the range of magnifications available, with the six objectives and three eyepieces mentioned above, would be from ×18 to ×1500.

When setting up a microscope, the correct alignment of all the optical components, from the lamp to the eyepiece, is of the greatest importance; modern microscopes have the most useful arrangement built-in.

Similar principles apply if the instrument is used with reflected light, for example, when studying metal surfaces. Here, intense lighting is required because so much is lost during reflection. Special accessories are required if higher powers are to be used with reflected light, and some instruments are designed solely for this purpose.

Other methods In addition to the straightforward types of illumination mentioned above, which have not changed in essence since the instrument was first invented, the twentieth century has seen the addition of three new techniques. *Phase-contrast* illumination allows specimens which appear almost invisible by normal illumination to be seen clearly, and is thus of major importance for direct investigation of living cells. The system was invented by Zernicke in the 1930s, and relies on condenser and objective pairs arranged in such a manner that light going through part of the condenser is slowed down relative to the rest, and is thus put out of *phase*, causing differences in refractive power to show

up as differences in light and shade.

Interference microscopy also relies on the interference of light beams with each other, and was developed after the phase-contrast microscope had shown the advantages of the method. The beam of light is split into two, one of which goes through the specimen and is modified by it before being recombined with the first beam. Differences in the light-retarding properties of the specimen show up as different colours. The advantage of the method is that actual precise measurements of the thickness of objects and other dimensions can be made.

Fluorescence microscopy requires the minimum of specialized equipment, but does need staining techniques, developed for this method. The principle is that blue or ultra-violet illumination is used to make parts of a specially-treated specimen emit light of various colours. Barrier *filters* absorb the light used to illuminate the specimen, which shows up only in the colours generated by the technique. This technique is of great use in *histochemistry*, which is the chemistry of living tissues.

Mounting specimens The modern microscope is versatile and relatively simple to use; many of the accessories thought essential in the last century are now being superseded. By itself, though, the instrument may be of little use: preparation of specimens for viewing through it is also most important, for most things are far too large to be viewed directly. The most usual technique is to make a slice of a specimen so that it is thin enough to be transparent. It will then need staining with coloured dyes to make the structure clear, and it will need mounting on a protective *slide* surrounded by a resin such as Canada balsam to give it the correct characteristics for viewing. The standard size of slide is 3×1 inches, and all kinds of specimens are mounted on these.

Other means of illuminating specimens show to advantage with some older slides. Two techniques have come down from Victorian times, *darkground* illumination and *polarized* illumination, and these can still give useful results after more than a century of application. In darkground work all direct light is cut off by a disc under the substage condenser, so that only very oblique rays are bent upwards by the specimen to enter the objective. This gives the effect of the specimen showing up brilliant white on a black background, and is especially useful for small living things. Polarized lighting requires polarizing filters, one below the specimen and one somewhere above, and relies on one of the properties of light as a wave motion. This is that the waves occur at all angles to the line of direction of the beam, and a polarizing filter cuts off all angles except a very few. If two such filters are used at right angles to each other no direct light can pass, but if the specimen is a crystal it may be able to alter some of the waves to enable them to pass the second filter, causing brilliant colours to be generated.

Electron microscope The electron microscope was developed to examine specimens in much greater detail than had been previously possible using an ordinary microscope, correctly known as the light microscope. It has proved extremely useful in studying metal as well as biological samples, such as viruses and cancerous tissue.

In 1873, a German physicist, Ernst Abbe (1840–1905), proved that in order to clearly distinguish between two particles situated closely together, the light source must have a wavelength no more than twice the distance between the particles. This therefore applies to adjacent points in a specimen.

The ability to clearly distinguish two particles is called *resolution*, which should not be confused with magnification. No matter how many times something is magnified if its image is blurred it will always be so. The wavelength of visible light is approximately 0.00005 cm (or 5000 angstrom units,

one angstrom being 10^{-8} cm). Thus the minimum details resolvable under a light microscope would be 0.000025 cm (2500 angstrom units) apart.

In the search for a new type of microscope, X-rays, which have a much shorter wavelength than ordinary light, were considered, but a 'lens' to control X-rays could not be produced. Scientists began to study the electron, finding that accelerated electrons travel with a wave motion similar to that of light but over 100,000 times shorter. Researchers found that either electrostatic or electromagnetic fields could be used to control an electron beam, these 'lenses' behaving in much the same way as the glass lens does in focusing a beam of light. Gradually during the 1930s the electron beam and its control by magnetic lenses was developed to produce shadow pictures of specimens until in 1939 the first commercial electron microscope became available, capable of resolving 0.00000024 cm (24 angstrom units). Since this first commercial instrument, designers have worked towards higher resolutions until today's instruments are capable of resolving 0.00000002 cm (2 angstrom units) as a matter of routine.

The electron microscope is an example of accelerated twentieth-century technology reaching a stage of development in less than twenty years, whereas it had taken 300 years to perfect the light microscope.

Transmission electron microscope The first electron microscope was known as a transmission electron microscope because the electron beam was passed through an ultra thin sample. The variation in density of the specimen resulted in a variation in the brightness of the corresponding area of the shadow image. The transmission electron microscope consists of a vacuum column which is essential for the free passage of the electrons. A tungsten hairpin filament, the cathode, is heated to a point at which it emits electrons. By

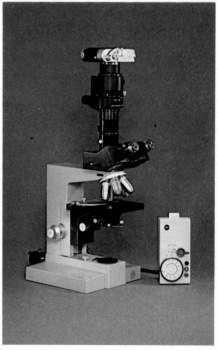

Above: a Dialux binocular microscope, of German manufacture, with 35 mm camera attached. A mirror allows the viewer to see exactly what will be in the photograph. Left: a revolving nosepiece.

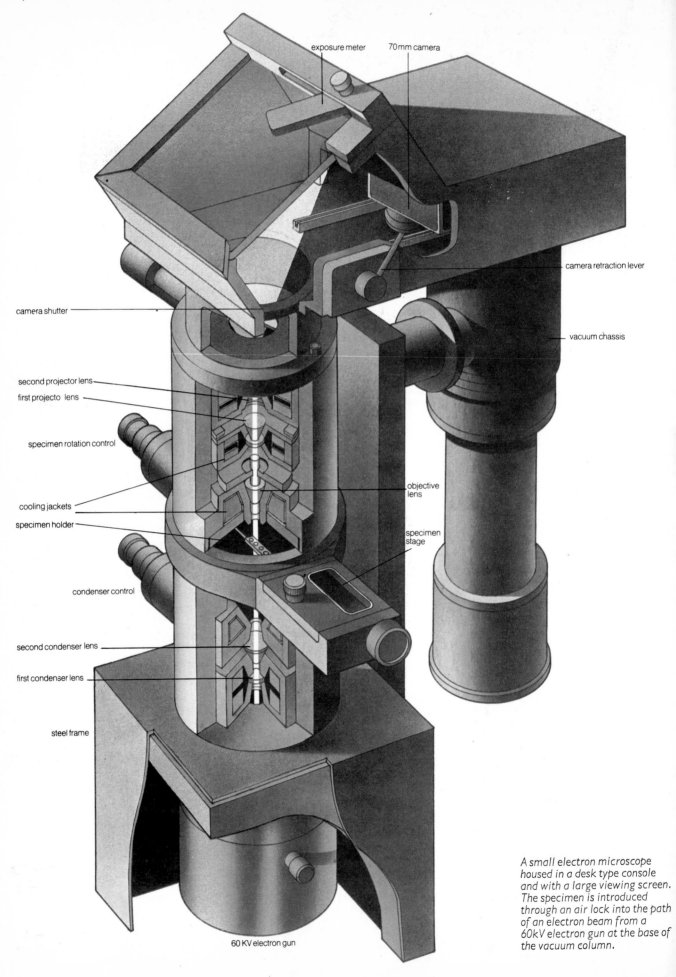

exposure meter

70mm camera

camera retraction lever

vacuum chassis

camera shutter

second projector lens

first projecto lens

specimen rotation control

objective lens

cooling jackets

specimen holder

specimen stage

condenser control

second condenser lens

first condenser lens

steel frame

60 KV electron gun

A small electron microscope housed in a desk type console and with a large viewing screen. The specimen is introduced through an air lock into the path of an electron beam from a 60kV electron gun at the base of the vacuum column.

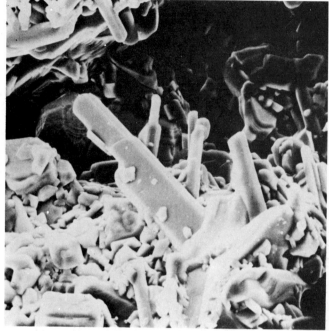

Left: 1,000,000 volt electron microscope at British Steel Corporation's laboratory. Below: a scanning electron microscope photograph of a nest of crystals in a cavity in a fragment of moon rock. They are about 3.9 billion years old, and were formed when the rock was cooling.

applying 20,000 to 100,000 volts between the cathode and the anode, the electron beam is accelerated down the column. (This system is called an electron gun.) Condenser 'lenses' control the beam size and brightness before it strikes the specimen, which is mounted on a 3 mm diameter copper mesh grid. The electron beam is focused and magnified by the objective lens before being further magnified and transferred on to a viewing screen by the intermediate and projector lenses. Most instruments cover the magnification range of $\times 50$ to $\times 800,000$. The screen is made of phosphorescent material (zinc-phosphide) which glows when when struck by the electron beam, and beneath the screen is located a camera for recording the image. Micrographs are not usually recorded at the highest magnification of the particular electron microscope as it is easier to enlarge them later by normal photographic processes.

The limitation of the transmission electron microscope (as well as the conventional microscope) is that it can focus on only a limited depth of the specimen (depth of field).

Scanning electron microscope In 1965 a second type of electron microscope became available with a depth of field enabling the study of specimens in three dimensions. This new instrument was known as the scanning electron microscope. It employs a column very similar to that of the transmission instrument consisting of an electron gun and condenser lenses, which are used to 'bounce' the electron beam off the surface of the specimen. Situated in the condenser lenses are a pair of coils which deflect a small beam spot across the surface; linked to this scanning system is a cathode ray tube (CRT), its electron beam being scanned across the screen in sequence with the beam in the microscope. The electron beam hitting the surface of the specimen drives off secondary electrons, which are drawn towards a detector which, via an amplifier, sends a signal to the grid of the CRT. The greater the number of electrons leaving the specimen the brighter the corresponding spot on the CRT. The magnification of the image depends on the relationship between the size of the area scanned and the size of the CRT, varying between 10 and 200,000 times. The image can be processed by the operator for brightness, contrast and display of either a negative or positive image. A conventional Polaroid or roll film camera can be used to record the image.

The smaller the size of the scanned spot the higher the resolution achieved, but as the spot is decreased the energy that it contains decreases. A balance between the energy required to drive off the secondary electrons and the minimum spot size results in a resolution limit of 70 to 100 angstrom units in present day instruments.

Scanning transmission electron microscope A third type of microscope first developed in the 1960s and commercially available in 1973 is the scanning transmission electron microscope, or STEM. It combines the most prominent features of its predecessors. The STEM has a new type of electron gun called a *field emission source*. The instrument scans the electron beam across the specimen, the electrons are collected by a detector and the image is produced through a conventional scanning display system. The field emission source enables a high energy beam, as fine as a few angstroms in diameter, to be produced. Thus the instrument is able to provide resolution as high as the transmission electron system with the flexibility and image display of the scanning electron microscope.

The largest electron microscope is a three million volt transmission instrument, so large that it is housed in its own building three storeys high. This massive instrument produces electrons of enormously high energy, enabling scientists to study specimens many times thicker than those which can be studied in conventional instruments. The ultra high voltages and the recently developed STEM will enable even more information to be obtained about specimens, using microscopy.

Specimen preparation In electron microscopy specimen preparation is divided into two categories: transmission, including scanning transmission, and surface scanning. Biological transmission specimens usually undergo a complex preparation before being cut into very thin sections with an expensive instrument called an ultramicrotome; the conventional specimen thickness range being between 400 and

1000 angstroms. Metallurgical specimens are usually thinned down to less than 1000 angstroms by means of electro-chemical polishing. Sometimes metal specimens containing particles can be examined as suitably transparent replica films, formed by deposition on to a thin plastic film that has been previously coated on a specimen grid. On the other hand, specimens for surface scanning are often examined with little or no preparation, but if a sample is non-conducting, a thin layer of gold may be deposited upon its surface to provide good contrast with the scattering of electrons.

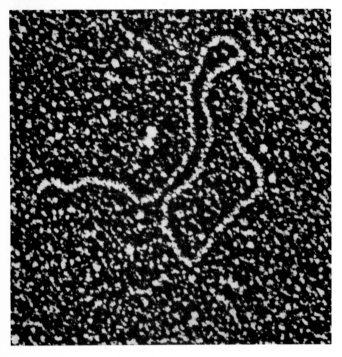

MICROTOME

A microtome is an instrument used to cut very thin *sections* (slices) of specimens for examination under a light micro-scope. Even thinner sections are needed when using an electron microscope, which has led to the development of the *ultra-microtome*.

The optical resolution of the light microscope is such that specimens are often best examined as thin sections, rather than whole. Nowadays medical and biological techniques require specimens of the order of 1 to 50 μm, the usual thick-ness being 4 to 5 μm (1μm, micrometre or micron is one-thousandth of a millimetre). Specimens are usually chemically preserved in a solution such as formalin (formaldehyde in water), dehydrated and embedded in wax. The wax block is firmly clamped on to a microtome—originally known as a 'cutting engine'—and sections cut with a specially prepared very sharp steel knife.

Types of microtome   There are three main types of micro-tome: the *rocking*, the *rotary* and the *sliding* microtome.

On the rocking design, the knife is clamped in a fixed horizontal position with the edge uppermost. The wax block is attached to the end of an arm pivoted near the knife and is moved or rocked in an arc past the knife edge. On the downward stroke the knife removes a thin section of the specimen. The block is advanced towards the knife by a ratchet mechanism with a micrometer thread for adjusting section thickness.

On a rotary microtome, the specimen block moves up and down in a vertical plane and the feed mechanism is actuated by a large hand wheel of which one rotation produces a complete cutting cycle. Larger and harder specimens can be sectioned on these machines and their rotary action is adapted easily to automatic power drive.

Sliding microtomes are the heaviest of all, enabling small and large sections to be cut of whole human lung and brain. One type of sliding microtome has a moving knife, drawn horizontally across the block, and is particularly useful for

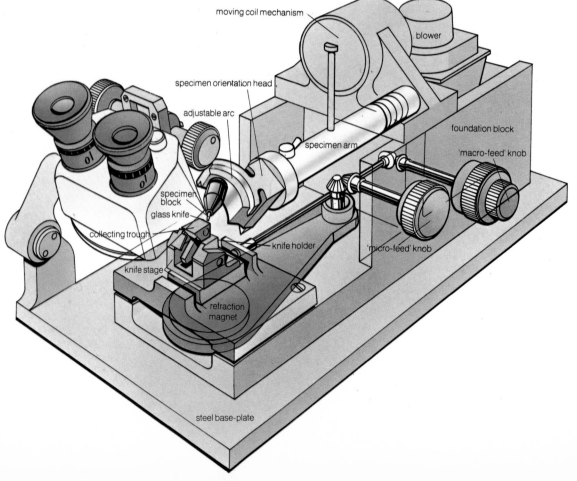

specimens which have been embedded in cellulose nitrate. The other most common and versatile type is the base sledge, where the specimen is mounted on a moving carriage or sledge and the knife is fixed. The micrometer feed mechanism is operated after each cutting stroke, either manually or by a trip device.

Ultramicrotomes The greatest resolution of the electron microscope over the light microscope requires even thinner sections. To obtain these, small 0.5 to 1.0 cubic millimetre biological specimens are embedded in very hard synthetic resins, for example methacrylate ('Perspex' or 'Plexiglas'), epoxy and polyester. Microtomes capable of cutting ultra-thin sections of these hard embedments—ultramicrotomes—have developed rapidly since 1950, allowing reliable and reproducible sections of even thickness to be obtained. A range of thickness for most machines is 5 to 150 nm (1 nm = one thousand millionth of a metre) with biological specimens being sectioned between 20 and 100 nm.

The embedded specimen is attached to a metal arm or tube which moves in a vertical plane past the knife edge. The knives are usually small pieces of plate glass broken in a controlled way to produce a fine edge 6 mm (0.24 inch) long and much sharper than any steel knife. Alternatively, expensive diamond knives are used with a long-lasting edge only 1.5 mm (0.06 inch) in length. As sections are cut they float out on to water contained in a trough, attached to the knife, and are often so small (0.1 mm square or less) that a stereoscopic binocular microscope is needed to see them. They are then collected on to small metal grids for insertion into the electron microscope.

The fine precision of the feed mechanism on these machines can be obtained either mechanically by a micrometer thread, reduced 250 times by a lever and leaf-spring system, or more often by electrically controlled thermal expansion of the rod or tube carrying the specimen. The rate of expansion and therefore section thickness will depend not only on the current applied but also on cutting speed.

During the cutting cycle the specimen block must not damage the knife edge on the return stroke, and this may be achieved by the specimen arm being displaced sideways on a D-ring movement after the section is cut. A less common but more advanced method is to retract the knife about 25 μm from its cutting position. This is done by means of a large electromagnet in the base of the machine, which is energized momentarily after a section is cut, while the specimen arm returns ready for another cutting stroke. Most ultramicrotomes can be automatically operated, the automatic models having complicated electrical circuitry.

Freezing microtomes Both the microtomes and ultra-microtomes described above have attachments capable of allowing sections to be cut of frozen unembedded specimens, thus avoiding damage to cells and tissue structures from chemical reagents and resins. Some microtomes are made specifically for this purpose, together with the cryostat — a refrigerated cabinet at —20°C in which the microtome is placed. With this machine, rapid frozen sections are produced of material removed from a patient during an operation allowing a diagnosis to be made within a few minutes and acted upon immediately.

MIRROR

All the objects around us reflect light. The surface of a mirror differs essentially from other surfaces only in being extremely smooth, so it does not jumble and diffuse the light rays that it reflects. Light rays diverging from an object will, after reflection by a mirror, continue to diverge, but from a new place. We judge an object to be at the place from which its light diverges. An observer will therefore judge light reflected from a mirror to come from a 'phantom' object—an *image*, which can be located behind or in front of a mirror surface. The study of image formation is part of the science of optics.

Above: if a mirror turns through a given angle, the image in it moves twice that angle, the principle of this infra-red line-scanning device for making heat images.
Left: a searchlight mirror, illustrating the image-forming properties of a curved surface. A magnifying shaving mirror works the same way.

Most mirrors for scientific use are aluminized on the front surface. This vacuum coating device is shown after being used to coat a reflector for use in a flight simulator.

The normal domestic mirror is flat and consists of a thin film of silver coated on the back of a sheet of glass. It forms images that are identical in appearance to their objects except that they are *laterally inverted*—for example, a left hand's mirror-image is a right hand. Other mirrors, such as car rear-view mirrors, are convex, and form reduced laterally inverted images. Concave mirrors, such as shaving mirrors, form enlarged laterally inverted images of objects that are close to them. The distorting mirrors of a funfair are concave in parts, and convex in others, thus stretching and squeezing different parts of the image.

Images can be formed by multiple reflection: double reflection removes inversion. Using two ordinary mirrors, it is easy to make a mirror that forms a 'correct' or non-inverted image.

So-called one-way mirrors consist of glass that has been coated with a layer of silver thinner than normal. This half-silvering reflects most of the light that falls on it (from either side), but permits some to pass through. On the observer's side of the mirror is a darkened room while on the other is a normally lit room. Most of the light from this side is reflected to form a bright image that masks the small amount passing through from the other side. Enough light passes through the mirror from the normally lit room to give the observer a clear view of it.

Mirror production   Metals have always been favoured for mirrors because they are highly reflective, able to be cast and polished to a high degree of smoothness, and hard wearing. Bronze hand mirrors, often highly ornate, were made by the Egyptians, Etruscans, Greeks and Romans in the centuries before Christ. Luxury mirrors were made of silver.

The Roman writer Pliny reports the existence of mirrors made of glass coated with tin or silver, but this principle was not in common use until the Middle Ages. It was first used on a large scale by Venetian craftsmen in the sixteenth century. They would lay a large sheet of tinfoil flat and coat it with mercury (which is liquid at normal temperatures), squeezing out excess mercury by laying a sheet of paper on top of this. Then a sheet of glass was gently laid on the paper, and the paper withdrawn. An *amalgam*, a chemical combination of tin and mercury, would coat the lower surface of the glass. It only remained to mount the whole on a protective backing.

This craft was supplanted in the nineteenth century by a chemical process invented by Justus von Liebig (1803–1873), who discovered how to deposit silver on to glass from a solution. This process is used today to make mirrors for everyday purposes. To prevent the delicate silver layer from being scratched, a coating of copper sulphate and other chemicals is added, finished with a layer of paint. One way mirrors dispense with these, and use transparent lacquer instead.

The glass of an ordinary mirror serves the purpose of providing a very flat surface, and of protecting the reflective silvering. It has the disadvantage of absorbing some of the light that passes through it, giving rise to multiple reflection, which in turn creates multiple images. These are too faint to be troublesome in everyday use, but in scientific work, mirrors are frequently front-silvered to avoid multiple images.

The concave mirror used in a modern reflecting astronomical telescope, for example, normally consists of an aluminium or chromium-plus-aluminium film coated on to the accurately shaped glass or ceramic surface. The coating is formed by *vacuum deposition*: a small piece of the desired metal is placed on a heating coil in a vacuum chamber. It vaporizes, and the vapour deposits on the uncoated glass, which is also placed in the vacuum chamber. The resulting film is only a few millionths of an inch thick, and can easily be removed chemically when necessary. It can then be renewed without disturbing the accurately ground glass surface. It is common to re-aluminize telescope and similar optics every couple of years or so, when the surface becomes less reflective, and large telescopes have their own built-in aluminizing plant, close by the observing floor.

MIXER, food

Food mixing processes include stirring, beating, whisking, kneading, chopping and blending, and each of these actions can be carried out with electric food mixers. The various types of mixers are often designed with their performance emphasized towards one or other of these operations.

One of the earliest electric mixers was the 'Universal Electric Mixer Beater' which was first produced about 1918 and consisted of the well-established kitchen whisk coupled to an electric motor. Another early mixer was the 'Peerless' industrial machine built in 1927, but it was not until after World War II that domestic food mixers started to become popular.

Large mixers   Large domestic mixers can be either the *orbiting action* or the *revolving bowl* type. Both use a *series wound* electric motor designed to give the correct speed-torque characteristics for the mixer. With this type of motor a fixed ratio reduction gearbox is used and the speed of the mixer is controlled by means of a governor attached to the end of the motor shaft. Under the action of centrifugal force the governor weights fly outwards. The centre of the governor arm moves out and bears on a pair of spring contacts which interrupt the electricity supply to the motor when the selected speed is exceeded. These contacts may control the motor directly or they may be connected to a semiconductor device which switches the main power to the motor. In the latter arrangement, the contacts only handle very light currents and their life is considerably extended. By moving the contacts away from or towards the governor, control of the speed can be obtained by a hand operated mechanism. Capacitors are connected across the contacts and across the supply connections to reduce the radio interference which would otherwise be caused by interruption of the mains supply.

The drive from the motor is taken to a gearbox by means of a toothed drive to avoid slip and to provide some speed reduction. In a typical design the motor is also brought to an outlet on the machine where a mechanical connection can be made to a liquidizer goblet or other attachment. The

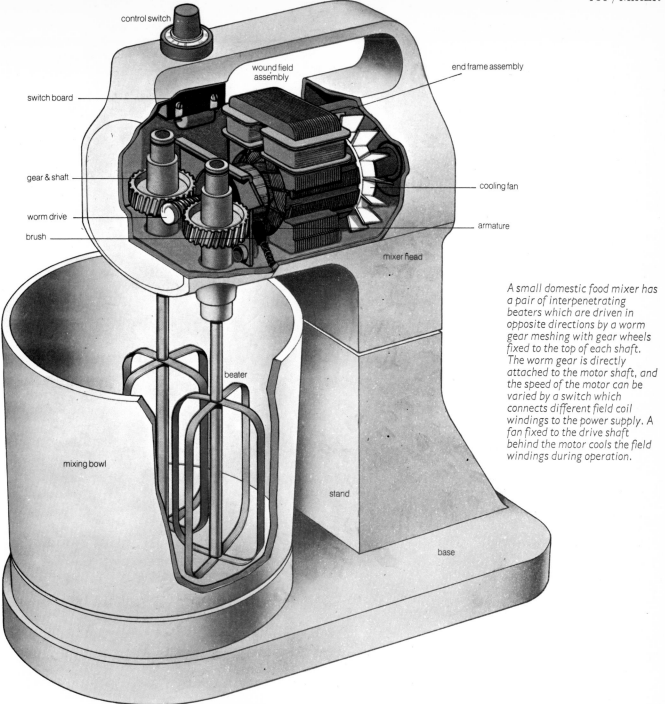

control switch

wound field assembly

end frame assembly

switch board

gear & shaft

worm drive

brush

cooling fan

armature

mixer head

beater

mixing bowl

stand

base

A small domestic food mixer has a pair of interpenetrating beaters which are driven in opposite directions by a worm gear meshing with gear wheels fixed to the top of each shaft. The worm gear is directly attached to the motor shaft, and the speed of the motor can be varied by a switch which connects different field coil windings to the power supply. A fan fixed to the drive shaft behind the motor cools the field windings during operation.

gearbox consists of a series of connecting spur gears which reduce the speed from the motor to a speed suitable for carrying out the various operations of the mixer. The reduction gearbox has additional outlets which provide power take-off points to additional attachments. A horizontal shaft is driven from the main vertical drive by a pair of bevel gears and, in orbiting action machines, this drives a hub situated above the mixing bowl into which the beater, whisk or kneader is inserted. The socket for the mixing tool is itself connected to a further gear, a planet gear, which meshes with a fixed ring gear mounted on the lower side of the hub gearbox. Rotation of the hub causes the planet gear to rotate about its own centre as its teeth engage the ring gear so that the mixing tool rotates about its own axis and also 'orbits' around the centre of the bowl. Each point of the tool has an epicyclic motion and all parts of the mixture are stirred. In revolving bowl machines, the mixing tool rotates on a fixed axis and the bowl itself rotates.

Industrial food mixers emphasize the kneading, beating and stirring actions. The mixer consists of an induction motor driving, by means of a change speed gearbox, a planetary gear combination inside a hub into which one or more beaters, dough hooks or wire whisks can be inserted. These tools are arranged to revolve on their own axes and also in a circle around the centre of the bowl. On the very largest of mixers the epicyclic action can become cumbersome, and in this case the beaters revolve on a fixed axis and the bowl itself is rotated with a separate drive.

Small mixers Small domestic mixers use a different principle. Their design emphasizes the whisking and beating actions and also allows lighter hand held appliances to be made. The essential components are a pair of whisks or beaters mounted vertically and spaced a short distance apart so that their blades interpenetrate each other. The beater shafts are driven in opposite direction by a pair of helical gears which lie on opposite sides of a worm gear. The worm gear is

Above: the 'Universal Electric Mixer Beater' an early type.
Above right: an industrial mixing machine in which the ingredients are weighed and fed to the mixing bowl automatically. The weigher dials are seen above the control panel.

mounted directly on the motor shaft. The contra-rotating motion draws material through the beaters, mixing it thoroughly in the process. Circulation of material in stand mounted mixers is achieved by turning the bowl which is mounted on a plate supported by a low friction bearing. The beaters are held to one side of the bowl and the viscous drag of the material being mixed turns the bowl automatically.

These mixers are usually fitted with universal motors, that is, electric motors which can be operated from a direct current source or a single-phase alternating current supply. In motors of this type the armature speed is inversely proportional to the magnetic field and this allows the speed of the mixer to be varied by the varying the numbers of turn on the field winding. Generally a number of coils are wound on top of each other and a simple three-position switch selects the coil required for the appropriate speed. An alternative method of controlling the speed is to use a thyristor regulator. Such a control permits infinitely variable speed adjustment and automatically maintains the speed irrespective of the output torque.

Liquidizers Another type of domestic mixer is the liquidizer or blender. These machines combine the actions of chopping and stirring, often by macerating solids in a liquid suspension. Blenders consist of two main parts, a goblet and a power unit. The goblet contains a number of small sharpened blades on a shaft which passes through a liquid-tight bearing in the base. The shaft is connected to the power unit by means of a plastic or rubber coupling designed to allow for any misalignment. The power unit consists of a series wound universal electric motor fitted inside a casing to which it is attached with rubber mountings. The blender or liquidizer blades are driven at high speed, usually about 12,000 rpm, and the rotating blades form a vortex or whirlpool in the liquid so that the solid particles are drawn down on to the cutting blades. Vertical ribs moulded inside the goblet control the vortex formation and assist vertical circulation of the particles. A series wound motor is used because it can be designed to give peak efficiency at high speed. Liquidizers have not always been fitted with speed controls, but some of the more recent models fit a thyristor control similar to that used on a hand-held mixer.

OPHTHALMOSCOPE

The opthalmoscope is an optical device used for examining the interior of the eye. The pupil is the tiny black 'window' of the eye; through it we can view a wide landscape, but no one could look in from the outside until the ophthalmoscope was invented.

It was believed that the blackness of the pupil was due to the total absorption of light rays by the eye, but Hermon von Helmholtz (1821-1894) discovered that most of the light entering the eye is reflected back and can be intercepted by an observer. Helmholtz was a physician and physicist who investigated the conservation of energy, the speed of nervous impulses, colour blindness, physiological acoustics and other subjects, but is most famous for his *Handbook of Physiological Optics* (1856-1866; complete edition 1867). In 1851 he hit upon the idea of directing a beam of light into the eye by means of a mirror in which there was a tiny aperture through which an observer could look.

For diagnostic purposes it is important to obtain a good view of the *fundus*, that part of the cavity of the eye which can be examined by looking through the pupil. This enables the observer to detect abnormalities and pathological changes in the eye; some diseases, such as diabetes, manifest themselves in the eye before symptoms appear elsewhere. The fundus, however, cannot be examined by a perforated mirror alone; this only gives a red reflex. Helmholtz found it was necessary to interpose a condenser lens, with about a four inch (10 cm) focal length in order to obtain an inverted image, magnified five times. This combination of mirror and hand-held condenser was called an indirect opthalmoscope and was in regular use for eye examinations until about 1920.

Nowadays ophthalmoscopy is carried out by a direct method with a hand-held instrument. From a tiny lamp powered by dry batteries which are located in the handle of the instrument, a narrow beam of light is directed through the pupil and into the eye of the patient by means of a prism

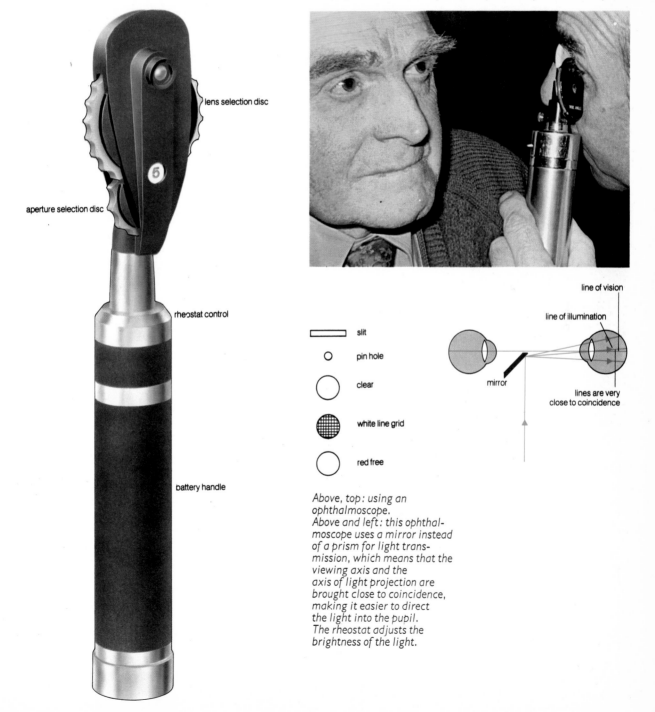

lens selection disc

aperture selection disc

rheostat control

battery handle

slit

pin hole

clear

white line grid

red free

line of vision

line of illumination

mirror

lines are very close to coincidence

Above, top: using an ophthalmoscope.
Above and left: this ophthalmoscope uses a mirror instead of a prism for light transmission, which means that the viewing axis and the axis of light projection are brought close to coincidence, making it easier to direct the light into the pupil.
The rheostat adjusts the brightness of the light.

or a perforated mirror of steel or glass. The image, magnified fifteen times, is viewed through the sight hole in the mirror and brought into focus by a revolving magazine of lenses, rotated by the index finger of the observer. It is interesting that the power required to focus the image represents the refractive error of the eye, and gives roughly the power of the spectacle lens required to correct the vision, but this is incidental to the real function of the instrument, which is the examination of the interior of the eye in order to detect abnormalities.

Many instruments are fitted with a variety of filters, such as red-free and polarizing screens, which show up conditions not visible with white light. Cross-line graticules can be projected on to the retina so that a particular point can be given a grid reference. There is also a new type of ophthalmoscope which projects a laser beam; this is used in eye surgery to coagulate the tissue around a detached retina. For clinical research and a more detailed study of the eye, there is a large binocular ophthalmoscope which stands on an instrument table. With this it is possible to obtain a large stereoscopic picture of the fundus magnified about fifteen times.

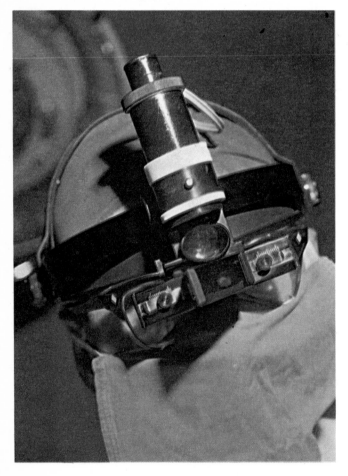

Above: the patient's view of a doctor using a binocular ophthalmoscope. Also called indirect ophthalmoscopy because it uses mirrors and results in an inverted image, the binocular system provides a three-dimensional image.

PACEMAKERS

The heart pacemaker maintains the steady pumping action of the heart by delivering a regular stimulation. Any muscle is sensitive to electrical excitation, but the heart is even more sensitive than most.

Heart pacemakers are divided into three categories: the natural pacemaker possessed by all animals, from the most primitive to the most complex; the external electronic pacemaker, located outside the body and used by hospitals in emergency and temporary situations; and the miniature pacemakers, driven by batteries, which can be surgically implanted in the body. The pacemaker adjusts the rate of stimulation according to the demands of the body; for example, during sleep, the adult human heart slows to approximately 55 to 65 beats per minute, and during hard work or exercise speeds up to more than 100 beats per minute. During temporary periods of great exertion the heartbeat may exceed 200 beats a minute. A failure of the pacemaker to communicate its signal to the heart results in the failure of the heart to pump blood, a condition known as *heart block*.

Natural pacemaker The natural pacemaker is a *sinoatrial node* (lump of small fibres) located near the top of the heart, which receives its instruction from outer masses of muscle tissue in the heart, according to the body's need for oxygen, its temperature, nervous excitation, and other factors. Biological experiments have shown that each cell of heart tissue seems to beat at its own rate, and that the instructions to the natural pacemaker are a consensus. In experiments with the salamander and similar creatures, grafted pacemakers from other animals can take over the function of the natural pacemaker without difficulty, but in man such a graft results only in scar tissue, so the function of the human natural pacemaker must be replaced when necessary by an electronic device.

External pacemaker The external pacemaker, used in emergency and temporary situations, works by means of a *catheter* (a tube inserted by a surgeon) extending through a vein to the right ventricle of the heart. The catheter contains electrical sensing and stimulating devices.

Implantable pacemaker Several types of 'portable' pacemakers have been developed since the 1950s. One model requires the implantation only of electrodes, with the device itself taped to the patient's chest. This device can be powered by a battery small enough to be worn on the chest or by batteries which are worn in a pocket or in a holder which clips to the patient's belt. Pacemakers which are entirely implanted in the body, battery and all, have been in use since the late 1950s and information on their long-term performance is still being collected.

Implantable pacemakers must meet certain design requirements. They must have long term reliability, measured in years. They must have long battery life (currently three years). The circuitry must be designed to operate at a low current drain, whose output pulse must remain unaffected by battery and load variations. The enclosure of the device must offer maximum protection from the ingress of body fluids, and the electrode system must be mechanically and electrolytically stable.

Batteries Pacemakers commonly used mercury cells, which are small enough and offer an excellent power to volume ratio and are capable of maintaining a constant output voltage for 90% of their life span. Other recent developments include an *isotope source generator* and the *metal-oxygen cell*. The isotope generator converts the heat generated by plutonium-238 into electrical energy using a *thermopile*, with the constant deep body temperature of 37°C (98°F) acting as a cold junction. Although a life of ten years is predicted, extensive testing is needed to ensure completely safe packaging. The metal-oxygen cell relies upon the oxygen content of the body tissue to supply it with energy; prob-

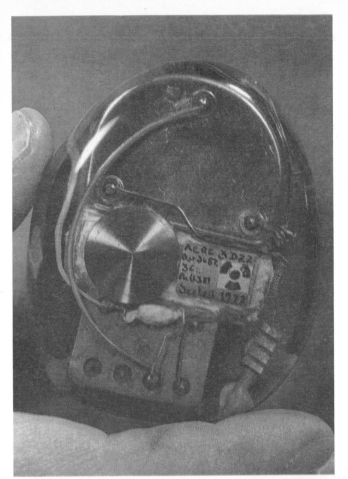

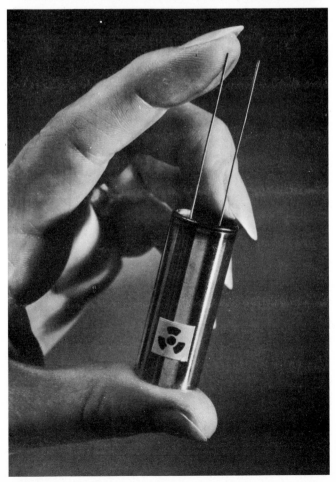

Left: an implantable pacemaker with a nuclear battery.
Below left: a close-up of
the nuclear battery.
Below: the isotope, 0.18 gm
of Plutonium 238, decays
radioactively, thus heating
the sides of the capsule.
The heat is turned into a
voltage.

lems to be overcome include the level of oxygen present and
the ability to maintain suitable rates of diffusion.

Circuitry The *blocking oscillator circuit* still in use in many
pacemakers is satisfactory, but because it needs a transformer
it is unacceptable for *thin film* circuit construction, which has
been found to be best for long term stability. *Complementary
pair* pulse generators have been tried; their advantage is that
they do not place a constant strain on the battery, but they
are particularly sensitive to line and load variation. *Mono-
stable* and *bistable* circuits have been adapted for pacemakers; difficulties have been overcome in achieving the
required pulse repetition rate and width without unacceptable differences in timing components.

The *demand* pacemaker possesses, in addition to the pulse
generator, a frequency selective amplifier which controls the
pulse generator *inhibitor*. The electrode that stimulates the
heart also monitors the ECG (electrocardiogram). The
amplifier inhibits the generator except when heart block
occurs, thus stimulating the heart 'on demand'.

The patient suffering from heart block is admitted to hospital, and his heart is assisted by an external pacemaker while
he undergoes minor surgery to have the subminiature device
implanted. He will be discharged from hospital, attend out-
patient clinics for check-ups, and be able to lead a normal life.

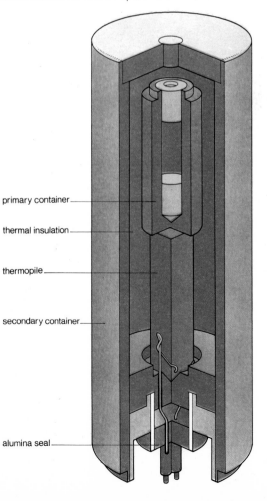

primary container

thermal insulation

thermopile

secondary container

alumina seal

PENS

The fountain pen is a writing device which consists of a reservoir to hold the ink, a nib (writing tip) made of gold or stainless steel, and a system of air holes and flow tubes for delivering the ink to the nib. The ink supply system takes advantage of the natural phenomena of *capillary action* and *surface tension*.

The earliest fountain pens, patented in Britain more than 150 years ago, did not use capillary action, but had a plunger device which had to be frequently operated by the user to keep the ink flowing. The tubes and air holes in modern pens are called the *feed*, and were first successfully used by an American, L E Waterman, in the 1880s. Waterman's fountain pens were immediately popular, and millions of them were sold each year.

The flow tube in a fountain pen must be exactly the right size; if the tube were too small, no ink would flow at all; if the tube were too big, all the ink would flow out of the pen and make a mess. Capillary action is the result of surface tension of a fluid inside a tube: it will cling to the walls of the tube, not flowing until some of it is removed from the bottom of the feed system, in this case by writing. The ink flows on to the nib, which has a veritcal slit in the end of it. The surface tension of the ink causes it to seek the slit but prevents it from running off the nib until the slight pressure of the nib on the paper causes the slit to widen. Then the ink flows on to the paper.

Surface tension is the property of a fluid which causes it to seek a spherical shape, like a free-falling raindrop. The molecules of a fluid attract each other in such a way that a drop of it wants to behave like an elastized container. This is why a pin will float on the surface of water. The pin is actually heavier than the water it displaces, but it is not heavy enough to break the surface tension.

There are a number of ways of filling the reservoir in a fountain pen. Some reservoirs are made of rubber or other elastic material and are compressed by means of a lever; then the tip of the pen is inserted in the ink and the lever released. The reservoir sucks up the ink in reverting to its natural shape. Other pens have a piston suction device, or a 'snorkel' (a tube which is extended from the tip and drinks up the ink as suction is applied by a piston). A cartridge pen uses a replaceable plastic container of ink; the cartridge itself is the reservoir. The body of the pen unscrews, the cartridge is inserted, and when the pen is screwed back together the mechanism opens one end of the cartridge.

Most fountain pens need an air hole to allow air to flow into the reservoir behind the ink, equalizing the air pressure; otherwise the ink would not flow at all. A variety of interchangeable nibs are available with tips of different width to suit the requirement of the user. Fountain pens come with screw-on or press-on caps to retard evaporation of the ink when the pen is not in use and to protect the clothing when the pen is carried in the pocket.

Ballpoint pens In the latter half of the 20th century, the fountain pen has largely been replaced by another writing instrument—the ballpoint pen. This relatively simple writing device uses a tiny ball bearing coated with a suitably viscous (thick) ink. The ball bearing is held in a socket at the tip, and this unit is connected to a thin tube ink reservoir. Ofter there is also a push-button spring mechanism for retracting the point.

Development In 1888, an American, John Loud, working independently on a pen that did not leak, invented the first ballpoint-type pen. More cumbersome than today's lightweight models, his pen consisted of a tube with a ball bearing at the tip, housed so that it could move freely in all directions. The tube was filled with a viscous ink and as the pen moved, the ink was drawn out by capillary action. Loud used this device to mark leather and fabrics, but the patent was allowed to expire.

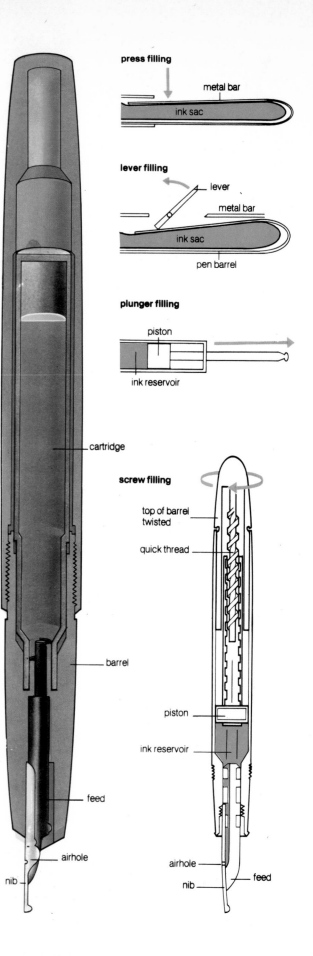

press filling
metal bar
ink sac

lever filling
lever
metal bar
ink sac
pen barrel

plunger filling
piston
ink reservoir

cartridge

barrel

feed

airhole

nib

screw filling
top of barrel twisted
quick thread
piston
ink reservoir
airhole
feed
nib

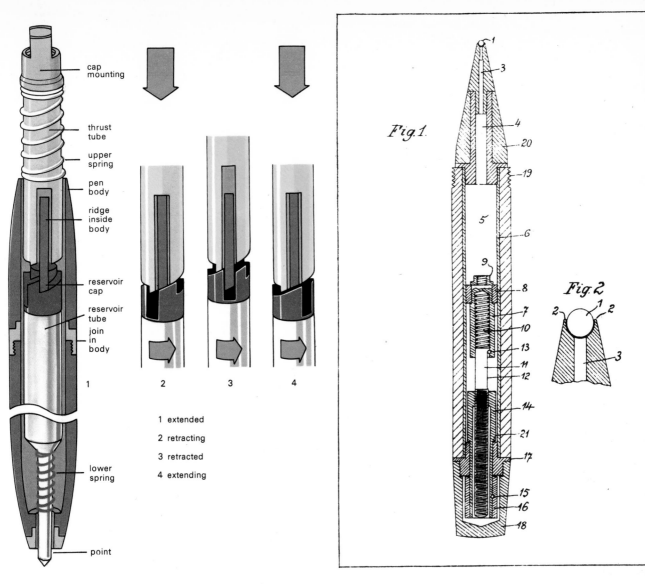

cap mounting

thrust tube

upper spring

pen body

ridge inside body

reservoir cap

reservoir tube

join in body

lower spring

point

1

2

3

4

1 extended
2 retracting
3 retracted
4 extending

Fig. 1.

Fig. 2.

In 1938, two Hungarian brothers, László Biró, a painter, journalist and sculptor, and his brother Georg, a chemist, patented a ballpoint pen which was more workable than previous inventions. At the outbreak of World War II they moved to Argentina, licensing their invention to companies in several countries. By 1939 the 'biro' was being widely used in Britain. The impetus for American acceptance came in 1942 from the US Army Quartermaster General who wanted a pen that did not leak for use in high-flying aircraft.
Today's designs The ball of the pen is usually made of steel, though synthetic sapphire and tungsten carbide have been used recently for expensive pens, and is normally one millimetre in diameter. It is fitted into a socket so that it rotates freely. Several internal ducts in the socket feed ink to the ball which therefore effectively rests on several metal ridges. The other end of the socket is drilled and fitted on to a metal or plastic tube which contains the ink. When the ball is pressed on paper and moved, the capillary action draws the ink from the reservoir, and impressions are made as the ink flows down the ducts. In effect, the ball functions as a valve to prevent overflow, and on rotation it acts as a suction pump drawing out the ink.

One problem was that as some of the ink ran out, a partial vacuum was formed between the back of the ball and the ink reservoir, which cut off the supply. This was solved by making a small hole at the far end of the reservoir. As the ink at the tip is sucked out, more ink from the tube is drawn into the socket to fill its place, the vacuum being prevented by air

Opposite page: on the left, a cartridge pen. On the right, a selection of other filling mechanisms. On all fountain pens, the ink runs down a groove into the slit in the nib.
Above, left: a retractable ball-point pen. The thrust tube and reservoir are both spring-loaded. Sloping ratchet teeth rotate the reservoir; slots and ridges inside the tube body keep it from revolving. Depending on whether a slot or a solid area lodges opposite the ridge, the reservoir is extended or retracted.
Above: the Birö's original patent drawing for a ball-point pen. It was designed to be refillable.

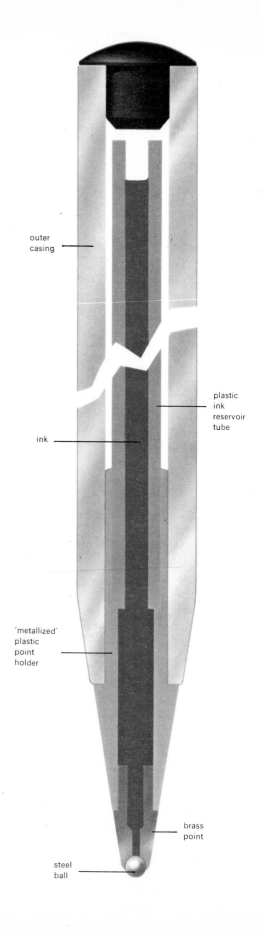

Right: cross section of a typical low priced disposable ballpoint. It has an 0.5 millilitre reservoir, which allows the use of a simple open-ended tube (wider tubes have to be sealed with a cap and a layer of viscous liquid). Only the very tip of the point is metal; for economy, the rest of it is made of matching gold coloured plastic.

outer casing

plastic ink reservoir tube

ink

'metallized' plastic point holder

brass point

steel ball

which is drawn in through the vent.

This method of inking the ballpoint pen depended on finding an ink that was susceptible to capillary action but which would not leak from the vent. At first, printer's ink was used but it was soon discovered that it was not viscous enough to prevent leakage. Therefore the principal research into making an efficient ballpoint pen in the last twenty years has been concerned with developing a suitable ink. The earliest ballpoint inks were of a heavy gelatinous type, but now there are two main kinds. The first, containing a dye soluble in oil, dries on the writing surface by absorption. The viscosity of the ink is high but the impressions formed tend to be less sharp than those created by spirit-soluble inks which dry on the writing surface by evaporation.

The capacity of the ballpoint reservoir varies from 0.5 to 1.5 millilitres. In the smaller reservoirs the tube is open at the far end, as the viscosity of the ink is sufficient to prevent leakage, but 1.5 ml reservoirs need a vented plug and the use of an even more viscous liquid known as a *follower*. The follower is solid enough not to leak but liquid enough to follow the ink as it is used up.

There are several types of mechanism for retracting the ink reservoir but basically they all include a spring wound round the lower half of the ink tube to retract it, and a thrust device which is pushed down against the top end of the reservoir tube when the button is pressed. In one type, when the thrust tube is pressed down it rotates, guided by a ratchet device. It is connected to slots (which look like alternate deep and shallow castellations) on the inside of the body and when the button is released it comes to rest slotted into the forward writing position or further back in the retracted position.

Another type of device consists of a spring mounted thrust tube attached to the top of the reservoir tube. The locking mechanism consists of a flexible pin which, guided by a simple vertical ratchet mechanism, rotates in a roughly heart-shaped guide slot coming to rest at the bottom and top of the 'heart' for retracted and writing positions respectively.

A much simpler device, used on some disposable ballpoints, consists of a thin plastic rod plugged into the open end of the reservoir, and with a small projection on one side. The top of the rod is pressed to extend the point until the projection clinks into a hole in the pen body. Pressing the projection releases it and the reservoir springs back.

The enormous reduction in the price of ballpoint pens compared with the fountain pen has practically ousted the nib pen from general use. But the fountain pen still serves a function since the lines made by a ballpoint pen can be seen to be discontinuous when examined under a microscope, so for stringent handwriting tests and signatures on some documents the use of a ballpoint pen is prohibited.

PINBALL MACHINE

The pinball machine, often called *pin table* in Britain, is an amusement device in which the player obtains the highest score possible on a display board by causing a steel ball to travel around a wooden playing board, negotiating obstacles.

Construction The working units are contained in a rectangular wooden cabinet measuring 22×54 inches and 16 inches deep (about $56 \times 137 \times 41$ cm). It is mounted on four steel legs, giving a sloping playing field 36 inches high by 38 inches (91×96 cm) wide with the low end where the player stands. The display board is at the other end of the cabinet facing the player and is 28 inches (71 cm) high.

The working components are electrically operated by 24 volts AC, and the lighting display by 6.8 volts AC, the power being obtained from a double-wound transformer. The basic working units fall into four categories: fractional hp motors, relays, solenoids and switches. The motors are used to drive cams which cause switches to pulse. Relays are used to transfer voltage links to different circuits, and solenoids are used to turn power into mechanical movement.

Operation Upon insertion of a coin, switches close, starting a motor which turns a wheel carrying painted numbers, and a certain credit is shown on the display board through a little window, according to the value of the coin. The player pushes a start button, which delivers a ball to the playing board, causes the machine to light up, causes all the electrical devices to home to zero and again operates the credit device, this time removing a credit.

Next the player pulls a spring-loaded plunger at the right-hand corner of the table, which sends the ball up to the high end of the sloping board, and the ball then rolls back down the board, negotiating obstacles. Runways and obstacles have rubber rings around them to give rebound to the ball. The chief mechanical obstacle is the *bumper*, a round device several of which are mounted on the playing board. When the ball rolls against a thin plastic ring round the base of the

bumper, a vertical rod in the centre of the ring, extending below the surface of the table into the well of a contact blade, moves in the well, causing a switch to make contact. A solenoid creates a magnetic field which pulls down sharply on an assembly of plunger and yoke around the top of the bumper, and the ball is pushed sharply away from the obstacle careering across the table to hit other obstacles. Each time the ball strikes an obstacle a score is registered on the board.

The other important device is the *flipper*. There are two, at the low end of the table, operated by buttons on each side. If the ball rolls between them, it goes down an *exit hole*, but the player can use the flippers (pivoted arms) to bat the ball back up into the obstacle course, and he can shake (*tilt*) the table to make the ball roll towards the flipper. If he shakes the table too much, it is sensed and the game is over.

There are also two runways at each side of the table, one of which is called the *out-hole*. These are provided because otherwise a skilful player might be able to keep a ball in play for a long time, and the table wouldn't make much money.

Normally, a player gets five balls, and bonus scores for

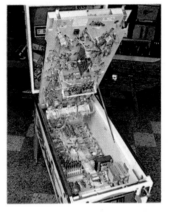

Left: a pinball machine, with the glass top removed and the works open.
Below: the bumper, one of the scoring devices on a pin table. When a ball strikes the plastic ring, the vertical rod in the centre moves in the well, causing an electrical contact. A magnetic field pulls sharply downward on the yoke at the top of the bumper, pushing the ball away, and the bulb lights up.

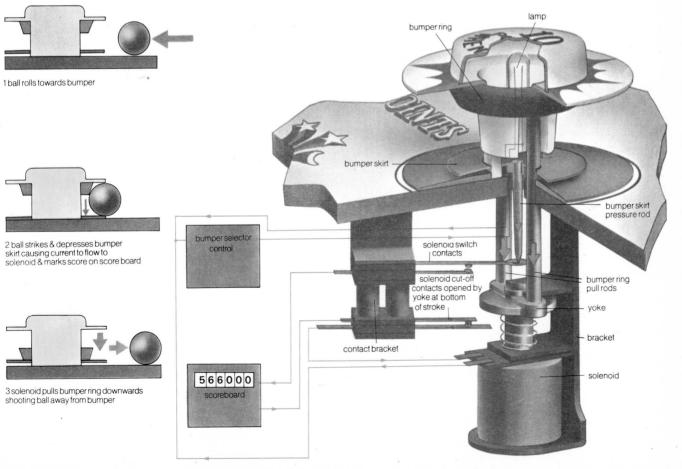

1 ball rolls towards bumper

2 ball strikes & depresses bumper skirt causing current to flow to solenoid & marks score on score board

3 solenoid pulls bumper ring downwards shooting ball away from bumper

bumper selector control

5 6 6 0 0 0
scoreboard

lamp

bumper ring

bumper skirt

bumper skirt pressure rod

solenoid switch contacts

solenoid cut-off contacts opened by yoke at bottom of stroke

bumper ring pull rods

yoke

bracket

solenoid

contact bracket

Below: the advance unit, which causes various units to light up when the balls strike them. Bells are also supposed to ring, but they are often disconnected.

achieving a certain number of points; if he gets above a certain score, he may receive an extra ball or a free play (five more balls). As each ball disappears down the hole, a new ball becomes available and the player uses the spring loaded plunger again to bring it into play. Normally also, bells ring each time a ball strikes an obstacle but the bells are often disconnected because of the noise factor.

The display board is screened (painted) with a garish design, and the games are called such things as *Delta Queen*, *Skylab*, *Lucky Ace* and so on, although they are all essentially the same. Hidden within the design are 'ghosted' areas which appear only when illuminated, giving a message to the player such as 'game over', 'ball to play' or 'tilt'. If the tilt limit has been exceeded the game cannot continue unless the player inserts a new coin.

PITOT TUBE

A pitot tube [impact tube] is primarily used for the measurement of speed or velocity of a fluid, either gas or liquid. The instrument is named after Henri Pitot, who recorded work in 1732 on the measurement of water flow. In principle, if an open-ended tube faces the flow of the fluid, a pressure will build up which will increase with the velocity of flow; thus, if a pressure gauge is attached to the outlet of the tube, the gauge will indicate fluid velocity.

The tube facing the fluid is known as the *head* and is required to sense a *total pressure*, which is the addition of the surrounding (*static*) pressure and the pressure due to flow (this is called *dynamic* pressure). The face of the tube creates an obstacle to the flow; this is known as a *stagnation point* which creates a *stagnation pressure*, and it requires very careful design to make the stagnation point give the required *total pressure*.

To derive an indication of velocity a pressure difference must be measured between the total pressure and the surrounding static pressure, which is sensed by *static vents*. In a simple system the vents are holes pierced in the side of the outer tube assembly, but in large installations such as aircraft, the airflow is complicated and therefore static vents are placed at a number of positions and the average pressure is taken.

Total pressure is directly dependent upon the density of the fluid, which is directly dependent upon *absolute temperature* (temperature measured in Kelvins, where 0°C is 273 Kelvins—273°K). The error on measured velocity due to

temperature change is approximately 0.2% per °C in air and therefore compensation must be included for high accuracy. Aircraft air speed indicator (ASI) systems do not usually have temperature compensation because first the error is not serious in this application and second the aerodynamic properties which control flight are all affected in the same manner.

Design features A typical design is the L shaped tube, which is satisfactory where the fluid flow is straight and not swirling, that is, there is no turbulence. This design will begin to give significant error if the flow deviates more than about ± 10°, but by shaping the inlet hole, flow angles of up to ± 60° are acceptable. A general purpose pitot static tube would be about 8 mm (0.3 inch) diameter and have a smooth surface. The nose must be carefully tapered or shaped to give both minimum disturbance to flow and achieve a good total pressure. Smaller and larger versions are made for special purposes, often with built-in temperature-sensing elements; also, very special shapes are necessary for supersonic conditions. A number of pitots may be grouped to measure angles of incidence or turbulent flow and if icing is probable, electrical heating may be added.

The basic tube is usually metallic, either non-ferrous or stainless steel and the mounting and surrounding assembly is often plastic or of a fibrous material. In conditions where damage may occur a protective shield may be added but this must not interfere with the flow; additionally, the nose may be hardened to prevent abrasive action disturbing the shape.

Aerodynamic properties In subsonic conditions, the pressure difference sensed by the pitot-static system is proportional to velocity squared, that is, every time the velocity is doubled the pressure difference increases by a factor of four. As velocity increases towards the speed of sound in the fluid, shock waves begin to build up until supersonic conditions are reached, where the pressure can in some circumstances decrease with increased velocity. At these speeds,

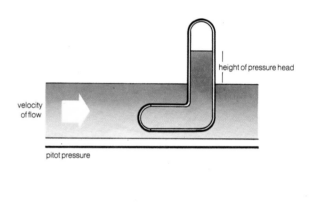

velocity of flow

height of pressure head

pitot pressure

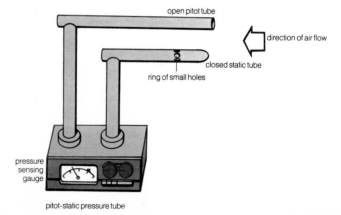

open pitot tube

direction of air flow

closed static tube

ring of small holes

pressure sensing gauge

pitot-static pressure tube

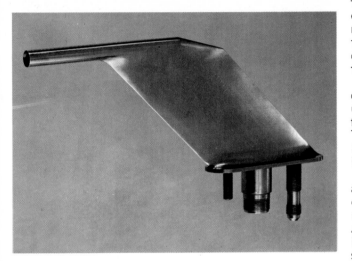

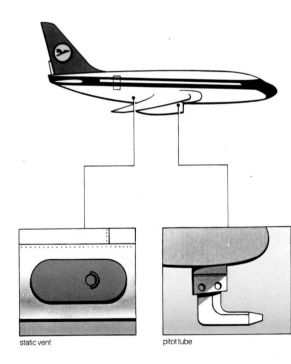

static vent   pitot tube

there is a need for special tube shapes.

At low air velocities, the pressures are extremely small, for example, at 2 mph (about 3ft/sec, 1 m/s) the pressure is only about 0.0001 psi; at 200 mph (100 m/s) this increases to about 1 psi. Thus for faster moving vehicles such as aircraft, the pitot becomes a practicable method of air speed indication. Complications arise, however, because, due to air density changes with altitude and temperature, the indicated air speed (IAS) is not the true air speed (TAS). At 3000 ft, the TAS is approximately twice the IAS.

Calibration To obtain accurate calibrations, the pitot system must be placed in a fluid flow which is directed accurately along the axis of the tube, without disturbance or turbulence. This is known as *laminar flow*. A common method is to use a *whirling arm rig*, which consists of a rotating arm, perhaps 10 to 20 ft (3 to 6 m) long, whose outer end travels in a circular trough. The pitot is mounted on the outer end of the arm and although the travel is circular, the radius is large enough for the fluid flow to be considered laminar. This method is useful for the lower velocity calibrations and has the advantage that the fluid velocity is known directly from calculation of the arm speed and does not require reference to any other fluid flow instruments.

The wind tunnel is another method for calibration and can be more useful for higher speed air flow. The tunnel is usually rectangular and can range from a few inches to many feet across with a length which is many times the aperture. This, plus careful design of the air driving fans, creates laminar flow. In a wind tunnel, the velocity calibrations must be compared with an instrument of known calibration.

Both of the above rigs are means of generating fluid velocity and, apart from calibration, they are frequently used for conducting flow experiments where pitots would be the means of determining the resultant velocities.

Associated equipment Some sort of pressure measuring instrument is required to indicate the pressure difference sensed by the pitot-static system. The *liquid manometer* is widely used for experimental purposes, where the pressure may be conveniently fed to each end of the classic U tube configuration.

Differential pressure gauges are more practicable for many purposes, for example, the aircraft air speed indicator; in this case the linkage within the gauge is mechanically designed to compensate for the square law (where pressure is proportional to the *square* of velocity of the fluid) and provide readings on a linear scale. Where an electrical signal is required for control or computation, a *differential pressure transducer* would be used. In this case, the compensation for the square law would be done electrically.

Typical uses One of the greatest uses is for the measurement of velocity of airborne vehicles, aircraft, missiles, and so on, where the higher speeds and the requirement for aerodynamic information make the pitot ideal. Considerable use is found in research and development where 'clusters' or 'rakes' are often fitted to rigs, to test pumps, turbines, compressors and so on, and establish both velocity and direction of flow.

Opposite page: principles of measuring pitot pressure.
Top left: static and pitot-static transducer. Signals from this unit are fed to an air data computer.
Centre left: fuselage mounted pitot tube with built-in de-icer.
Left: the position and appearance of tubes on an aircraft.

POSTAL SORTING MACHINERY

Mail was first franked in 1844, in Britain with a combined place-date and number rubber stamp, the parts of which could be used separately. Initially this was done by hand, but in 1857 it was tried, unsuccessfully, with a steam-powered franking (cancelling) machine. In 1949–50 the first franking and facing machines were introduced in Britain; they operated on a light-dark scanning principle, a dark-coloured stamp being the trigger. However, the scanner also reacted to prominent writing and to labels.

Research turned to stamps themselves. Graphite lines were printed on the backs of the stamps, but these were easily broken, causing loss of conductivity; problems were also caused by paper clips and other metal objects triggering the machinery. The successful system finally devised uses *phosphorescent* materials which only become visible after exposure to ultra-violet light. These materials are either incorporated into the stamp paper or printed on to it when the stamps are made. They do not actually contain phosphor, which is toxic; British stamps, most of which are printed by Harrison and Sons, use organic resins with an activator in a unique formula. In some countries stamps include zinc sulphide with a copper activator as the luminescent agent. *Fluorescent* materials are also used, which glow under the ultra-violet light, but do not have an afterglow, as do the phosphorescents.

Segregator Mixed postal consignments are emptied into a hopper. Conveyor belts carry it to a slowly rotating drum $1\frac{1}{2}$ m (about $4\frac{1}{2}$ feet) in diameter and as long as a room. The inside of the drum is fitted with hinged plates and slots through which letters can slip to another belt; newspapers, small parcels, and letters thicker than 6 mm ($\frac{1}{4}$ inch) or wider than 140 mm (about $5\frac{1}{2}$ inches) pass through the drum to another belt. The infeed conveyor can handle about 30,000 items per hour; vibrating rollers beneath the belt help to provide a smooth flow. About one-fifth of the input will be separated as non-machineable.

Automatic letter facing Nowadays the automatic letter facer (ALF) is linked to the segregator and can handle its machineable output. The letters have been aligned on their long edges and are carried through on a conveyer belt. An optical scanner looks for the stamp and if necessary the letter is turned around by an electro-mechanical device.

In Britain, first class stamps have two phosphorescent bars printed on them and second class stamps have only one. The scanner on the ALF recognizes the afterglow excited by the

ultra-violet light source and switches off immediately. It opens again after 4.5 milliseconds, and if the eye still sees an afterglow the letter is channelled to the franking device for first class mail. If there is no further reaction the letter goes to the second class franking head. Each ALF is fitted with four franking heads and four boxes for receiving cancelled mail; there is a fifth box for unrecognized letters which must be dealt with manually.

An optical recognition device is also fitted for 'recognizing' official mail, which bears two dark bars and the words 'official paid', as well as pre-paid reply cards. This mail is franked by the same four franking heads as the other mail.

The coding desk At the coding desk the letters are fed one at a time in front of an operator at eye level. He has two viewing positions; he reads the address in the first position and keys the address of the envelope in the second. If a mistake is made or if the machine fails to respond to the keyed code he can check the address again. He types the letter-figure combination on the keyboard of the coding unit, which operates the *translator* which prints a phos-

The segregator (above right) has slots allowing letters of the correct size for automatic sorting to fall through. The rest of the mail passes through the tunnel to be sorted by hand. The letters are mechanically faced and franked; then they go past the coding desks (right) where the operator keys the address code and the machine prints corresponding phosphorescent dots on the front of the envelope. The newest designs of coding desks are lower and more compact, so that postal employees are not so isolated behind the machines.

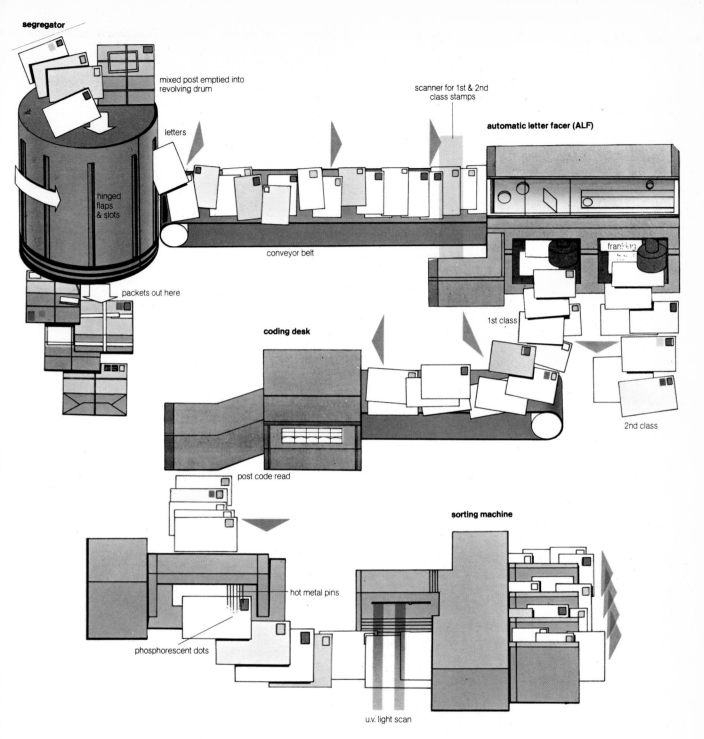

segregator

mixed post emptied into revolving drum

letters

hinged flaps & slots

conveyor belt

packets out here

scanner for 1st & 2nd class stamps

automatic letter facer (ALF)

franking

1st class

2nd class

coding desk

post code read

sorting machine

hot metal pins

phosphorescent dots

u.v. light scan

phorescent code on to the letter face; this is done by means of heated metal pins which strike a coated paper or tape. Thus the address code is translated into a language which can be read by machines. The translator is also called the *electronic dictionary*. The *binary* code consists of two rows of 14 dots and lines which could provide over 16 million possible combinations; it is called a fourteen-bit pattern. Each row consists of a starting beam, twelve dots and a parity dot.

The phosphorescent materials used in the stamps and for printing the code are different, so that the one will not trip the scanning head looking for the other. The signal of the postage stamp will only give an afterglow when excited by a short-wave ultra-violet radiation (256 nm), while the dots respond only to a long-wave (365 nm). In some countries phosphorescent materials of different colours are used.

Above: Incoming mail is tumbled in the segregator; letters of machineable size pass through the slots. The ALF machine faces the letters and cancels the stamps, reading the phosphorescent bars on them to determine first from second class. At the coding desk, an operator reads the postal code and types a code in phosphorescent dots on the front of each envelope; this is read by the sorting machinery.

Right: the letter coding device which prints the phosphorescent dots on the envelopes. The coated paper can be seen in the photo. The dots are printed in a computer code dictated by an 'electronic dictionary'.

Letter sorting machine There are about a dozen large post offices in Britain using the automatic high speed letter sorting machines developed by the M S Thrissel and GEC-Elliott companies. They forward 7000 to 8000 letters an hour into 150 stacking boxes, which are arranged in five rows of thirty boxes each, one row above another. The sorting machines are each served by five or six coding desks. The phosphorescent dots of the code are irradiated with ultra-violet light, identified and converted into routing instructions for the appropriate collecting box.

The Post Office and the Elliott company have experimentally operated one of these machines at a speed of 20,000 letters an hour, but there is only one such machine and at present it is not fully operational. There is a space of two inches or so between the ends of the envelopes as they go through the machine, and the exact throughput of the machine depends on the length of the letters. A machine handling Christmas cards will be able to handle mail faster than a machine in a post office which handles a lot of long commercial envelopes.

Several machines in parallel are needed to provide the hundreds of groupings in a modern postal system; each machine can be changed to another sorting system by electronic switching within the translator. The translator itself used to be a threaded core device in which the codes were represented by wires threaded through a magnetic core, but is now a computer with a magnetic drum using a paper tape input. It is programmed to serve each sorting machine in turn at regular intervals according to a *time slot* or *timing pulse*. During the pulse the machine asks a question in the form of a binary address and receives an answer from the appropriate sorting plan, and operates the mechanical part of the machine accordingly.

Character recognition Optical character recognition is also used in postal sorting. Machines can read typed or printed characters, although over two hundred different typefaces are encountered, many of them not very well printed. The OCR machinery can be asked to search into the body of the

address or into the postal code, which should contain all the necessary information.

The first option requires large, expensive equipment, because fifty or sixty characters may have to be identified; the advantage is that more information is carried in the address than is actually needed. For example, the word BIRMINGHAM can be correctly identified if six out of the ten letters are masked. The rest is called redundant information. The postal code of five to seven characters is less demanding, but requires every character to be recognized. In practice there is a small amount of redundancy built into the system, since not all combinations of letters and numbers are used.

OCR equipment, attached to automatic machinery or separate and operating at speeds of up to 30,000 items per hour, is being tried out in a few locations, mainly in Japan and the United States.

Postal codes In Britain, the code consists of letters and numbers (hence is *alphanumeric*) and has five to seven characters in two parts. The first part indicates the town and the second part the street or a large individual user of the service. In the USA, the code is called the ZIP code (for Zone Improvement Plan). It is a five-digit number, of which the first three digits indicate the section of the country and the last two the local post office or postal zone of the addressee. Both plans went into use in the 1960s.

Future development The long term aim is to achieve completely automatic mail handling, so that the letter can be fed into the machinery and not be handled until it gets to the postman's bag. At present only 10% or less of the letters handled in Great Britain are sorted by means of the phosphorescent dots. The technology is mostly already developed; the problem is to get people to use the postal codes, and the greater use of OCR equipment would require the international standardization of addressing envelopes. In the Soviet Union, envelopes are printed with boxes on the front for letters and characters, and the user is supposed to print the address in computer style.

Above: the automatic high-speed letter-sorting machine. It reads the code on the envelope printed in the form of phosphorescent dots and sorts 7,000 to 8,000 letters per hour into 150 boxes. Such a machine has sorted 20,000 letters per hour experimentally.
Left: the conveyor belts inside the letter sorting machine.

RADAR

Radar, an acronym derived from *Radio Detection and Ranging*, means the detection and location of remote objects by the reflection of radio waves. Radar proper came into existence in the 1930s principally as a result of recognizing the need to have warning of air attack long before the hostile aircraft could be seen or heard, and was made possible by the existence of sufficiently advanced radio techniques. That objects were capable of reflecting radio waves had long been known, and indeed the principle of reflection was the means used for studying the ionized layers in the upper atmosphere, which are important in long distance radio communication.

Sir Robert Watson-Watt, the 'father' of British radar, had engaged in such research. Asked to comment on the possibility of a 'death ray' in the form of powerful radio waves as a defence against air attack, he pointed out that while this was totally impracticable, it was quite feasible to use radio waves to detect an aircraft long before it could be seen. Out of this suggestion arose the British chain of Early Warning radar stations which were operational at the outbreak of war in 1939 and were a vital factor in winning the Battle of Britain. While most of the major powers had discovered radar principles before the war and had made efforts to develop them for military uses, the British defence chain was probably the most advanced operationally in 1939.

Basic principles The phenomenon of acoustic echoes is familiar; sound waves reflected from a building or cliff are received back at the observer after a lapse of a short interval. If the initial sound is a short sharp one such as a hand-clap and if the speed at which sound waves travel is known, the interval between the initial clap and its echo is a measure of the distance of the reflecting object or surface. Radar uses exactly the same principle except that the waves involved are radio waves, not sound waves. These travel very much faster than sound waves (186,000 miles/sec, 300,000 km/sec) and are capable of covering very much longer distances. If, in addition to timing the interval between the initial transmitted signal and the echo, the direction of arrival of the latter is observed, the actual position of the remote reflector is obtained.

While it is possible to use both *continuous* waves such as are used in broadcasting and interrupted or *pulsed* signals in which the radio waves are emitted in short bursts or impulses, the way radar works is probably most easily understood in terms of pulsed radio signals. A radio transmitter connected to a directional aerial [antenna] (an aerial which concentrates its radiation in a beam along a particular direction) sends out a stream of short pulses of radio waves. Each pulse will normally be a few millionths of a second long but may be even shorter, and the pulses are separated by a time interval which is substantially longer than the time it takes for radio waves to travel to any object (or 'target') of interest and back.

Any object in the path of the transmitted beam reflects some of the energy falling on it back to a radio receiver located near the transmitter. In the receiver there is thus a stream of reflected pulses slightly displaced in time, with respect to the stream of transmitted pulses, by a short interval corresponding to the time any one pulse takes to travel from the transmitter to the target and back. This is a measure of the range of the target. If the transmitting and receiving aerials are both beamed in the same direction, only targets lying in the beam reflect any signals and so the direction in which the target lies is obtained.

In practice, a single aerial is usually used for both transmission and reception, the receiver being momentarily 'suppressed' during the brief period of the transmitted pulse but reactivated in time to receive any echo pulses. The device used for this purpose is a form of switch triggered by the transmitter pulse and called a *TR cell* (Transmit-Receive cell). The single beamed aerial will normally be rotated in *azimuth*

(horizontally) at a steady rate and is then said to be scanning. This is the common arrangement for *search* or *surveillance* radars. In practice the aerial beam is one or two degrees wide in azimuth but extends in elevation from the horizon to perhaps 15° or 20° elevation. This is the so-called *fan-beam*. It produces two-dimensional information only (range and bearing).

It is possible to devise more elaborate systems which give three-dimensional information (range, bearing and elevation). In one form the 15° wide vertical 'fan' is divided up into a number of individual narrow beams (1° to 2° wide), this 'stack' of beams being simultaneously scanned in azimuth as before. In more advanced systems the elevation information is obtained by scanning a narrow pencil beam up and down in elevation at a rapid rate while rotating it in azimuth at a much slower rate. In this way a volume of sky is scanned by the narrow cone. The azimuthal scanning is usually carried out mechanically while the elevation scanning is done by varying the radio frequency, the aerial having been designed so that the direction of its beam in elevation is controlled by the radio frequency used.

In search radars with a rotating beamed aerial, the direction of a target is obtained from the direction of maximum received signal. When greater angular precision is required other arrangements are possible. For example two directional beams can be arranged so as to partially overlap. In the overlapping area, very accurate measurement of direction is possible by taking the ratio of the signals received in the two aerials. The two partially overlapping beams may be made to rotate or 'spin' about the equi-signal line (along the centre of the overlapping area), producing what is known as a *conical scanning aerial* commonly used in many anti-aircraft fire-control tracking radars.

Radars for fire-control purposes must follow a target continuously once it has been picked up or *acquired*. Such a radar is called a *tracking* radar and usually has an aerial beam which is a narrow cone or pencil shape. Once the radar has

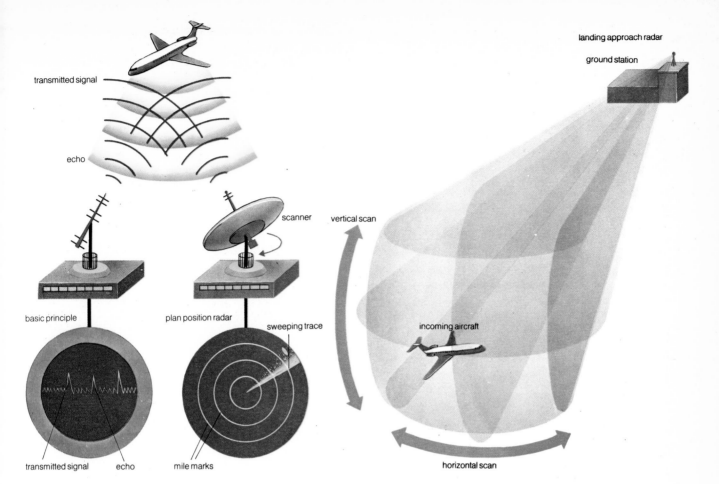

landing approach radar

ground station

transmitted signal

echo

scanner

vertical scan

basic principle

plan position radar

sweeping trace

incoming aircraft

transmitted signal   echo

mile marks

horizontal scan

Above: the plan position indicator radar scans horizontally to give range and bearing. The landing approach radar scans both horizontally and vertically giving range, bearing and elevation.
Far left: a photograph taken with side-looking airborne radar. The signal is transmitted to one side of the aircraft and the reflected signals are electronically modified to give a vertical view of the terrain, then recorded on photographic film. This is part of western Nicaragua, showing the San Cristobal volcano, the town of Chinandega, the Pan American Highway, areas of woodland on the slopes of the volcano, and areas of cultivated land.
Left: the aerial of an Air Traffic Control radar.

Right: an Air Traffic
Control display screen.

found a target the beam can be made to follow it automatically; it is said to be 'locked-on' to the target.

The transmitter of a pulse radar must be capable of producing short pulses of radio energy of high intensity but at substantial intervals of time. For example, typically the pulses may each be 4 microseconds long spaced at intervals of 4 milliseconds. With these figures, the transmitter is actually operating for only 1/1000 of the total time. This is called the *duty cycle*. The 'mean' power of the transmitter will be only 1/1000 of the 'peak' power. The two most important types of transmitting tubes used in radar are the *cavity magnetron*, which is a self oscillator and the *klystron*, which is an amplifier.

Displays The receiver detects and amplifies the received pulses without undue distortion of their shape, and it embodies a display system to present both the range and directional information available. The timing of the interval between transmitted and received pulses can be effected in several ways. The most common means uses a cathode-ray tube similar to that used in a TV set. The cathode-ray beam traverses the tube face horizontally at a steady rate. Where the beam strikes the tube face a bright spot is produced which, as the beam moves across, traces out a bright line.

The start of this line coincides with the moment of emission of the transmitter pulse. When an echo is received the spot is momentarily deflected transversely back to the original point, producing a trace. The length of this trace is a measure of the time interval or *delay time* of the echo pulse. When the spot reaches the extreme right hand end of its travel it 'flies back' to the left ready to commence the cycle over again when the next pulse is emitted. This was one of the earliest forms of display used in early warning radars and is still used for certain purposes.

In an alternative arrangement, the cathode-ray beam is adjusted so that in the absence of any received signal it produces only a very faint trace on the tube face but is intensified to give a bright spot at the point where an echo is received. The beginning of the trace coincides with the centre of the tube face and the line rotates in synchronism with the scanning aerial. The tube face presents a map-like picture of the space around the radar, bright spots appearing at positions corresponding to the range and bearing of any targets. This is the display known as the *plan position indicator* or PPI and is commonly used in many radars.

Doppler radar Another familiar acoustic phenomenon is the Doppler effect, in which the sound heard by an observer is raised or lowered in pitch if the sound source is moving towards or away from him. The same effect occurs with radio waves, so that the radio frequency of an echo signal will be above or below the frequency of the emitted pulse if the target is approaching or receding. This principle can be put to use in many ways. In a search radar, for example, since objects like buildings, trees or hills are stationary their echoes show no Doppler frequency shift and this may enable them to be cancelled out on the display, only moving targets such as aircraft being presented. Moreover an aircraft can use radar to measure its ground speed by noting the frequency shift of echoes from the ground below and ahead of it. These principles are also put to considerable military use, in missile guidance systems for example.

Range and wavelengths Radar waves, like light waves, travel more or less in straight lines when in free space. In the Earth's atmosphere, however, a small amount of bending takes place because the atmosphere decreases in density with height, so radar can see marginally beyond the horizon, but in designing a radar set this atmospheric bending can be ignored for all practical purposes. The range of a search radar is thus fundamentally limited by the curvature of the Earth, a serious matter so far as the detection of low targets

is concerned. Raising the radar extends the horizon but to obtain a worthwhile extension, for example to detect very low level aircraft, it is necessary to carry the radar aloft in an aircraft. This arrangement is known as an *Airborne Early Warning* or AEW system.

All objects are capable of reflecting radio waves to some extent. The reflecting power, often called the *scattering coefficient*, depends on the shape and size of the object and the wavelength used. Large objects like ships and aircraft are good reflectors at all wavelengths up to 10 metres or more in length (the first British early warning radars used a wavelength of between 10 and 15 m). Smaller objects are more easily detected, in general, at shorter wavelengths in the centimetric or microwave region. Most radars now use wavelengths in the range 3 to 25 cm.

A typical modern long-range surveillance radar for early warning of approach of aircraft or for long-range air traffic control (ATC) purposes might have the following characteristics: a wavelength of 25 cm; peak power 2 MW, mean power 2 kW; aerial size 12 m (39 ft) wide and 5 m (16 ft) high, scanning at 10 rpm; and a range on aircraft (assumed above horizon) of 300 miles (483 km).

Secondary radar The form of radar thus far described, which relies on the passive reflection of radio waves, is now called *primary* radar to distinguish it from a more complex form known as *secondary* radar or sometimes *secondary surveillance radar* (SSR). In this the target (usually an aircraft) carries a small device which is both a receiver and transmitter, called a *transponder*. This receives the pulses from the ground radar and retransmits them to ground on a slightly different wavelength. At the same time it adds to the retransmitted signal a few additional pulses which are coded to convey such information as the identity (call-sign) and height of the target. The basic radar functions of measurement of range and direction are thus carried out while in addition the identity of the target and possibly its height are obtained. This is of great value in ATC and it is now mandatory for aircraft in certain categories flying in particular regions of the sky known as *controlled air space* to carry appropriate transponders. The system also forms the basis of the military identification system known as IFF (identification, friend or foe).

Applications Some important applications of radar such as long-range early warning for defence and the longer range aspects of ATC have already been mentioned. Both defence and ATC also make use of medium-range search radars (100–150 miles range), the former for the control of defence weapons such as fighter aircraft and anti-aircraft missiles, and the latter for the *airfield approach* and *terminal control* phases of ATC. Still smaller radars may be used for local *airfield control* purposes, that is, management of traffic within a few miles of the field.

Search radar is used on ships for both air defence purposes and for navigation in poor visibility. Its use by merchant ships for the latter purpose is now very extensive. Tracking radars are primarily of value in such operations as the control and guidance of anti-aircraft weapons and for space exploration.

Radar carried in aircraft has many uses. It can be used by fighter aircraft to locate and intercept enemy aircraft. A form of surveillance radar can be used to produce a 'map' of the ground over which the aircraft is flying so that it can locate targets on the ground which it is seeking. Another similar type of radar known as *cloud and collision warning* radar can detect the presence of high ground or intense rain storms ahead of the aircraft and thus improve air safety.

Portable radars are used by ground troops to locate moving vehicles and even men up to ranges of a mile or two. To distinguish moving objects from the large number of stationary objects that tend to confuse the radar picture, the Doppler principle is usually employed. One such form of radar is also used by the police in 'speed traps' to detect speeding motorists. It may also be used as a perimeter defence to detect illegal entry into, or exit from, a defended area such as a prison.

Radar has applications in surveying, particularly in terrain where access is difficult. A device known as a *radar tellurometer* is in fact a type of distance measuring instrument. One use of radar in meteorology is the location of rain storms. This is important, particularly, in providing warning of the occurrence and movement of tropical storms. The use of tracking radar to follow meteorological atmospheric sounding balloons and thus enable upper atmospheric wind to be measured is another important use.

Safety Living tissues can absorb radio waves to a certain extent and this results in some rise in temperature which may have harmful effects. The intensity of signals required to produce such effects is, however, very great. Only the most powerful radars such as those used to detect intercontinental missiles at long range are really dangerous, and even so only within comparatively short distances of the radar. All radars for more normal applications are, to all intents and purposes, harmless unless the living tissues are placed directly in the aerial beam and very close to it (within a few feet).

RADIO

Radio is the name given to the system of transmission and reception of information by the propagation of *electromagnetic radiation* as radio waves through space. It is the most significant contemporary technique for the transmission of information over distances.

A single signal source, after amplification, is used to *modulate* a carrier wave and after further amplification is fed to an aerial for transmission. At the receiver, the radio wave is selected (to the exclusion of all other radio waves), demodulated, amplified and fed to the speaker which reproduces the original sound.

Modulation With radio reception, if some means were not available for distinguishing between the desired programme and all other programmes, the result would be very poor reception. Furthermore, the situation would deteriorate

Below: a Marconi high-frequency curtain array.

with the number of transmitters in a given area, especially if the desired signal was from a less powerful or more distant transmitter.

To overcome this problem, modulation is used. This effectively provides each transmission with a 'signature tone' —this is called the *carrier frequency*. Each transmitter has its own signature tone so that any transmission can be selected. The signal is in some way superimposed on this carrier wave before transmission—this is *modulation*. The receiver is 'tuned into' this carrier frequency and, with suitable electronic circuitry segregates the carrier wave from the signal, amplifies the latter and feeds it to the speaker.

In transmission, as with reception, a number of different techniques is possible. With one particular type of transmitter or receiver the circuit design can have even greater variety and so the descriptions below will refer only to the basic types, the major classifications being on the type of modulation employed.

AM radio transmitters The first requirement of an amplitude modulated (AM) transmitter is a stable carrier frequency. If it is not stable, the reception at the receiver will not be consistent in quality and likely to fade and distort. Stability is provided by a crystal oscillator. Piezoelectric crystals are used, of which the quartz crystal has by far the best characteristics. The design of quartz oscillators for these applications is similar to those used in quartz clocks where stability is of the utmost importance.

The voltage output from the crystal oscillator is a sine wave and this is amplified to a high power level by several amplifiers in series. Such amplifiers require special design

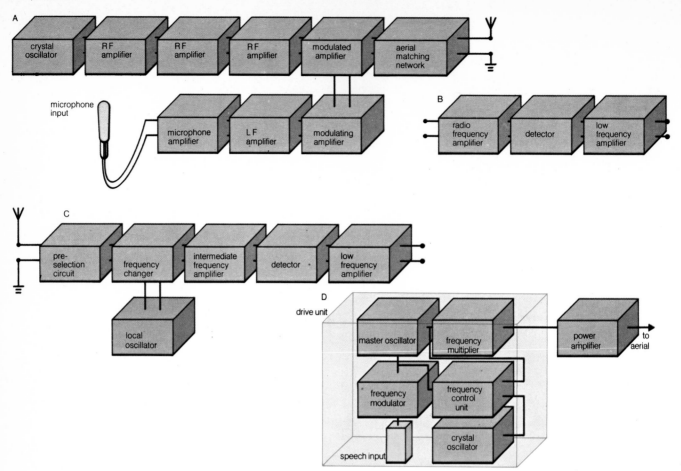

Above: block diagrams of basic units in radio transmitters and receivers. A and B show simple AM transmitter and receiver. An alternative and more sophisticated AM receiver is shown in C. This is the supersonic heterodyne (or superhet for short). Here the carrier frequency of received signal is altered, using a 'frequency changer', into a (fixed) intermediate frequency which matches a fixed tuned circuit. An FM transmitter is shown in D—here the carrier frequency is altered according to the amplitude of (speech) signal.

because of the high frequencies involved (from 30 kHz in the low frequency, LF, band to upwards of 30 MHz in the very high frequency, VHF, band). They are known as radio frequency, RF, amplifiers. The greatly amplified sine wave then passes to a *modulated amplifier*.

The signal to be transmitted is first amplified using a low frequency, LF, amplifier and then passes to the *modulating amplifier*. The output from this alters the amplitude of the high power carrier sine wave in the modulated amplifier according to the instantaneous magnitude of the signal— this is amplitude modulation. The AM signal then passes through a *matching network* and on to the aerial for transmission (see 'aerials and matching networks' below).

The type of circuit described above produces what is called a double-sideband, DSB, AM transmission. This follows from the way in which the signal and carrier wave are combined—essentially by *multiplication*. In general, the multiplication of one sine wave (frequency f_1) by another frequency (f_2) produces a waveform containing two frequencies ($f_1 + f_2$) and ($f_1 - f_2$). Where the signal contains a range of frequencies (as in speech) the resulting AM signal contains the original carrier frequency, f_c, with two *sidebands* about this frequency. If, for example, the signal has frequencies up to 4 kHz (typical speech) and the carrier frequency is 100 kHz then the total frequency bandwidth of the AM signal is 8 kHz about the 100 kHz mark, that is, from 96 kHz to 104 kHz. The portions on either side of the 100 kHz mark are known as the *sidebands*.

There is a certain *redundancy* in this situation because the information about the original signal is contained separately in both sidebands. Also, the transmission bandwidth is twice that which is really necessary. Where the radio spectrum is crowded (as it generally is between LF and VHF) this presents an unnecessary waste of frequency 'space'.

Single-sideband, SSB, transmission is therefore used in

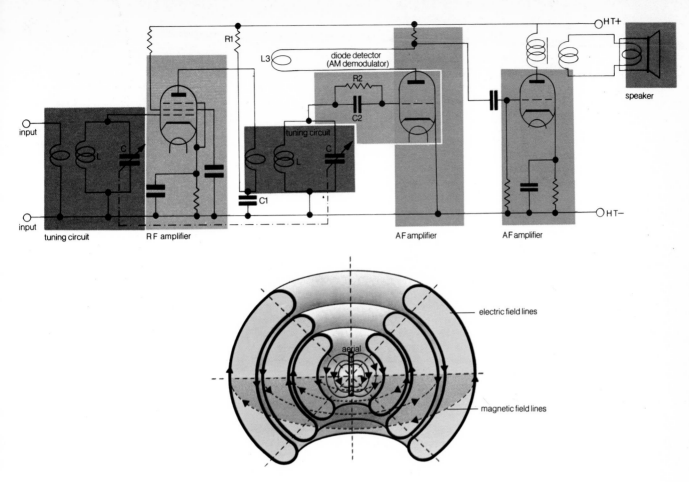

Above: circuit diagram of
simple AM receiver showing the
functions of the various
components. The input from the
aerial is transformer coupled
to the first of two tuning stages
which reject neighbouring
broadcasts. The 'accepted'
signal goes to an RF amplifier
and then on to second tuning
circuit. The signal now has
sufficient amplitude to be
demodulated by detector
circuit using a diode detector
(R2, C2 and grid circuit of
second valve). This valve also
acts as AF amplifier and after
more amplification (third
valve), the audio signal goes to
speaker. Distribution of electric
and magnetic fields from an
aerial (bottom) shows that they
are everywhere at right angles.

some situations by filtering out one of the sidebands. Furthermore, using two such circuits with one common carrier frequency, two independent sidebands (representing two independent signals) can be superimposed about the carrier frequency for transmission. Such systems are used, for example, in remote control models—providing two separate control signals in one transmission—and for stereo transmissions.

AM receivers The receiver must be able to receive any program from the broadcasting spectrum—this means able to select a particular carrier frequency and its sidebands but excluding everything else.

Before 'disentangling' signal from carrier it is usually necessary to amplify the aerial signals. This requires an RF amplifier. Next, the signal is demodulated by the *detector circuit* and the resulting audio signal amplified, using an LF amplifier. Frequency selection is carried out in the pre-amplification stage before demodulation. This is achieved with a *resonant circuit* whose resonant frequency is adjustable by means of a variable capacitor.

Because of the density of broadcasting, especially in the lower frequency part of the spectrum, the selectivity of the tuning circuit is important when considering high fidelity reproduction. In some systems based on the above scheme, two or more resonant circuits are coupled together through amplifier stages. They are individually tuned with variable capacitors, but these are 'ganged' together with a common control knob for ease of tuning.

Another method, used extensively, is to take the frequencies in the region required (a rough selection) and transfer them to another part of the frequency spectrum—called the *intermediate frequency* (IF). This is achieved using the heterodyne principle—where two sine waves are mixed to produce beats—and this type of receiver is known as a *supersonic heterodyne* or 'superhet' for short. The superhet principle is

oscillator

mixer

automatic
volume
control

radio
frequency
amplifier

aerial
socket

D/F aerial
socket

audio
frequency
amplifier

intermediate
frequency

tone filter

power supply 5.4V

loudspeaker/phone
socket

simply that rather than tune the circuit to the carrier frequency it is better to change the carrier frequency to suit a
fixed tuned circuit. This way the tuned circuit can be designed
to have the best possible characteristics. The filtered signal
from this stage passes to an intermediate frequency, IF,
amplifier and from here on to the detector as before.

Distortion, noise and FM The reason for turning a simple
radio principle into a complex arrangement of LF, IF, and RF
amplifiers, tuned circuits and detectors is to make sure that
at every stage the signal or carrier or both are being handled
in the best possible way. For example, it is virtually impossible to design an amplifier which will handle low frequency
(audio) and high (radio) frequency signals equally well. It
must be designed to do one or the other.

The factors that an engineer looks for in an electrical
circuit and in the final sounds from the speaker are *distortion*
and *noise*. Distortion implies unfaithful reproduction and is
mainly caused by non-linearities in the amplifier. For
example, if a 10 volt signal is amplified to 20 volts, but a

20 volt signal is amplified to 35 volts, then the gain is not the
same for all input signals and distortion results.

Noise is all unwanted signals. This could be interference
from an electrical machine or closely packed broadcasting
stations straying into the fields of reception. AM systems do
not cope with noise as well as frequency modulation (FM)
systems. Here, rather than superimpose the signal amplitude
on a constant frequency carrier wave, the carrier frequency
is made to change (deviate) from the central (carrier) frequency according to the instantaneous value of the signal.
Any clicks or noises that become mingled with an FM signal
affect the magnitude, not the frequency, and the signal
continues virtually unchanged.

The one major disadvantage with FM is that it requires a
much larger frequency bandwidth to transmit a given signal
than with the equivalent DSB AM method. In the very high
frequency (VHF) and ultra high frequency (UHF) broadcasting bands, however, there is much more room and FM is
used extensively.

Aerials and matching networks When a stone is thrown into
a pond the wave ripples spread out in concentric rings. A
similar process occurs when an alternating current travels
along a conductor except that the ripples are electromagnetic
in nature. The aerial is the 'conductor' in this case and the
alternating current is the AM or FM signal from the last
amplifier stage in the transmitter.

Using a simple straight-wire aerial, the radio waves produced would, if they could be seen, appear like ever-
expanding doughnuts. Such an aerial is *omnidirectional*. By
careful design of the aerial, radio waves can be directed or
beamed and at microwave frequencies they begin to look like
the reflectors used in a flashlight.

*Above (with key): receiver for
use on small marine craft. This
is a superheterodyne receiver
operating between 160 and
4150 kHz. It can be adapted for
reception of world-wide
standard time signals and for
radio compass.*
*Above right: an 11 GHz
broadband radio link for the
transmission of colour TV
signals or 960 simultaneous
telephone calls.*

RECORD PLAYER

In 1857, a Frenchman called Léon Scott was the first to record sound. His invention, called the *phonautograph*, drew the waveform of the sound on a smoke-blackened cylinder. It was claimed that the voices of famous people could be recorded for posterity, but the device was not very practical, since there was no means of playing the recordings back.

Thomas Alva Edison was the inventor of the first practical recording machine, which was also a playback machine and was called the *phonograph*. In 1877, Edison wrapped tinfoil around a brass drum; when he cranked the drum around and spoke into a diaphragm with a stylus attached, the stylus made indentations in the tinfoil. A separate stylus and diaphragm was used for playback. It was impractical to unwrap the tinfoil to make copies of it; Alexănder Graham Bell developed a means of recording in wax, in 1885, calling his invention the *graphophone*.

Emil Berliner, in Germany, began working with Scott's invention, using photoengraving to make a playable recording from the tracing in smoke. To eliminate the difficulty of unwrapping and rewrapping the photoengraving on to a cylinder, he turned to disc recording. (Bell had also experimented with discs.) The phonograph and the graphophone both used a *hill-and-dale* or *phono-cut* method of modulation, which meant that the stylus moved in and out of the surface of the cylinder; Berliner, with his *gramophone*, patented the *lateral cut*, or *needle cut*, in which the modulation was from side to side at a constant depth.

In 1894, Berliner began production of his gramophones in America, in partnership with the operator of a machine shop in Camden, New Jersey. The firm was incorporated as the Victor Talking Machine Company and eventually became a part of the Radio Corporation of America (RCA). Until 1913, both cylinders and discs were manufactured, and both hill-and-dale and lateral cut discs were available which could not be played on the same record player. Playing speed of recordings varied (with a maximum of 100 rpm) and was often inaccurate. Eventually the 78.26 rpm lateral cut disc became standard, although Edison continued to manufacture cylinders, for owners of his phonograph, until 1929. Today the record player is still called the phonograph in the United States and the gramophone in Europe.

Acoustic players   Records were played by means of a stylus and diaphragm assembly called a *soundbox*, and the sound was amplified by means of a horn. The arm of the record player was hollow and of increasing cross-sectional size from the soundbox to where the arm opened into the horn. A great deal of research was done into the exact shape of the horn for best results; on more expensive models, the horn was made larger or longer for more volume. If the record player included an elaborate cabinet, the horn could be built into it; the user 'turned down' the volume by closing the cabinet door or by stuffing a rag into the horn.

Spring-driven motors were used, after some early experimentation by Edison with electric motors. The records were made of powdered slate mixed with shellac; the grooves were abrasive, and the tracking pressure of the soundbox was as much as six ounces, so that the playback stylus (improperly called the 'needle') had to be changed after each record. Styli were made of steel, sometimes chrome plated, and various kinds of thorns, cactus needles and bamboo were also used; these gave a softer tone and reduced record wear, but would barely last one play without being resharpened.

Electric players   Beginning about 1915 and interrupted by World War I, research was done into electrical methods of recording and playback using technology developed for the broadcasting and telephone services. By 1925, most recording companies had changed over to electrical recording, and electric record players were becoming widely available. Recording techniques dramatically changed; with microphones, it was possible to record under fairly natural circum-

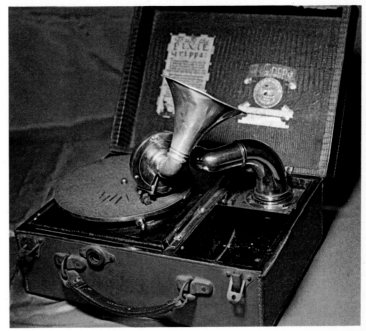

Above, top: a drawing of Berliner's 'Gramophone, or speaking machine'. This 1891 instrument was the ancestor of modern record players.
Above: a portable acoustic record player, with the horn attached directly to the soundbox. Note the selection of 'needles'.

stances, instead of grouping musicians or singers around a horn. In playback, the modulation of the groove on the record was changed by the *pick-up* (or *cartridge*) into an electrical impulse, which was then sent through an electronic amplifier to a loudspeaker.

Pick-ups The first pick-ups were magnetic. The stylus was connected to a piece of steel called the *armature*, whose vibrations excited a magnetic field to produce current in a coil. Electrical recording necessitated an attenuation of low frequencies during the recording process, in order to limit the side-to-side movement of the groove on the record; the bass frequencies were restored in playback by the mass of the pick-up head resonating with the resistance of the stylus to motion.

Piezoelectric materials began to be used in the mid-1930s. A slice of crystal, such as Rochelle salt, was coupled to the stylus; in a piezoelectric material a bending or twisting force is converted to an electrical voltage. Early crystal pick-ups were more efficient than the primitive magnetic types, giving much higher output, but Rochelle salt had the disadvantage of being soluble in water and melting in hot climates. Later, ceramic pick-ups were developed, and these are still used for their piezoelectric properties on low-priced record players today.

Development of magnetic pick-ups continued. The first lightweight pick-up appeared in Germany in 1939. Its mass was greatly reduced; the stylus tracking pressure was brought down to just over an ounce (28 g); the stylus shank was more compliant and the stylus tip itself could be made of a jewel or a precious metal, giving hundreds of plays before wearing out. These factors meant a much higher quality of the signal sent to the amplifier; low frequencies now had to be restored by circuitry in the amplifier, called *compensation*. While the output of the light-weight pick-up was of high quality, it was also greatly reduced in power, so the *pre-amplifier* became necessary, which is nowadays usually built into the amplifier. After World War II, Decca in Britain and General Electric in the USA continued to improve the lightweight magnetic cartridge.

Modern records During the 1930s, RCA in the USA had introduced a long-playing record which was made of the same materials and had the same size groove as a 78, but was played at $33\frac{1}{3}$ rpm. It was a failure, partly because the slower speed without improvement in other factors meant much higher surface noise.

After World War II, research into thermoplastic materials and the high state of pick-up development made possible new types of records. In 1948, CBS in America introduced a new long-playing record made of polyvinylchloride (PVC) and having a *microgroove*. RCA introduced an entirely new system, comprising a 45 rpm record, 7 inches (16.5 cm) in diameter, with a large centre hole and a new record changer especially designed for it. RCA's record was also made of vinyl and had a microgroove. During the early 1950s, the 'battle of the speeds' resulted in the 78 becoming obsolete and the 45 rpm record becoming the medium for popular singles.

Subject to quality control in record manufacture, surface noise from the record itself has disappeared; the magnetic pick-ups of today operate at a tracking pressure of well under two grammes, less than one hundredth of tracking pressures in the acoustic era. With the introduction of *variable pitch* records in the 1950s by CBS, a method of cutting the groove spirals closer together where the dynamic level of the recording is low, the playing time of the 12 inch (30 cm) $33\frac{1}{3}$ rpm record rose to more than thirty minutes per side, compared to about four minutes for a 12 inch 78.

Record changers Record changers were developed because of the short playing time of 78 rpm records, which required several sides for one symphonic movement.

The first successful record changers were designed by

E W Mortimer. At the end of the record is a wide spiral, called the *lead-out* groove. When the pick-up enters this groove, its position causes a lever attached to the pick-up arm, or tone arm, to engage with a pawl attached to a cam under the deck, or motor board. This causes the cam, which has teeth on it, to rotate a few degrees and engage with teeth on the main spindle carrying the turntable. The cam makes one revolution, during which various cam profiles actuate levers which lift the arm from the record, swing it clear of the stack of records on the spindle, operate the record changing mechanism itself, and deposit the pick-up in the lead-in groove of the next record.

The earliest record changing mechanism consisted of a pair of posts with blades or shelves supporting the stack of records on opposite sides. During the record changing cycle, the posts rotated a few degrees, forcing the lower blades between the records, separating the lowest record from the rest of the stack and allowing it to drop to the turntable. This often broke large chips from the edge of the records.

A better system was the *pusher-platform*. The edge of the stack of records rested on the platform; the spindle was inclined to balance the stack, and indented to the thickness of one record. The platform pushed the bottom record off the indentation. Then the 'pusher' was installed in the spindle itself; in most record changers today, the spindle is indented, a metal finger in the spindle pushes the bottom record off the indentation, and a stabilizing arm balances the stack. When the stabilizing arm falls as far as it can after the

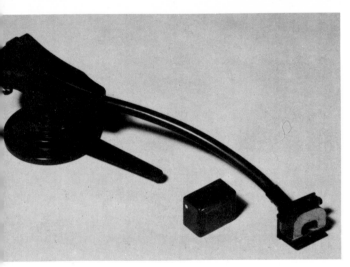

Far left, top: a spring-driven gramophone motor, from 1919. The centrifugal governor with its flyweights is on the right.
Far left, below: test dept. of a gramophone factory, photographed in 1905.
Left, top: the first British record changer which could be obtained as a separate component, in 1932. There are three posts with adjustable blades for ten-inch and twelve-inch records; the curved arm at the rear sensed the record size and also when the last record had been played. The turntable speed was slightly adjustable for variation in record speed.
Left, below: an HMV pick-up arm, showing the size of the magnet in an early magnetic pick-up.
Below: a simplified drawing of a record changer. The record-dropping mechanism is driven by a cog on the turntable, which turns a large cog one revolution until a gap in its teeth causes it to stop. The falling record knocks the size-feeler, moving the stop-lever to one of three points, depending on the size of the record.

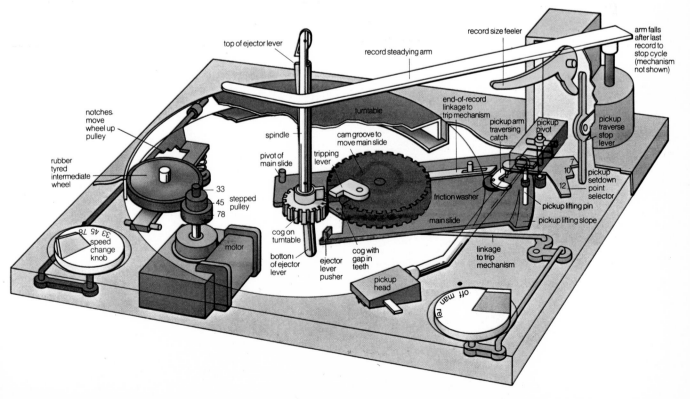

last record drops, the record changer shuts itself off at the end of that record.

Record changers can be well made nowadays, but automatically changing records is still hard on them, particularly on the centre holes. With the longer playing time of records today, the single-play turntable or transcription deck is preferred by many people.

Motors and drives The first electric motors for driving record players had centrifugal governors and drove the turntable through a worm gear. After World War II the method of driving the turntable by means of a rubber *intermediate* wheel or *idler* wheel was developed. The wheel itself

Above, top: a modern stereo amplifier, with five separate controls for different sections of the frequency range, as well as conventional tone controls. Above: a belt-driven turntable, which isolates motor noise from the record.

is driven by an ungrooved pulley on the motor shaft. This system helped to isolate the turntable from motor noise as far as was necessary then, and is still used on most low-priced record changers today. It is also easy to design with three speeds, requiring only steps of different diameter on the pulley.

With the high quality of records and pick-ups today, it is essential for best results to further isolate the motor from the record. The most common method in use is *belt drive*, in which a belt from the motor drives a pulley and a longer belt made of soft rubber runs from the pulley around an inside rim under the table. This system is used in the majority of single-play decks.

The popular record changer and the higher quality deck both use the shaded pole type of induction motor. The speed is controlled by the mains frequency, but also by the load, so the turntable platter is often made heavy to have a flywheel effect, in order to maintain constant speed. Some decks have small *synchronous* motors which are locked to the mains frequency; the flywheel effect is supplied by the huge flywheel in the power station. Recently some 'direct drive' models have appeared on the market, with the turntable mounted directly on the motor shaft; keeping motor noise and vibration out of such a system can only be done by means of extremely high quality engineering at an equally high price, and the benefits are questionable.

Stereo Before World War I research had already begun into multi-channel recording. It was originally proposed to cut two-channel records by combining a lateral and a hill-and-dale cut in the same groove, but this would have meant limiting the quality of stereo records in order to keep them compatible with existing mono cutting and playback techniques. When stereo records were finally marketed in the late 1950s, the record companies had adopted the 45/45 technique, with each wall of the groove (45° from a line perpendicular to the surface of the record, but 90° from each other) carrying a separate lateral-cut signal. The stereo pick-up is actually two magnetic (or ceramic) assemblies in one. This advancement spurred development of even lighter tracking pressures, smaller stylus tip radius, and more compliance. For about ten years the record companies manufactured separate mono and stereo editions of each record.

Now the industry is marketing quadrophonic, or four-channel sound, which is said to duplicate more completely the ambience of the room in which the recording was made. Public reaction so far has been sceptical. The object, after all, is not technical sophistication, but simply to hear the music. The theory of stereo is logical, but nobody has four ears.

REFLECTIVE MATERIALS

Reflection is the result of light rays striking a white or coloured surface and 'bouncing' off it. There are three basic types of reflection: *diffuse*, *mirror* (or *specular*) and *retro* (or *reflex*) reflection. Although the first two are often natural phenomena, retro reflection is engineered and retro reflective materials are now being used extensively on traffic signs, car number plates and safety garments because of their unique ability to return a large amount of the light shone at them directly back to the source. When illuminated by car headlights at night, a reflective material will reflect the light back to the driver's eyes and thus can appear to be several times brighter than any ordinary white or coloured surface.

Construction Reflective materials make use of the optical properties of glass beads. A beam of light entering a spherical glass bead will be affected in one of two ways; either it will be reflected back to the source by *total internal reflection* at the back surface of the bead, or the bead will act as a strong converging lens. If the latter is the case, a conventional reflecting surface must be positioned behind the glass bead to enable it to act as a retro reflector.

A typical reflective material will have about ten thousand glass beads to the square centimetre, each bead having a diameter of about 0.1 mm (0.004 inch). It is not possible to ensure that all the beads are the same size, and this means that the light reflected back is slightly spread out and does not all travel directly back to the source. This allows a driver whose line of vision, on average, is two to three feet (0.6 to 0.9 m) above his head lights to see the full effect of the retro reflection.

It has taken many years since the original concept to develop a practical and usable reflective material. A reflective sheet consists essentially of three plastic layers: a transparent front layer, a middle layer in which the glass beads are embedded, and a reflecting back layer. Sometimes there is a layer of contact adhesive behind the back reflecting layer so that the reflective sheet can easily be fixed in position.

Uses When used to make up traffic signs, the sheets of reflective material are cut into letters or shapes and stuck on to coloured sheets of aluminium. The large blue and green signs on roads normally have reflective legends and symbols, and it is these that appear to glow at night. Of course, what the driver is seeing is a reflection of his own headlights in the shape of the letters which make up the destination. Many other smaller traffic signs, such as red triangles and speed

limit signs, are put together in a similar way. In Britain and many European countries all new cars must now be fitted with reflective number plates by law. These are manufactured by sticking sheets of white or yellow reflective material on to aluminium and then applying the registration numbers. Many people who are involved in work on roads at night, such as road workers or policemen, now wear bright orange jackets on to which have been sewn strips of retro reflective material, and it is these strips which are visible in the headlights of oncoming cars.

Another type of reflective material is the surface of a modern cinema screen. Although a matt white surface makes an adequate screen, much of the reflected light is lost by scattering at the screen surface. Reflective cinema screens have a surface which is impressed with a pattern of shaped indentations so that most of the light is reflected back only towards the audience. The permitted angle of reflection with this type of screen is greater in the horizontal plane than in the vertical plane, to correspond with the area occupied by the audience in a cinema.

Reflective sheet materials have the property of retro reflection; in other words, they reflect light back in the direction of its source. The reflective sheets are made from a reflecting base layer, a clear plastic middle layer in which thousands of tiny glass beads are embedded, and a top layer which is transparent and usually coloured.

SCREWS

The origins of the screw are not known exactly, but some types of screw devices such as the screw auger (a kind of drill) were in use in Greece and Egypt before the 3rd century BC. By the 1st century BC, heavy wooden screws were used in presses for making wine and olive oil, and the character of the screw had thus been given a new dimension and was used to exert a force, its modern counterpart being called a *power screw* or *screw jack*. Metal screws and nuts first appeared in the 15th century, and can be seen in some mediaeval armour.

A screw consists of a circular cylindrical barrel on to which is formed a spiral ridge which is generally of roughly triangular cross-section. There are two broad categories, *machine screws* and *wood screws*. Both are generally made of metal but whereas the machine screw is of constant diameter and mate with a threaded *nut* or hole, the wood screw tapers to a point and the wood into which it is turned is deformed into a mating thread. The latter must usually be started in a hole made by an awl or a drill. The principle of operation in both cases is that as the screw is turned the spiral thread translates the rotation into an axial movement. This ability in fact leads to other uses of the screw than just fasteners. The lead screw on a lathe or the screw micrometer are examples of the screw being used to transfer power or provide axial measurement.

There are many varieties of machine screw or *bolt* used to clamp machine parts together, either in conjunction with a nut or with a threaded hole. These screws stretch when tightened, so that the axial tensile load thereby created holds the parts together. The heads of smaller screws are generally made with screwdriver slots, Phillips type heads, or hexagonal recesses to take an Allen key. The larger types almost invariably have hexagonal heads to which large torques can be applied with open-ended or box-end wrenches.

Threads There are an enormous number of different profiles for machine screw threads. The main ones in use in Britain were the British Standard Fine (BSF) and British Standard Whitworth (BSW). These were developed during the Industrial Revolution and they all have the same included angle at the bottom of each groove, namely 55°. The difference between them is the 'pitch' of the thread, which is measured in terms of the number of threads (or grooves) per unit of axial length, BSW being the coarsest of those mentioned (that with the fewest threads per inch). During World War II the Unified system of threads was adopted by Britain and Canada to be compatible with the American National Standard, and this resulted in the Unified Course (UNC) and Unified Fine (UNF) thread types, having an included angle of 60°. The International Organization for Standardisation (ISO) urges industry to adopt the ISO metric system in future designs, with ISO Unified as second choice. The internationally agreed metric system is now being used around the world, but it will be a very long time before all machines using the old systems are obsolete.

Self-tapping screw Besides the wood screw and the machine screw there is a third class of screw fastener, namely the *self-tapping* screw. This forms or cuts its own mating thread in such materials as metals, plastics, glass reinforced plastics, asbestos and resin impregnated plywood, when driven into a drilled hole of which the diameter is less than the overall diameter of the screw. The threads are formed in the hole by the displacement of material adjacent to the hole so that it flows around the screw.

Screw jacks Screws that modify force and motion are known as power screws, one form being the screw jack which converts torque (turning movement) to thrust. In one version, the thrust, used to lift a heavy object, is created by turning the screw in a stationary nut. By using a long bar to turn the screw, a small force at the end of the bar can create a large thrust force, and the screw jack can then be said to have a high *mechanical advantage*.

Above: an adjustable prop used as a temporary support during building work. It is in two parts, the lower one carrying the threaded section. When the threaded collar is turned, it moves up or down the threaded section and adjusts the height of the upper part of the prop. Top: an old Roman screw press which was used for pressing grapes or olives, found at Pompeii in Italy.

SEWING MACHINE

The sewing machine works on a different principle from that of hand sewing. In hand sewing, the needle and the free end of the thread are pressed right through the fabric and pulled through on the other side. No normal sewing machine could do this, since its needle is attached to the mechanism and cannot be released. Furthermore, hand sewing is done with a limited length of thread, the whole of which is pulled through the fabric at each stitch (except for the part that has already formed stitches). This would not be practical for a high-speed machine, which must be able to draw thread continuously from a reel [spool] or bobbin.

There are now a few highly complex machines which can imitate hand sewing, thanks to a free 'floating' double-pointed needle and other devices, but the vast majority of machines use either chain-stitch or lock-stitch.

Chain-stitch In the simplest form of chain-stitch, only one thread is used. This is pulled off a reel above the fabric and threaded through the eye of the needle—all sewing machine needles have the eye at the same end as the point.

The needle enters the fabric, pulling a loop of thread through with it. It then withdraws slightly, but friction against the fabric prevents the thread from withdrawing, so that it broadens out into a loop under the fabric. A looper—basically an oscillating hook—then comes across and catches the loop, after which the needle withdraws fully and the fabric moves on one stitch length.

The looper holds the loop under the fabric in such a position that when the needle descends again, it passes through

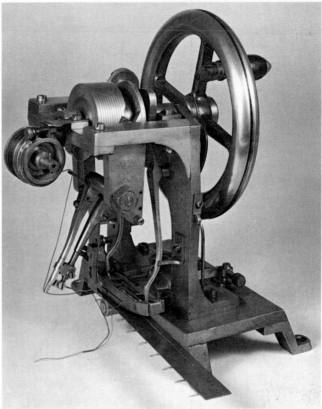

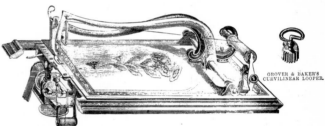

Left: in Thimonnier's 1830 chain-stitch machine, the fabric was fed through by hand. The wheel on the right is a flywheel, turned by the treadle.
Above, top: Howe's 1845 machine used lock-stitch. The shuttle, loaded with red thread, is at the bottom, just above the spiked baster plate which held the fabric as it passed through the machine.
Above: the Grover and Baker machine, patented in 1851, was the first to use two-thread chain-stitch, which did not come unravelled like the single-thread type, but used much more thread.

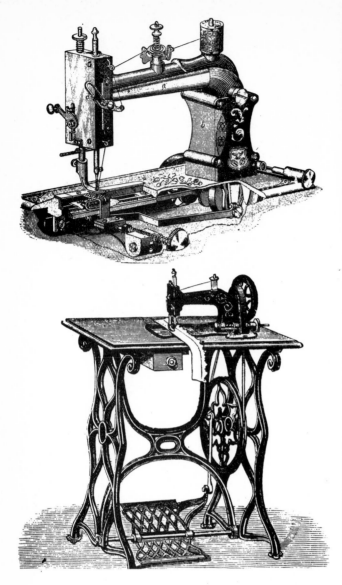

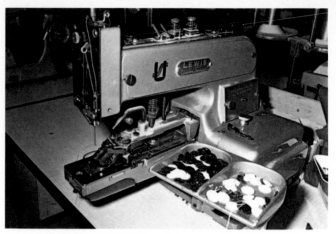

Above, top: Many 19th-century lock-stitch machines still had a straight-line reciprocating shuttle like that of a loom. This is a machine by W G Wilson, dating from 1881.

Above: before the adoption of the electric motor, most machines were driven by a treadle, working a flywheel and belt drive.

Above, right: Modern industrial machines are specialized. This is a button sewer, one of the few types using single-thread chain-stitch.

Opposite page: a modern domestic rotary lock-stitch sewing machine.

the held loop before forming a new loop, which is caught in turn by the looper. Thus a succession of loops is formed under the fabric, each one laced through the previous one.

Lock-stitch Two threads are used in basic lock-stitch, one above the fabric, pulled continuously off a reel, and the other below it, taken from a small bobbin mounted in a bobbin case or shuttle. The upper thread, or needle thread, is carried down through the fabric as before, but the loop it forms is caught by a hook travelling on a curved path (either oscillating or revolving fully) which passes the loop right around the bobbin, thus looping it around the bobbin thread.

The needle then withdraws, pulling the intersection of the threads into the fabric. In this way a stitch is created which looks the same from above and below; each thread runs across the surface of the fabric and dips into it at intervals to loop around the other thread halfway through the fabric.

Relative merits Chain-stitch can be executed very quickly, typically at 7000 stitches per minute, corresponding to 8 ft (2.4 m) per second at 12 stitches to the inch (5 per cm). Lock-stitch cannot match this because it puts a greater strain on the thread. But single-thread chain-stitch can be un-ravelled by simply pulling apart the end of an unfinished seam, or cutting one stitch and pulling. It is therefore only used on certain industrial machines where this does not matter, such as button sewers and tacking machines. It was also used on the earliest machines. Other chain-stitch machines use *two-thread* chain-stitch, where a seperate under-

Modern machines Most modern household machines have a *swing needle* which produces a *zig-zag stitch*. There is an elongated hole in the needle plate under the fabric, and the needle moves left and right at alternate stitches. The width of zig-zag can be set from zero (straight stitch) to 5 or 7 mm (0.2 or 0.3 inch). Adjusting the stitch length can give a long, loose zig-zag for sewing stretch fabrics or finishing raw edges to stop fraying, or a very short, tight one, for making button-holes or doing *satin stitch*, which is a simple embroidery stitch. If the forward motion is completely stopped, the machine will sew on buttons or make *bar tacks* (reinforce-ments at points of strain). The zig-zag can also be used for *blind hemming* (turning up the bottom of a garment so that stitches do not show from the outside) by folding the fabric, right sides (outsides) together, and stitching so that the point of each zig-zag just catches the fold without passing right through the fabric.

Fancy stitches are produced by cam boxes which vary the width of the zig-zag and the direction of the feed in a set pattern.

Industrial machines perform all these functions and more, but are larger and heavier, and tend to be specialized one-function machines, such as buttonholes. They also nearly all use chain-stitch. There are many complex multiple versions of chain-stitch for extra-strong stitching on heavy fabrics; up to nine threads may be used at once.

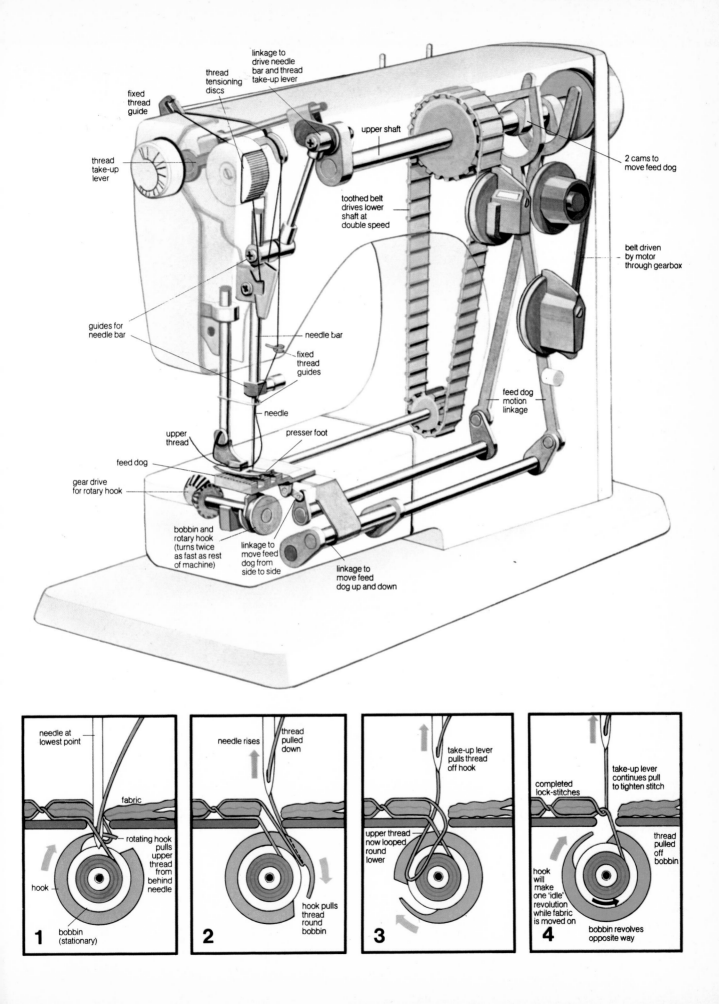

thread
take-up
lever

fixed
thread
guide

thread
tensioning
discs

linkage to
drive needle
bar and thread
take-up lever

upper shaft

2 cams to
move feed dog

toothed belt
drives lower
shaft at
double speed

belt driven
by motor
through gearbox

guides for
needle bar

needle bar

fixed
thread
guides

needle

feed dog
motion
linkage

upper
thread

presser foot

feed dog

gear drive
for rotary hook

bobbin and
rotary hook
(turns twice
as fast as rest
of machine)

linkage to
move feed
dog from
side to side

linkage to
move feed
dog up and down

1 needle at lowest point / fabric / rotating hook pulls upper thread from behind needle / hook / bobbin (stationary)

2 needle rises / thread pulled down / hook pulls thread round bobbin

3 take-up lever pulls thread off hook / upper thread now looped round lower

4 take-up lever continues pull to tighten stitch / completed lock-stitches / hook will make one 'idle' revolution while fabric is moved on / thread pulled off bobbin / bobbin revolves opposite way

Automatic control has begun to enter the clothing industry, and many machines are programmed to do complete operations by themselves, for example making shirt collars. There have been experiments with threadless machines with hollow needles which inject a fluid into the material to be 'sewn'—this is hardened by heat treatment to form a 'seam'. But these are really no longer sewing machines. thread is interlooped with the ordinary loops under the fabric, making the stitch secure. This, however, results in very high thread consumption; about 5 inches of thread are used for every inch of seam.

Lock-stitch is secure and uses about half as much thread, but its use is restricted not only by its maximum speed, but also by the fact that the needle thread has to pass right around the bobbin at each stitch. This makes it impossible for the bobbin thread to be drawn from a large fixed reel, and since the size of the bobbin is restricted by the size of the loop that can be drawn from the needle thread, its capacity is severely limited. (The underthread for two-thread chain-stitch *can* be drawn from a fixed reel.) For this reason, lock-stitch is most suitable for home sewing machines, where speed and thread capacity are less important than in industry, where chain-stitch is generally preferred. Chain-stitch was also used in many home machines during the 19th century.

Feed The fabric is advanced at each stitch by a *feed* mechanism, which consists of a toothed bar under the fabric, gripping it by pressing it up against a smooth spring-loaded *presser foot*. The feed bar moves in a *four motion*: up, forward, down, back and so on. This can normally be reversed to move the fabric the other way (thus securing the end of the seam by back-stitching) and its travel can be altered to change the length of the stitch.

History The first patent for a sewing machine was taken out in 1790 by Thomas Saint, a London cabinet maker. It used single-thread chain-stitch, and had a forked needle which went through a hole previously made by an awl. It was probably never built, since minor design faults in the patent specification would have made it unworkable.

In 1810, a German, B Krems, invented the eye-pointed needle, again without commercial success, but the first commercial machine was built in France by Barthélémy Thimonnier, a tailor, and had a barbed needle which tended to catch in the cloth. It also had no feed and the fabric was moved by hand, but nevertheless, Thimonnier set up a business making military uniforms, and operated successfully until his machines were destroyed by hand workers who feared they would lose their jobs. After this setback. Thimonnier abandoned any further development of his design.

The first lock-stitch machine was invented in the early 1830s in New York by Walter Hunt. Again, it was not commercially exploited, but in 1846 Elias Howe patented a fairly similar machine. It had an inconvenient feed mechanism whereby the edge of the cloth was held vertically on spikes on a *baster plate*, which then carried it through the machine. At the end of its travel this had to be moved back and the next length of cloth put on the spikes. The machine also needed a curved needle.

The first machine to have the general form of a modern one was produced by the American Isaac M Singer in 1851. This was a lock-stitch machine with a straight, vertically sliding needle and a spring loaded presser foot. It was the first machine to be foot-powered by a treadle.

Another American, Allen B Wilson, invented in 1852 the rotary hook, which took the needle thread around a stationary bobbin (Singer's machine had had a straight-line oscillating shuffle.) In 1854 he invented the four-motion feed. This completed the basic equipment of the modern machine, and later improvements were mainly in detail. The double-thread chain-stitch had been invented in 1851 by William O Grover.

SEXTANT

Since the earliest ocean voyages navigators have had to fix their position at sea by means of measuring the angles above the horizon of heavenly bodies, the Sun, Moon and stars. The simplest, and least accurate, of these is to measure the *altitude* (angle of the horizon) of the Pole Star (in the northern hemisphere). More accurately, the altitude of the Sun at local noon, or of bright stars whose position is known at their highest point, will give the latitude after simple calculations from tables.

The first instruments to be used for measuring these angles were astrolabes, cross staffs, back staffs and quadrants—all variations on the theme of a sighting bar moved along a scale of degrees. In most cases it was necessary to view both the star and the horizon at the same time from the deck of the ship, and it is not surprising that the observations were inaccurate. In the case of the quadrant the reference point was not the horizon but a plumb line attached to the scale. This made it possible to concentrate on the star only, but the plumb line could easily swing about, leading to further errors.

The device which replaced these, the forerunner of the sextant, was the *octant* invented by James Hadley in 1731. The principle and design of the octant was the same as that of any sextant in daily use nowadays: the main difference is that the octant had a scale which was one eighth of a circle, 45°, while a sextant has a scale of one sixth of a circle, 60°. Because both devices measure an angle which is reflected by a mirror, the octant will measure angles up to 90° and the

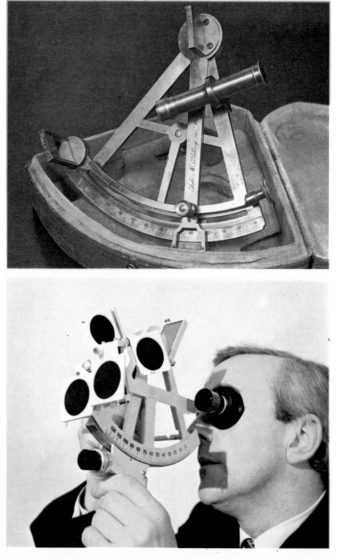

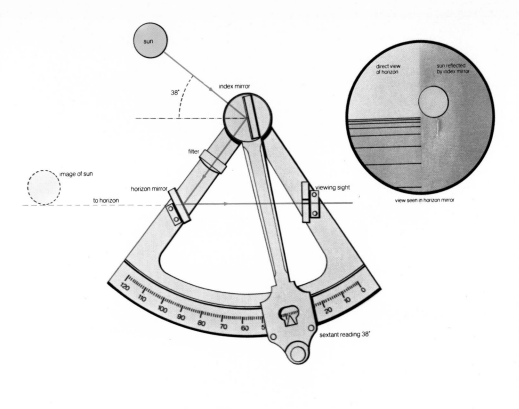

sun

index mirror

38°

filter

image of sun

horizon mirror

to horizon

viewing sight

120 110 100 90 80 70 60 50 40 30 20 10 0

sextant reading 38°

direct view of horizon

sun reflected by index mirror

view seen in horizon mirror

Left: the basic layout of a sextant. The procedure is to move the index mirror until the bottom of the Sun's image just touches the line of the horizon as shown in the inset. This reading must be corrected to give the position of the centre of the Sun.
Opposite page: the general design of the sextant has not changed since its invention. This brass instrument, made in about 1800, compares with the modern type at bottom. The older sextant has a clamp to fix the arm on to a bar which can be adjusted by a fine screw at one end, while the modern instrument has a vernier screw gauge for the same purpose. Modern optical design has shortened the telescope.

sextant angles up to 120°. The octant was often referred to as a quadrant because it had the same range as one, though its scale was only 45°. In practice, few sextants measure exactly 120°, but the name is retained.

Hadley's invention was to use a pair of small mirrors to reflect the image of the star to be observed so that it appeared to be on the horizon. The navigator could keep both in view at the same time, and as the ship rolled both would move together. The movable mirror was attached to the pivot of the movable index bar, at the radius of the scale or *arc*, so that as the angle was changed so the mirror would move. This mirror reflected the star's image to a second mirror, permanently set to view the first one. The navigator looked through a sight to the second mirror, past which he could see the horizon. He moved the first by moving the index bar until the star's image reflected by both mirrors exactly touched the horizon. The angle was then read off the arc, which was graduated in degrees (but twice as closely as a true scale of degrees, to allow for the mirror's reflection).

Hadley's octant was immediately accepted by navigators. In 1757, John Campbell introduced the true sextant, which was capable of measuring a greater angle. Captain Cook was probably the first to fully apply the potential of the sextant for measuring not only vertical angles but also angles at any inclination. By measuring the angle between the Moon and a given star, he could calculate the precise time, using tables of the Moon's motion, enabling him to find his longitude as well as the latitude—the method used for the charting of New Zealand during the voyage of 1768-1771. The invention of accurate timekeepers made the procedure unnecessary, and the sextant was then used to measure the altitude of stars or the Sun at precise times, thus giving the longitude.

Although the earliest sextants and octants had simple sighting devices, the accuracy was much improved by the use of a small telescope instead. The second mirror, the horizon mirror, would only be silvered across half its width so that the telescope would show both the horizon and the star side

by side. Dark filters could be moved into the light paths to cut down the brightness of the Sun or horizon.

The sextant has remained basically unchanged from 1800 to the present day, but there have been some changes to the way in which the arm of the sextant is made to travel along the arc. In the early days there was no fine adjustment screw and the navigator merely moved the arm along the arc and clamped it to the frame so that the reading could be taken. On a moving deck this was rather hard to do, so in the 1760s a fine adjustment tangent screw was added. This meant that the operator could quickly take his sight to the nearest degree and then, by using the tangent screw, make the final close adjustment. The only drawback with the clamping variety of sextant was that the tangent screw frequently had to be returned to its starting position, otherwise it would come to the limit of its thread as a sighting was being taken. The problem was solved in the 1920s. A toothed rack was cut into the sextant frame and the tangent screw was now meshed into this. The arm could be moved along the arc by pressing a quick release catch, and the tangent screw could travel the full length of the arc without being reset.

The sextants of this period were still using the finely engraved scale, which had to be read with a magnifier, as they had been over the previous hundred years. Around 1933 the micrometer sextant was evolved which is still in use today. Instead of engraving the fine divisions on the arc, they were transferred to an enlarged tangent screw head, doing away with the magnifier and making the sextant easier to read.

Other types The sextant may still be used for air navigation far from the busy air corridors. In this case the horizon cannot be used as it is below the true horizontal, and a system which reflects the image of a bubble level into the field of view is used. *Astrodomes*, small transparent domes into which the sextant will fit, may be set into the top of the aircraft, or in the case of the faster aircraft a periscope system will be fitted. Land sextants, in which the horizon is provided by a small trough of mercury in the sextant, have been made.

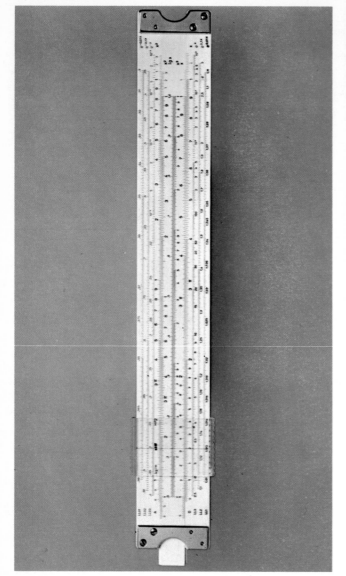

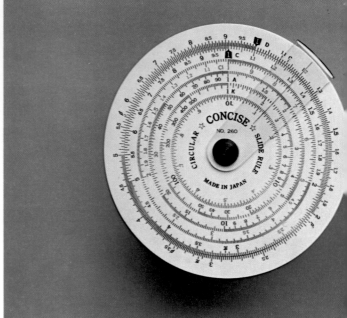

Above: a general purpose linear slide rule. The C and D scales show the division of 1.2 by 1.32—the answer 0.91 is found on D scale adjacent to '1' mark on C scale. Other scales are also marked. There are two ways of labelling these scales: the alphabetic labels marked on the left end and mathematical notation on the right. The C and D scales are marked x. The AB scales (x²) are used for obtaining the square of a number (and conversely, square roots). They have two cycles of the basic CD scales but reduced in size. The side shown also has log log scales (six in all) marked on the white fixed sections. Positioning cursor on D scale at x = 2 gives value of e² on LL3 scale (eˣ)—e is the 'natural number', 2.718. Log log scales can be used to find any number to any power.

SLIDE RULE

The slide rule is an extremely versatile instrument for performing calculations quickly and with reasonable accuracy. It has certain advantages over the pocket-sized calculator; however, with mass-production of electronic calculators, quality slide rules are no longer relatively cheap.

The electronic calculator has the advantage of handling longer numbers, upwards of six digits, whereas the slide rule can only handle up to about four digits. The slide rule is therefore limited in range. Furthermore, with the floating decimal point system, electronic calculators can simultaneously cope with numbers less than a millionth and greater than a million. Slide rules depend on the user to remember the 'scale' of the numbers involved and position the decimal point by reason and skill at the end of a calculation.

Slide rules, however, have the advantage that they can handle easily some calculations that only the most expensive calculators otherwise could cope with. Also, they require no battery to power them and they have few parts that can go wrong.

Linear slide rule The most common slide rule is the straight, or linear, version. This consists of two rigidly connected scale sections (like rulers) with an inner scale that can slide. Surrounding this arrangement is a transparent sliding plate with a fine hair line engraved on it perpendicularly across the three sections. This plate—called the *cursor*—can move independently of both the fixed and the sliding parts.

The type of calculation that can be performed on this instrument will ultimately depend on the scales provided but, in general, the following procedure is used where two separate numbers are involved. Firstly, one number is located on the fixed scale and the cursor moved along until the hair line lies directly over this number. Secondly, the reference point on the sliding scale—in multiplication, this will be the 1 (unity) mark—is moved until it also lies directly under the cursor line. Lastly, the second number in the calculation is located on the sliding scale and the cursor moved until its lies directly over this point. The answer is found by determining where this same line crosses the fixed scale.

The most common scales found on slide rules are those used for multiplication and division. These are *logarithmic* in character and multiplication (or division) is performed by effectively adding (or subtracting) along these scales. Other

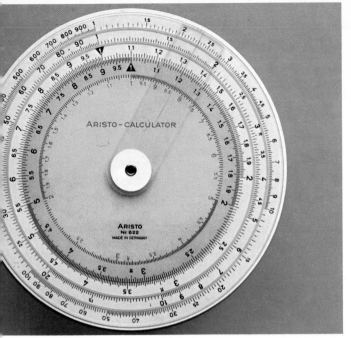

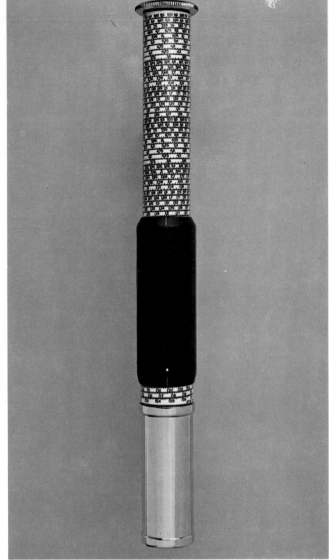

scales will produce the square of a number, square root, sine, cosine and tangent and the logarithm of a number. Some scales give the constant multiple of a number, and one commonly found is the π (pi) multiple used in calculations involving circles. Another important scale is the log log scale used for finding the result of any number taken to any power.

Some slide rules will contain all these scales (and more), and this is achieved by making the slide rule two-sided with the cursor line precisely marked on both sides.

Apart from the precision of machining and constructing these instruments (and this is important in all types of slide rule) the accuracy of linear slide rules is limited by their length. The longer the scales, the finer the divisions (graduations) that can be marked. There is, however, a practical limit to the length of linear rules because they are primarily designed to be portable and are usually no more than a foot (30 cm) long. One way of overcoming this limitation is to use a cylindrical slide rule.

Cylindrical slide rule In this, rather than placing the scale in a straight line, it is wrapped in a helix around a tube. In this way, a five foot (1.5 m) scale can be accommodated on a tube two inches (5 cm) long and one inch (2.54 cm) in diameter. The principle behind its design and use is identical to the linear slide rule accept that it is 'circularized'. They are, however, more limited because only one type of scale can be incorporated (usually the logarithmic scale for multiplication and division).

The fixed scale is the tubular scale, attached to a handle (for ease of operation). Over this fits a metal tube and at the other end is a sliding scale—a second tube which fits inside and can slide independently of the other two. The metal tube is the cursor with marks at the two rims where the scales project. The technique of cursor-scale manipulations involved in calculations is identical to that of the linear slide rule.

Circular slide rule The logarithmic scale (to the base ten) used in multiplication and division need only be presented between 1 and 10 because other decades (for example, 10 to 100) are identical in form. Because of this cyclical character of the logarithmic scale the slide rule can cope with any number with only one range of scale numbers (1 to 10). For example, to use the number 635 in a calculation one locates instead the number 6.35 (this exists on the slide rule scale) remembering that it is one hundredth of the desired figure.

This cyclical nature of the logarithmic scale makes it ideal

Above: cylindrical slide rule. This can only be used for multiplication and division. The lower end of the black cylindrical cursor is shown set against the number 1.32 on the fixed scale and the top cursor mark against 1.2 on the sliding scale.

Above, left: two types of circular slide rule. These are light, compact and have the advantage over linear slide rules that calculations cannot 'run off' the end of the scales. Left type shows division of 1.2 (on outermost, D scale) by 1.32 (on adjacent C scale). The result (0.91) is found on D scale opposite '1' mark on C scale. Radial cursor line shown over 1.2 and 1.32 markings.

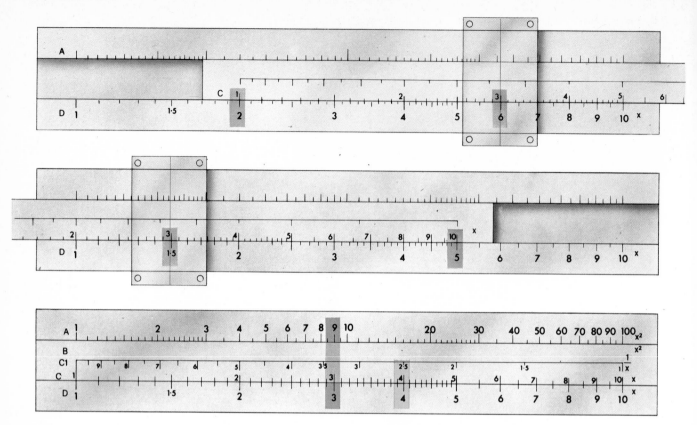

Above: the top situation shows how to multiply 3 by 2. The unity mark on the C scale is set against the 2 mark on the D scale. By placing the cursor over the 3 mark on the C scale the answer (6) is found on the D scale. The middle situation shows that the unity mark on the other end of the C scale must sometimes be used. Shown here set against the 5 mark on the D scale, the multiplication of 5 by 3 is found under the cursor on the D scale. The bottom situation demonstrates squares and square roots (using A and D scales); the reciprocal of a number (located on D scale) is found on the C1 scale, showing that the reciprocal of 4 is 0.25.

for circular slide rules. The scale between 1 and 10 is made to fit exactly into a circle so that the 1 and 10 marks coincide. With circular slide rules, the base disc has the fixed scale marked on it. A second, and smaller, disc turns within the base disc and contains the sliding scale. Over both lies a transparent arm with a radial line engraved—this is the cursor. Unlike a cylindrical slide rule, a circular one may have several concentric scales.

Construction Several factors are taken into account in the design of slide rules. The sliding surfaces must be hard and smooth to prevent wear and permit a smooth and even movement. The material must retain its dimensions—that is, not expand or contract through changes in temperature or humidity. Also, it must be designed to withstand rough handling (as opposed to abuse). The components are commonly made of injection-moulded thermoplastic materials. Some low-priced rules are wooden, with printed paper labels covered in plastic film to protect them.

SOLENOID

A solenoid is an electrical device consisting simply of a coil of wire, and can be made, for example, by wrapping wire around a cylinder. When a current passes through the wire a magnetic field is set up, and this moves a ferrous core to actuate valves, switches and other devices. The solenoid is therefore a direct application of an electromagnet.

Outside the solenoid the lines of magnetic flux behave in a similar fashion to those of a bar magnet. A solenoid freely suspended horizontally in the Earth's magnetic field will set itself along a North-South line. Its ends behave like the poles of a bar magnet, their polarity depending on the direction of the current in the spiral. Any ferrous material brought into the vicinity of the solenoid will be attracted to the poles along the lines of the magnetic field.

The strength of the magnetic field within the solenoid is uniform for most of its length but near the ends, known as the poles, the field diverges. At the poles the field strength dies rapidly to about one half of the strength in the centre.

Inside the solenoid, at distances from one end of greater than about $3\frac{1}{2}$ times the diameter of the coil cross section, the field strength is 99% of the calculated value for an infinitely long solenoid. Hence in practice a 'long' solenoid should have a length at least seven times its diameter.

Applications The ability of the solenoid to produce a magnetizing force leads to its use in starting devices and power operated valves, as only a switch need be turned to energize it. For example, solenoid switches are widely used to engage starter motors in cars. Here two solenoids, the 'draw-in' coil and the 'holding' coil, are mounted on top of the starter motor with a plunger running through the inside of both (thus operating in the region where the field strength is uniform and at a maximum). One end of the plunger is attached to a lever which engages and disengages the starter motor pinion with the flywheel. The other end of the plunger is connected to a switch.

When the ignition switch is turned, the 'draw-in' coil is energized and the plunger is drawn to the right, thus engaging the starter motor with the flywheel. When the plunger makes contact with the switch the 'holding' coil and the starter motor are energized and the 'draw-in' coil is short circuited. This is because the 'draw-in' coil drains more power from the battery than is needed just to hold the plunger in position, and this power is now required to turn the starter motor. After the engine has started, the ignition switch is released, the 'holding' coil is de-energized and a spring returns the plunger to its original position, thus disengaging the starter motor from the flywheel.

In a simple power operated switch, such as one would find in a domestic central heating system, the solenoid provides the power to open and close the valve disc. The disc is con-

Left: solenoid operated rotary switch with multiple contacts. The electromagnet operates a ratchet mechanism to the contacts.

nected by a rod to the core of the solenoid. When the solenoid is de-energized, the disc is held against the aperture by a spring and the valve is off. Switching on a current through the solenoid produces a magnetic field which draws the rod, and therefore the valve disc, away from the aperture, so that the valve is on. When the current is turned off the spring returns the disc, thus shutting the aperture and the valve.

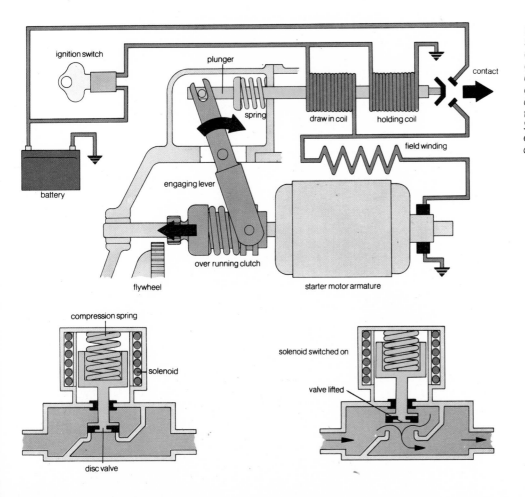

Left: two applications of a solenoid. The starter motor arrangement for an internal combustion engine can be operated by two solenoids—one draws in the clutch and the other holds this position while the motor turns the engine. Solenoid valves are used extensively in industry for control purposes.

SPRAY GUN

The origin of the spray gun can be attributed to Dr DeVilbiss, a medical practitioner of Toledo, Ohio, USA, who in the early 1800s searched for the ideal method of applying medication to oral and nasal passages. As an alternative to swabbing, the atomizer was devised as a means of introducing a liquid from a container into an air stream which was directed on to the area to be treated, and after a period the atomizer evolved into the paint spray gun which is now widely used throughout industry. The spray gun is used to apply coatings such as paints, lacquers, and glazes to every imaginable mass produced article which requires a finish for decorative or protective purposes.

A spray gun is a precision tool which relies on compressed air for its energy or power. Paint and compressed air are directed separately into the spray gun and by a system of air channels and a paint passageway the air is aimed at the paint stream at the head of the gun. This stream of compressed air imparts additional motion to the material being applied, and since every material has a critical speed at which it will break up or *atomize*, the material is therefore broken up into small spheres or globules—this action being known as *atomization*.

The force of the compressed air performs two functions. First it atomizes the paint and second it vaporizes the solvents in the paint. By correct adjustment of air pressure and the amount of fluid fed to the gun, it is possible to accurately control the degree of atomization, thus ensuring maximum atomization with the least vaporization of the solvents.

Operation The mechanical action of the spray gun may be described as follows. Fluid enters the spray gun through the fluid inlet immediately below the spray head of the gun, and is forced past a needle valve (the *fluid needle*) and out through the *fluid tip* or nozzle in the form of a jet or stream. The pressure driving the fluid may be created by gravity (by mounting the fluid container on top of the gun), by suction from the flow of air through the gun, or by use of a pump or a pressurized fluid container. A spring-loaded valve, operated by the gun trigger, controls the supply. From the compressed air supply, air enters the gun through the air inlet at the bottom of the spray gun handle. It passes through the gun body and is projected through a series of small holes in a *baffle ring* behind the *air cap* which surrounds the fluid nozzle. The air supply is regulated by another spring-loaded valve, also under trigger control.

When the trigger is depressed both valves act simultaneously, bringing the flows of fluid and air together, and the air directs the fluid stream into a conical spray pattern. This spray pattern can be altered to suit the work in hand by means of a spreader width adjustment valve mounted at the rear of the gun body. This valve alters the setting of an opening inside the gun, which regulates the flow of air through a second series of holes in the baffle ring to air ports contained in the two 'horns' of the air cap. These horns direct jets of air into the spray pattern from above and below.

The emission of air from these 'horn holes' impinges upon the conical spray, spreading it into an elongated shape, and by this means a horizontal fan shaped pattern can be achieved.

Types of gun Spray guns are made in various sizes and are classified according to their capacity to atomize and apply amounts of paint. They are rated for high, medium and low production requirements. The three principal parts of a spray gun are the air cap, the fluid tip, and the fluid needle. These are the parts which may be changed in a gun to make it more suitable for spraying one kind of material or another. The type of cap, tip or needle used depends on the type of paint to be sprayed and on the method of supplying paint to the gun (by suction, gravity or by pressure feed).

The amount or volume of air used by a spray gun is

Right: a cutaway diagram of a spray gun. The compressed air supply enters the gun through the air inlet at the bottom of the spray gun handle and passes through the lower spring loaded valve, which is controlled by the trigger of the gun. It then travels through the body of the gun to the fluid nozzle where it meets the paint flow which is leaving the gun under the control of the needle valve, which is also operated by the trigger, and adjusted by the upper spring loaded valve. The air and paint flows leave the nozzle in the air cap as a conical spray, and the cone can be flattened and widened into a horizontal, fan-shaped pattern by allowing air from the air cap's horns to impinge upon it. This air is controlled by the rear adjustment valve.

adjustment valve

spring loaded valve

baffle ring

nozzle

spring loaded valve

air cap horn

fluid inlet

spray gun handle

air inlet

determined by the size and design of the air cap, and this is measured in terms of cubic feet (or cubic metres) per minute. Two types of air caps are generally available. One is the suction feed cap. This type is designed so that the air passing through the centre hole of the cap creates a vacuum in front of the fluid tip, thus allowing atmospheric pressure to force paint up to the spray gun from the container beneath it. The pressure feed cap is not basically designed to create a vacuum in front of the tip, and these caps are used in fast production where the fluid is fed to the spray gun by a pump or from a pressurized container.

The fluid tip is the part of the spray gun which meters the amount of paint flowing from the gun, and acts as a valve seat for the fluid needle. Fluid tips and needles are made in various sizes according to the viscosity and type of paint to be used.

The fluid needle is normally held in the closed position against the fluid tip by means of a spring, until withdrawn by the trigger action. The distance the needle is withdrawn, and thus the amount of fluid delivered, is controlled by the fluid adjusting screw.

Spray painting is a mechanical means of applying paint. The benefits derived by spray painting are economical application of finishing materials and high quality finishes at high rate of production. To enjoy these benefits it is important that the spray gun be regarded as a precision instrument and must be properly maintained and skilfully used in order to obtain maximum efficiency.

Electrostatic painting Many industrial painting processes, such as the application of primer paint to car bodies, now use electrostatic painting systems. The object to be painted is given a negative electrostatic charge and passed through a mist of positively charged paint sprayed into the painting chamber. Electrostatic forces between the object and the paint attract an even layer of paint on to the surface of the object, including crevices that would not normally be reached. Finishing coats may subsequently be applied by conventional painting methods.

STETHOSCOPE

The invention of the stethoscope is usually attributed to R T H Laënnec who, in 1816, used a wooden cylinder to transmit heart sounds to the ear. In 1828 Poirry modified this very basic instrument by adding an earpiece and a trumpet-shaped chest piece, but it was not until about 1850, with the invention of the binaural stethoscope (two earpieces) that the instrument took on the shape we know today.

Acoustic stethoscopes These instruments are used to investigate the condition of the respiratory, circulatory and digestive systems. Most sounds of interest to a doctor, especially heart sounds, have frequencies in the range 60 to 600 Hz, but some *mitral diastolic murmurs* (irregular sounds heard over the heart during expansion of the heart and indicating an abnormality in the mitral valve) have frequencies below 60 Hz, and a few sounds, such as *crepitations* (crackling sounds heard over the chest in some diseases of the lungs), have frequencies of up to 1400 Hz. Acoustic stethoscopes do not amplify sound, they merely convey it to the ear in as efficient a manner as possible.

A typical modern stethoscope has a combined *bell and diaphragm* chestpiece made of stainless steel. The bell section of the chestpiece is open, has a diameter of 1.13 inch (2.87 cm) and is 0.25 inch (0.64 cm) deep. When this side of the stethoscope is in use, the patient's skin acts as a flexible diaphragm across the mouth of the bell to transmit the sound. The diaphragm section has a diameter of 1.72 inch (4.37 cm) but is only 0.13 inch (0.33 cm) deep. It is covered by a rigid linen-Bakelite diaphragm. The doctor can select either side of the chestpiece by rotating it relative to the *collecting tube* which

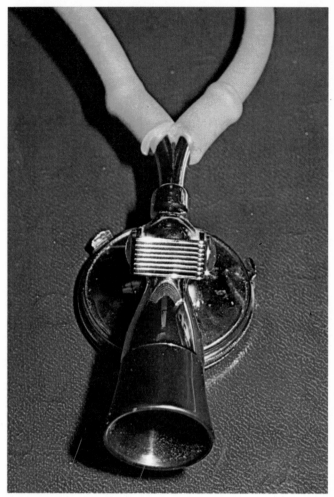

Below: the bell-and-diaphragm chestpiece of a modern stethoscope. At the bottom of the picture is the bell side of the chestpiece which is used for listening to relatively low frequency sounds.

connects it to the earpiece tubing. The bell side of the instrument will normally be used by the doctor to listen to relatively low pitched sounds in the range of about 30 to 500 Hz, whereas the diaphragm side of the chestpiece is designed to filter out the lowest sounds but pass the highest frequencies in the range 200 to 1400 Hz.

The stethoscope tubing is made of a flexible plastic material with a very smooth interior surface, and the earpiece tubes are generally made of stainless steel. The design of the plastic tips for the earpieces is very important, they must be as large as possible, usually with a diameter of from 0.50 to 0.63 inch (1.27 to 1.60 cm), so that external sounds are blocked from the ears; a leak only five times the diameter of a human hair has a very marked effect on the performance of the stethoscope, particularly at low frequencies.

Phonocardiographs These are devices for monitoring heart sounds electronically. A phonocardiograph gathers no more information than a simple acoustic stethoscope, but it displays the sound in visual form on a cathode ray tube or a pen recorder, and this often makes diagnosis simpler. Phonocardiograms are frequently used in conjuction with electrocardiograms to diagnose heart valve defects. The input to a phonocardiograph is from one or more microphones, which are usually attached to the patient's chest by suction.

Top: Valdemar Poulsen's
telegraphone, invented in 1898.
It recorded telephone speech on
steel piano wire at a speed of
84 inches (213 cm) per second.
Poulsen foresaw the use of tape
coated with metallic powder.
Above: a Marconi-Stille
tape recorder, which used steel
tape. This one was used by the
BBC in 1935.

TAPE RECORDER

Magnetic recording was first demonstrated in 1898 by Valdemar Poulsen, a Danish scientist, but it took another 50 years to perfect the tape recorder. Poulsen used steel wire, and later machines used a flat·steel ribbon, but these materials were awkward to handle and store. Research in Germany before World War II produced a more versatile medium for magnetic recording. This consisted of a thin coating of iron oxide on paper strip; the paper was replaced by plastic film a few years later, to make the first magnetic tape.

In the basic design of tape recorder, the *supply spool* of 6.25 mm ($\frac{1}{4}$ inch) wide tape is loaded on the left hand spindle and the loose end of tape is threaded across the face of the *magnetic heads* on to the empty *take-up spool*. The speed at which the tape winds from one spool to the other is not determined by the revolutions of the take-up spool. Instead, a precision *capstan* is rotated at constant speed by an electric motor and drives the tape, which is pinched between the capstan and a *pressure roller*.

All magnetic recorders rely on the principle of electro-magnetism, that is the interaction of electric currents and magnetic fields. During the recording process, the signal received from a microphone, record player or radio, is in the form of a tiny electric current whose strength varies with the intensity of the original sound, and whose frequency corresponds to the pitch of the sound. This signal passes through an amplifier and then through a coil in a specially shaped electromagnet known as the *record head* and produces a magnetic field which also imitates the sound in strength and frequency. As the tape is drawn past the front gap of the head, where the magnetic field is concentrated, the magnetic state at each instant is effectively recorded by alignment of the iron oxide particles, each of which behaves like a tiny magnet.

During the playback process, the tape is drawn past the *playback head*, which is an electromagnet resembling the record head (indeed the same head is used for both recording and playback in most home recorders). The changing magnetic field pattern on the tape induces a current in the playback head coil and, when this is amplified and fed to a loudspeaker or headphones, the signal is heard as originally recorded.

A new reel of tape is supplied in completely demagnetized state, that is the magnetic coating particles are arranged in a random manner. It is essential that the tape should be in this demagnetized state when it reaches the record head. Therefore, to allow tapes to be used over and over again, all recorders incorporate an *erase head*, which scans the tape at a point just before the record head. Originally a small permanent magnet was used for this job of demagnetizing, but this resulted in noisy tapes. Modern erase heads are fed with a current at an ultrasonic frequency which magnetizes the tape first in one direction, then the other, leaving it more effectively demagnetized.

The oscillator used to generate the high frequency erase current has a second function in most tape recorders. This is to supply high frequency *bias*. A small amount of the ultra-sonic signal is mixed with the programme signal going to the record head. This has the effect of reducing non-linearity in the *hysteresis loop* of recorded flux and so reducing distortion. The erase bias frequency is usually chosen to be at least three times the highest audio frequency which the machine is intended to reproduce, therefore values between about 50 to 80 kHz are usual.

Because the recorded *waveform* is more cramped at lower speeds of tape travel, a limit is set to the ability of the machine to reproduce high frequency sounds at slower tape speeds. On the other hand, higher tape speeds use more tape for a given duration of recording, and so a compromise is usually sought. The common speeds are 15, $7\frac{1}{2}$, $3\frac{3}{4}$ and $1\frac{7}{8}$

Left: one of the first modern tape recorders, made in Germany in 1936. One use of tape soon became common: opera recordings made on machines like this and featuring the famous artists of the pre-war years are still bootlegged around the world. Tape was also a convenient means of broadcasting the speeches of Adolf Hitler.
Below: a tape-to-disc transfer studio. The operator adjusts the sound level from the master tape, using the mixing console, as the disc-cutting lathe cuts the master disc in the background.

inches per second (38, 19, 9.5 and 4.75 cm sec), with the higher speeds preferred for high quality and the lower speeds for economy. In recording studios, 15 and even 30 ips speeds are used, while on home recorders, the technology is now so highly developed that satisfactory recordings can be made at $3\frac{3}{4}$ ips, and adequate ones at $1\frac{7}{8}$ ips, the standard speed of a cassette.

Economy in the use of tape is further achieved by making separate recordings occupy only part of the tape width. Thus the heads of a *half-track monophonic* recorder, for example, would have a vertical gap designed to scan only the top half of the tape. The user can then turn the tape over and make a second recording on what was previously the lower half of the tape. *Stereophonic* recording needs two tracks simultaneously for the left and right channel signals and so, with economy in mind, stereo recorders today are based on a *quarter-track* head configuration. The tracks used in either direction are interleaved, so that tracks 1 and 3 are used

for left and right in one direction and tracks 4 and 2 when the tape is turned over.

Special effects On very simple recorders, the only way to obtain sound-over-sound or *superimposing* of one recording on another (for instance to sing or play a duet with oneself) was to disable the erase function temporarily during the second recording. This is sometimes done by means of a 'trick' button which simply switches off the erase current (or physically holds the tape away from the erase head). Unfortunately this method inevitably weakens the first

recording, partial erasure taking place due to the presence of high frequency bias in the second signal. Also, there is no way of listening to the previous recording while making the new one unless the machine has a separate playback head—and then the two recordings are out of *synchronism* because of the time lag between the two heads.

Most quarter-track machines, and some stereo half-track ones, permit more polished trick recordings. For multiplay recordings, for example, the procedure is as follows: record the signal A on track 1: rewind the tape: play back track 1 and simultaneously re-record it plus signal B on track 2. The *mixed* recording now on track 2 can then be fed back to track 1, while adding a further signal C and so on. Note that it is important to be able to *monitor* (listen to) the previous track, preferably on headphones, while adding new sounds. This is possible only on recorders which permit one track of the record head to be switched to playback while the other is in the record mode. This is called *sync playback*.

Machines with a separate playback head often provide a so-called 'echo' effect. For this, a controlled amount of the playback signal can be fed back to the record head during a recording. Because of the head spacing, this return signal is delayed: when it in turn is recorded and again reaches the playback head, it produces a second echo and so on. The effect is of many echoes fading out more or less gradually, depending on the level of signal fed back. The frequency of the echoes will be lower at lower tape speeds, and this may sound less natural.

In professional recording, while a single tape head delay may be used for the initial repetition, the random dying away which constitutes *reverberation* in a real concert hall or church is more truly simulated by sending the signal to a reverberation chamber, suspended steel plate or spring and then mixing this reverberant signal with the original.

Editing One feature of tape recording which appeals strongly to many users is the ease with which tape can be edited. In the splicing technique usually adopted, the two ends of tape to be joined are each cut across with a razor blade or scissors (brass scissors are obtainable which avoid the clicks which steel scissors may put on the tape if they have become accidentally magnetized). A slanting 45° cut is used, and this helps to give smooth continuity of the sound background through the join, instead of a noticeable break. After a little practice, it is possible to edit musical passages to single note accuracy and speech can be cut to within a single syllable. When recording concert music, engineers routinely edit mistakes from the master tape, inserting re-makes so that the finished recording is closer to perfection than a live performance could possibly be, a practice of which some music lovers disapprove. The editing can some-times be heard, when listening carefully to a recording using headphones, and often shows as a change in the dynamic level of the recording.

Above left: this photo shows the recording head with the capstan and pinch roller which pull the tape past the head. Tension on the take-up reel is relatively small.
Centre: the heads on this tape recorder at EMI's Abbey Road studio have twelve tracks.
Above: a close-up of the two-track stereo recording head seen on the opposite page. Professional recording tape and the gaps in the heads are wider than on home machines.
Below left: tape recorders have many uses. This one is being used to detect bats.

TELEPHONE SYSTEM

Alexander Graham Bell stated the principle of telephone transmission as follows: 'If I could make a current of electricity vary in intensity precisely as the air varies in density during the production of sound, I should be able to transmit speech telegraphically'. Bell, a Scotsman who had emigrated to Canada, then set about inventing such a device, transmitting the sound of a twanging clock spring by wire in 1875. On 10 March 1876 he transmitted a complete sentence to his assistant in another room. From that grew today's worldwide network of over 300 million telephones.

Bell was the first to utilize a continuous current intensified and diminished in proportion to the sound waves projected into the transmitter, but his early telephones failed because the electrical resistance of the wires through which the current passed soon made the current ineffectual.

Early telephoning was done over a single iron wire connecting two telephones, with an earthed [grounded] return circuit (that is, using the earth as the return conductor). In the USA and in the UK development was bedevilled by a plethora of private telephone companies rushing to set up non-interconnecting systems, often using incompatible equipment. In Britain the Postmaster General, in 1880, won a lawsuit giving him rights to 'acquire, maintain and operate' telephones, while private companies were allowed to operate under licence from the Government, the licences expiring in 1911. Thus the companies, which were largely American owned, held the patents needed to build up a good service, but lacked the incentive to develop services under the threat of a Post Office take-over. Growth was slow until 1912, when the Post Office took over the National Telephone Company and began working towards a compatible national network.

Engineering developments had not been standing still. Edison invented the carbon granule transmitter (or microphone), which is still used in most telephones today, and Ericsson combined the receiver and handset in the same instrument.

The UK public telephone system consists, basically, of a large number of single telephones unevenly distributed throughout the country, each one connected to a switching centre, called a *telephone exchange*. Each exchange has lines to other exchanges and also to a suitably positioned *main switching centre*—which has access to all the other main switching centres in the system.

Although the shape of the telephone instrument has varied greatly over the years, the principles involved have remained constant. It contains: a device to change sound waves into electrical waves and devices to reverse the process, to call the exchange, to indicate the required number, and to attract the subscriber's attention.

Transmitter and receiver Sound waves are changed into electrical waves by a carbon granule transmitter. This consists of a thin metal disc—the *diaphragm*—and a box containing particles of carbon which are in light contact with the centre of the diaphragm. Sound waves cause the diaphragm to vibrate, which produces changes of pressure on the carbon granules with a corresponding change in the overall resistance through the granules. An electric current is passed through the carbon, and each change in pressure produces a similar change in the flow of current.

The receiver, which reverses the process, is a metal diaphragm held tightly around its circumference very close to the two poles of an electromagnet. When a changing current passes through the electromagnet coils, the pull on the diaphragm varies accordingly. The diaphragm thus vibrates, and completes the process of converting the current changes back into sound waves.

The problem inherent in Bell's early telephones was that the changes in resistance of the carbon granule transmitter with changes in sound pressure were significantly smaller than the overall resistance of the wires connecting it to the

exchange. Consequently, the varying 'signal' current component was inherently small—a problem which was exacerbated the longer the connecting wires. Edison overcame this using a *matching network* consisting essentially of a transformer. The circuit permits direct (DC) current to flow through the transmitter (necessary for its operation) while boosting the signal component, which is superimposed on this. This essential device is used in all modern telephones today.

Signalling the exchange Direct electrical current, supplied from a battery at the exchange, is used for signalling between the subscriber and the exchange. A circuit is completed between caller and exchange when the receiver is lifted, which closes a small set of contacts. In early manual systems, before the advent of automatic exchanges, the operator's attention was obtained by turning the handle of a hand generator which lit a lamp, rang a bell or operated an indicator at the exchange. The hand generator was also used to indicate the end of the conversation, this being achieved in modern telephones by simply replacing the receiver and separating the circuit contacts.

A dial is used to send the required number of signal pulses to the automatic processing equipment. The function of the dial is to interrupt the subscriber-exchange circuit a number of times corresponding to the figure dialled. As the dial spins back from the finger stop a toothed pulse wheel—attached to a common spindle—operates a pulse contact unit.

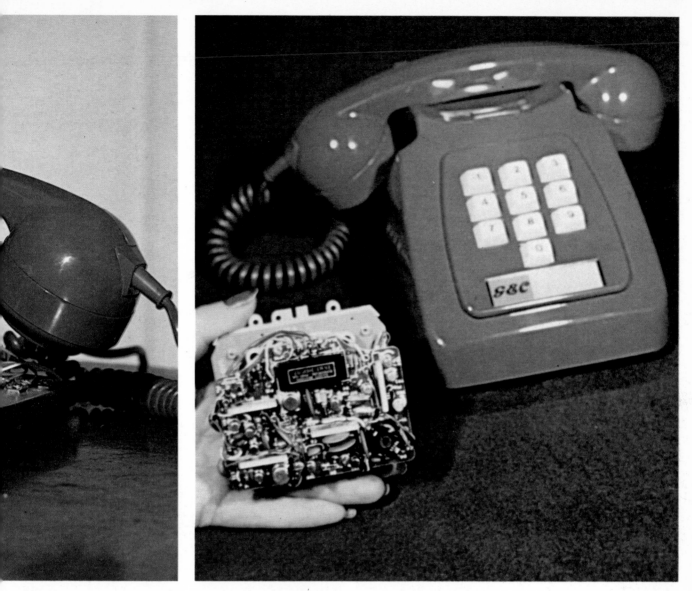

There are ten interruptions when '0' is dialled. The return speed of the finger plate—and consequently the rate at which the pulses occur—is controlled by a centrifugal type of governor which is incorporated in the dial mechanism.

In many countries, a variety of push button telephones have been developed to replace dial phones. Different approaches are used, but in general, tiny electronic circuits, oscillators and—in some types—silicon chip number storage units (needed to retain the numbers while an electro-mechanical exchange—which is much slower—goes through its line selection processes) replace the dial and pulse contact units.

The final main component of a telephone is the bell or tone-caller. Both are operated by an alternating current from the exchange. A bell is operated by pivoted armature with a hammer-end rocking in sympathy with the alternating current passing through two adjacent electromagnet coils. The 'warbler' type of telephone uses an oscillator circuit with amplifier and small speaker.

Distribution systems Originally all telephones were connected to the exchange by bare copper, bronze, or galvanized iron wires carried on insulators fixed to poles. Overhead wires have numerous disadvantages: the effects of the weather necessitate using strong wires many times larger than is needed just to carry the current, and overhead wires are exposed to inductive action from nearby high voltage equipment and from atmospheric electricity. They are also

Above left: a typical telephone handset. The two wires connecting this to the exchange handle both the speech signals and the pulses necessary to control the exchange switching equipment.
Above: a modern pushbutton telephone. The caller keys in the required number as fast as he likes. Integrated circuits containing 1200 semiconductor devices in about 0.01 square inch (0.065 cm²) can store an 18 digit number and transmit it to the exchange at a speed suitable for the system. Strowger equipment is very slow.

unsightly and expensive.

Nowadays little overhead wiring is installed. Current practice is to use large, multi-pair cables running underground from the exchange. The cables have either paper or polyethylene insulation around aluminium or cadmium copper conductors. The cables may be sheathed in lead or polyethylene covered, and may have from one to 4800 pairs of wires in them.

The large cables split into successively smaller cables, culminating in *distribution points*. From these, single pair cables run to the subscribers' telephones. In areas where the telephone density is high, the distribution points are poles fed by an underground cable carrying, usually, 10 to 15 pairs of wires; overhead wires radiate from the top of the pole. These wires may be either a bare cadmium-copper alloy or *drop wires*, that is, a pair of copper-plated steel conductors insulated with PVC. In other areas each telephone is served directly by a one-pair underground cable from the distribution point.

Exchanges The major cables run underground into exchanges and are terminated on distribution frames which split the cables into their component pairs of wires, for connection to the exchange equipment. Each national telephone system varies in its details compared with others, though most exchanges in developed countries are now automatic—Britain has no manual exchanges, for example. At manual exchanges the telephone wires are connected to switchboards. By means of cords with plugs at each end, the operator can link any two telephone circuits together. She is alerted by a warning light, illuminated when the circuit is completed by the caller lifting his receiver.

The first important step in replacing the operator was the invention in 1889 of an electromechanical selector switch by Almon B Strowger, a Kansas City undertaker. He conceived the idea of a many-positioned switch at the exchange, operated by pulses produced remotely by the subscriber's telephone dial. Switches using Strowger's principles, known as *selectors*, are used in the vast majority of the UK's 6000 exchanges, and Strowger-type exchanges formed the basis of automatic systems in other countries for many years. The following description refers particularly to the British system. American terminology is different: for example, Strowger's system is called 'step by step'.

A selector in its simplest form consists of a movable set of contacting arms, known as a *wiper assembly*, associated with

Right: the frames in the Bahrain telephone exchange. Every telephone has its own pair of wires and can be connected to any other telephone or exchange. Opposite page: telephones are connected either directly to a local exchange or via a private exchange. The local exchanges are designed to suit local conditions and may vary in size from 600 to 20,000 subscribers. Long distance telephone links combine up to 2700 conversations in one coaxial cable using multiplexing equipment or can be transmitted over a microwave system.

a fixed set of contacts known as a *contact bank*. The next step was the invention of *trunk-hunting switches*, a modification of the Strowger switch. When assigned to a particular call it automatically steps along a bank of trunk terminations until it finds a free one; that trunk line is then included in the call path until the end of the conversation.

When a subscriber lifts his receiver, completing the subscriber-exchange circuit, a rotating switch (called a *uniselector*) associated with that line starts searching for an unoccupied Strowger switch, the *first selector*. When a free first selector is found, the uniselector stops and a dialling tone is connected automatically to the caller's telephone. After the uniselector, *two-motion selectors* are used. The wipers of this type of selector can be moved in two planes. The selectors can have more than one contact bank, each of which is composed of ten horizontal arc-shaped layers, known as *levels*, with ten equally spaced sets of contacts on each level. The banks are positioned so that the levels are

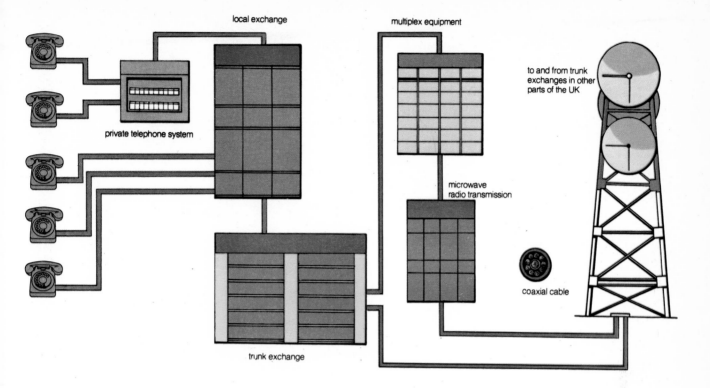

local exchange

multiplex equipment

private telephone system

microwave
radio transmission

to and from trunk
exchanges in other
parts of the UK

coaxial cable

trunk exchange

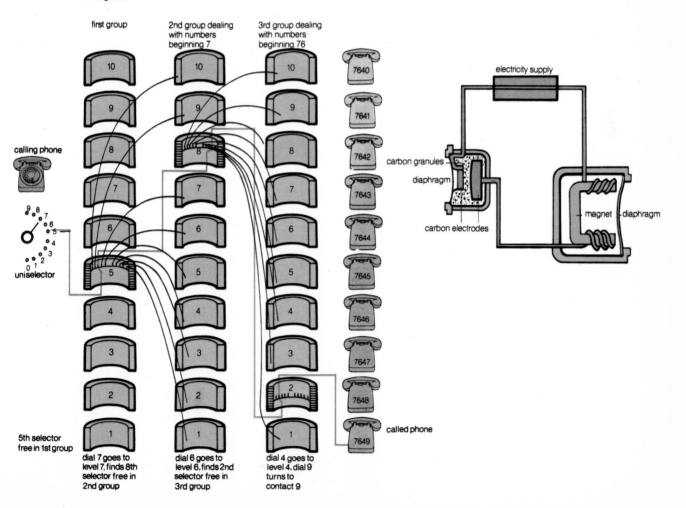

dialling 7649

first group

2nd group dealing
with numbers
beginning 7

3rd group dealing
with numbers
beginning 76

calling phone

uniselector

electricity supply

carbon granules

diaphragm

carbon electrodes

magnet   diaphragm

7640

7641

7642

7643

7644

7645

7646

7647

7648

called phone

7649

5th selector
free in 1st group

dial 7 goes to
level 7, finds 8th
selector free in
2nd group

dial 6 goes to
level 6, finds 2nd
selector free in
3rd group

dial 4 goes to
level 4, dial 9
turns to
contact 9

concentric with the shaft which carries the wipers, so the sets of wipers can be positioned on any set of contacts by two movements of the wiper-shaft—a vertical stepping motion and a rotary stepping motion. The shaft is motivated by a ratchet and pawl system. If, say, 3487 is the number to be called, the first digit dialled—3—lifts the contact arm of the first selector switch to the third row of contacts. The contacts in this row are connected, by trunk lines (these refer to lines within the exchange and not those between main exchanges), to the second selectors. The first selector moves along the third row until it reaches a contact point that is linked to an available second selector. If no contact is made, an 'engaged' tone is sent to the caller. If a free trunk is found, control passes to the wiper of the second selector, and it rises to the fourth row (4 was the second number dialled) and goes through the same process as in the first selector.

The process moves on to the final selector, which carries the connections to the subscribers' lines. The third number dialled lifts the wiper by the dialled number of steps (8) and the fourth turns it clockwise (by 7) to reach the contact corresponding to the number 3487. The final selector control circuit is arranged to test the called subscriber's line. If the line is not engaged, an alternating ringing current is sent to work the bell, and a ringing tone is returned to the caller. If the line is engaged, a busy tone is returned to the caller and the call is not recorded on the meter. When the caller replaces his receiver at the end of the call, the selectors restore to normal.

In theory an exchange like this should give a subscriber direct access to 9999 other lines (ten on the first selector times ten on the second times one hundred on the final, less the caller's own line), but in practice the capacity is reduced. The top level of the first selectors is connected to equipment allowing subscribers to dial their own long distance calls (subscriber trunk dialling, or STD). The code '100' is reserved for calls to the operator, and the necessary equipment to route the call is connected to level one of the first selectors. (In America, 'O' is reserved for calls to the operator, and subscriber trunk dialling is called direct distance dialling, or DDD.)

Another level (eight) of the first selectors is connected to second selectors giving access to nearby exchanges, while level nine of the first selectors leads to second selectors with access to exchanges a little further away than those off level eight. Sometimes third selectors, connected to levels of second selectors, are required to give extra outgoing routes if there are a lot of surrounding exchanges. Further levels are reserved for information service final selectors and the emergency service. The provision of these direct dialling and service facilities means that at least four levels of the first selectors cannot be used for subscribers' line switching, so the capacity is reduced to around 6000 lines.

The circuits coming in from other exchanges and from the operator's switchboard terminate directly on first selectors which have access to both the subscribers on that exchange and certain of the outgoing routes to other exchanges. Access to the outgoing routes is provided to allow the exchange to act as a switch on calls between other exchanges, thus removing the need to have direct circuits between every exchange. The dialling code for such a connection will then consist of a code to route the caller through his nearby exchange to the intermediate exchange, and another code to complete the routing from the intermediate exchange to the final exchange he needs.

The country is divided into groups of exchanges with one exchange nominated as the *main exchange* (switching centre) of the group. There are at present two group numbering schemes in use. The older one involves dialling a different code for the same exchange from different points in the group, but in the newer scheme, the same five or six digit number is dialled from anywhere in the group. Unwanted

digits are absorbed within the equipment. The exchanges close to the main exchange are usually called *satellite exchanges* and may be fully interconnected, as well as with the main exchange. To obtain subscribers on other exchanges with which there is no direct connection, the call is routed via the main exchange (it is used as a *tandem exchange*). A tandem exchange can be considered as a collecting point for small amounts of traffic from other exchanges which do not justify a direct route to a particular distant exchange. Exchanges further away from the main exchange— *minor exchanges*—have junctions to the main exchange and to nearby exchanges, but are not fully interconnected. Each main exchange will be, or will have access to, *a group switching centre* (GSC) which is part of the STD network. Each GSC is fully interconnected with all adjacent GSCs and either directly or via another GSC to almost all GSCs'.

Common control systems Although the majority of British exchanges work on the Strowger principle, there are two other types in use elsewhere, and which are replacing Strowger exchanges in the UK. They use *common control systems*, whereby information about the call is first passed to a central control point which processes it and selects the path through the exchange the call shall take. Not until the control equipment has found and reserved a free path do any of the switches operate.

One form of common control is the *crossbar system*. The crossbar switch consists of a matrix of relays, each having several springsets which are actuated when the relay armature operates. Each caller has access to a number of outlets and wired logic ensures that no two callers can be connected to the same outlet. The springsets form a network of fixed precious metal contacts from which a path is selected electrically. Crossbar has fewer working parts than Strowger, is faster and operates more quietly. It is widely used in Europe and the USA, but another, even more advanced, system is being developed which uses *reed relays* and in the

Above, left to right: 1. A typical multi-pair cable may contain up to 4800 pairs. This is laid underground. It is terminated on a distribution frame in the exchange and split into smaller cables en route which terminate at distribution points. From here single pairs connect private phones.
2. Testing uniselectors used in 'step-by-step' (or Strowger) exchange equipment. In a busy exchange, each incoming telephone line has its own uniselector. This 'hunts' for a free first selector (a two-motion selector). When one is found, a dial tone is sent to the caller and he can start dialling the required number.
3. A reed relay/electronic exchange. Electronic techniques are used to control the reed relay functions. Like crossbar exchanges, these use common control systems—a wired central program unit coordinates all the exchange functions. Such systems are reliable, fast and offer much greater flexibility than crossbar or Strowger. Their modular construction is adaptable to particular requirements.
4. A reed relay switching matrix. This particular matrix is used in a 600-line private exchange.

Below: an electromechanical switching matrix for a crossbar telephone exchange.

long term this is likely to replace both.

A reed relay consists of reed inserts operated by an electro-magnet; each reed insert consists of two flat springs—the reeds—made of ferromagnetic material and sealed in the ends of a glass tube. The glass tube is filled with an inert atmosphere to prevent contamination of the contact surfaces. When current flows through the electromagnet, the reed ends attract each other and an electrical path is made through them.

When a subscriber originates a call, the calling loop is detected by the subscriber's line unit, which signals to the control equipment that a call is being originated. The control equipment then sets up a path from the line unit through the reed relay switching network and a selected supervisory relay set to one of a number of registers. When the caller dials the number it is kept in storage reed relays which stay open in the register while the control equipment checks that the required line is not busy or out of order. If the line is free, the control equipment selects and switches a path from the called subscriber to the supervisory relay set and checks that the connection is properly established. The supervisory set applies ringing current to the called subscriber's line and returns a ring tone to the caller.

The register and control equipment are released to deal with other calls, the control of the connection being left to the supervisory relay set. The control equipment can deal with only one connection at a time, but it works so fast that this is no handicap. Reed relays also seem to produce fewer faults than Strowger or crossbar.

Other developments Whatever type of exchange is used, there are problems in sending electrical currents over long distances between them. In the early days of the telephone very thick—and unwieldly and expensive—cables were used, which created less resistance and allowed the current to flow more freely. Now long distance telephone circuits are passed through a series of repeaters, which contain amplifiers for each direction of conversation. By using amplitude modulation many different calls can be transmitted on different *carrier frequencies* within the same cable (which can be multipair or coaxial).

The current is generally amplified by transistors. In ordinary land cables, separate amplifiers are used to avoid 'leakage' of signals between wires, but on *submarine* cables a more complex two-way amplifier is used.

For short distances, a new system which can carry up to 30 telephone circuits simultaneously on two pairs of wires uses pulse code modulations. The principle of this system lies in converting the analog (continuously varying) signal into a series of pulses. The pulses represent a binary code of the amplitude of the original signal, sampled (that is, inspected) at a frequency twice the maximum frequency of the signal that needs be transmitted. Because intelligible speech does not require high fidelity, the maximum frequency that need be transmitted is only 4 kHz; consequently, the sampling rate need only be 8 kHz (8000 samples per second). Coaxial cables, however, have been designed which will handle frequencies in the megahertz (MHz) region and so handle many PCM signals simultaneously. At the receiver a PCM decoder converts the pulses back into an analog signal.

A further alternative is radio. A radio microwave network, in which high frequency radio waves are beamed between line-of-sight directional dish or horn shaped aerials links major centres. Microwaves are also used for long distance telephone calls which travel via communications satellites. Subscribers are able to contact almost every country in the world via satellites strategically sited over the Indian, Pacific and Atlantic Oceans.

TELESCOPE

The idea of enlarging the apparent size of distant objects to make the details more easily visible was achieved around the start of the 17th century, when several people independently discovered the principle of the telescope. By 1609 telescopes could be bought on the open market, and in the following year Galileo turned one to the sky and thereby inaugurated a revolution in astronomy. Modern applications of telescopes include gunsights and other military purposes, use in surveying instruments and laboratory spectroscopes, as well as their obvious role in astronomical research.

All telescopes have essentially two parts: the *objective*, which may be a curved mirror (in the *reflecting* telescopes, used only for astronomy) or a lens (in *refracting* telescopes, which can be used for all purposes), and the *eyepiece*, consisting of a lens or a group of lenses. The objective forms an image of the distant object at its focal point, and this image is then magnified by the eyepiece, which acts as a simple magnifying glass. For many astronomical purposes, a photograph of the object is required, and in this case the eyepiece is dispensed with. The film is placed at the focal point of the objective where the image is focused, so that the arrangement is exactly the same as that of an ordinary camera.

Magnification   When the telescope is in 'normal adjustment', the focal point of the eyepiece coincides with the intermediate image produced by the objective, and the final image seen by the eye is apparently at a very great distance ('at infinity'). The *magnification* of the telescope is the ratio of the apparent size of the final image to the apparent size of the object itself, and for normal adjustment this is equal to the focal length of the objective divided by that of the eyepiece. Different magnifications can be achieved with the same objective by using eyepieces of various focal lengths; for example, a small astronomical telescope with an objective of focal length 48 inches (122 cm) would give a magnification of 48 with a 1 inch (2.5 cm) focal length eyepiece, and a magnification of 96 with a $\frac{1}{2}$ inch (1.3 cm) eyepiece.

With any particular eyepiece the magnification can be increased by moving it nearer the objective, but as this brings the final image closer to the eye it can result in eyestrain when viewing for a length of time. The size, or *aperture*, of the objective determines the maximum magnification which can be usefully employed in a telescope. A small objective does not give as much detail in the intermediate image as a larger objective, because of the diffraction of light at the edge of the objective, so that a high-powered eyepiece will not show any more detail in the final image than a moderate powered one. A magnification of 200, for example, is perfectly adequate for a 6 inch (15 cm) aperture telescope, and the use of a higher power may magnify the image to such an extent that its dimness makes the details difficult to see.

The brightness of the final image also depends on the brightness of the intermediate image, and this decreases as the square of the *focal ratio*, which is the focal length of the objective divided by its aperture. The reason for this is that a longer focal length increases the size of the intermediate image and hence makes it dimmer, while a larger aperture collects more light and hence produces a brighter image. For observing stars, which appear more or less as points of light even in the largest telescopes, the focal length is not important, and the faintest star which can be seen depends only on the aperture. It is for this reason that astronomers have always built successively larger telescopes which can reach fainter and thus more distant objects.

Refracting telescopes   The earliest telescopes used a lens as the objective, but it was soon found that the images suffered from *aberrations* or imperfections. The most serious aberration of these *refracting* telescopes was the production of false colour around the images, and until the introduction of the achromatic lens by John Dollond in 1758, this was a

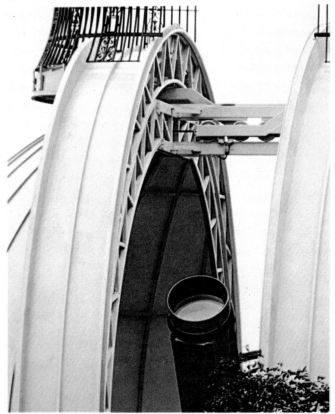

serious hindrance to astronomical research. An achromatic lens has two components of different kinds of glass, overcoming much of the false colour. The modern Keplerian, or astronomical, refracting telescope has an achromatic objective.

Simple refracting telescopes produce an inverted image of the object. When this image is magnified by the eyepiece it remains inverted, and so this type of telescope is not convenient for viewing terrestrial objects. One way of *erecting* the image (turning it right way up) is to incorporate a pair of lenses just behind the intermediate image, which erect it without producing any magnification; this arrangement leads to a longer and hence heavier tube. Alternatively, a pair of prisms can be used, as in prismatic binoculars. Both these arrangements, however, lead to an appreciable loss of light by reflection at the extra glass faces, and so they are only used for terrestrial purposes and not for astronomy.

Reflecting telescopes   Before the advent of achromatic refractors, the possibility of using a curved mirror as an objective had been investigated by James Gregory and Sir Isaac Newton. A major problem with reflecting telescopes is that the image is formed in front of the mirror, in the path of the incoming light. Only in the largest astronomical telescopes can an observer actually sit at the *prime focus* without blocking off a large amount of the incident light. Newton's solution was to place a small flat mirror in the tube at 45°, to reflect the light from the objective, or *primary* mirror, out of the side of the tube before it forms an image. This is then studied with an eyepiece in the usual way. The Newtonian reflector is popular with amateur astronomers because the flat secondary mirror is cheap, and this design does not

terrestrial telescope

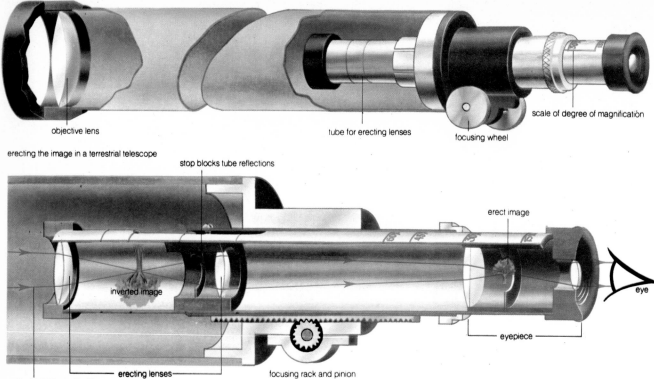

objective lens

tube for erecting lenses

focusing wheel

scale of degree of magnification

erecting the image in a terrestrial telescope

stop blocks tube reflections

erect image

inverted image

eye

eyepiece

path of light from objective

erecting lenses

focusing rack and pinion

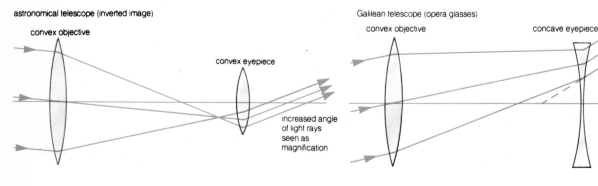

astronomical telescope (inverted image)

convex objective

convex eyepiece

increased angle
of light rays
seen as
magnification

Galilean telescope (opera glasses)

convex objective

concave eyepiece

image still erect

Above: at top is a common design of terrestrial telescope. It has an achromatic objective—essential for bright images free from false colour—and magnification variable from 15 to 60 times. A detail of the eyepiece end is shown at centre: the power is varied by altering the gap between the erecting lenses. As the telescope must be refocused as the magnification is changed, this is not a zoom lens.
At bottom, the principles of the astronomical and Galilean designs.

require a hole to be cut in the primary. A later development, the Cassegrain reflector, uses a convex secondary mirror to reflect the light back down the tube and through a hole in the centre of the primary mirror. The shape of the secondary gives the telescope a much longer effective focal length.

The mirrors for reflecting telescopes have to reflect from their front surfaces, to avoid double reflections, which makes them liable to tarnish. In addition, the two mirrors can get out of alignment much more easily than one lens, so reflecting telescopes are rather delicate and are used only for astronomy. Refractors, with tubes sealed at one end by the objective, are more robust and can therefore be used for everyday purposes, with erecting lenses, as well as for astronomy if necessary.

Galileo's telescope The telescope used by Galileo was of a rather different design from those described so far. The light from the objective lens is intercepted by the eyepiece, in this case a *concave* lens, before it forms an image, and this arrangement gives an upright final image. The field of view of a Galilean telescope is, however, small. Its principal use is in opera glasses, where the relatively short tube and erect image are important assets.

TELEVISION

The first successful television transmissions were carried out by John Logie Baird between 1928 and 1935, using the BBC's medium wave transmitters. In this system the pictures were composed of only 30 lines and small details could not be reproduced: this was known as a *low-definition* system. Several hundred lines are needed to give definition comparable with that of 16 mm or 35 mm film.

The world's first *high-definition* service (using 405 lines) was launched by the BBC in the autumn of 1936 from Alexandra Palace in North London. This service has remained in operation since then, except for an interruption between 1939 and 1946 caused by World War II. In 1964 the BBC introduced its second television programme using the 625 line system: this gave better definition than the 405 line system and the new line standards agreed with those used by many other broadcasting authorities, so facilitating the international exchange of television programmes.

The first regular colour television service began in the USA in January 1954, in Japan in 1960 and in Great Britain in 1967. In Great Britain in 1975 there were nearly 17.5 million television receivers (of which 7 million were colour) and colour transmissions were available to 95% of the population. In the USA in 1976, there were 130 million TV sets, more than the number of bathtubs. This development has taken place in less than 50 years since Baird's first successful transmission.

Black and white television   Television is a highly complex subject, and in the following description of the fundamental principles a number of simplifications have been made in the interests of clarity and brevity.

Close examination of a newspaper picture shows it to be composed of a large number of dots arranged in a regular geometric pattern. The dots are very small in almost-white areas of the picture but are large and almost touching in very dark areas, intermediate sizes of dot giving the various

Below: part of Baird's first television apparatus. In all the earliest systems, scanning was done by rotating apertured discs in front of the object to be televised. Baird's pictures were formed of 30 lines scanned ten times per second.

shades of grey between black and white. At normal viewing distances individual dots are too small to be seen and the picture appears continuous. An alternative way of reproducing a photograph would be to have dots of constant size but of various degrees of grey between black and white.

This technique of dividing the picture into very small *elements* is used in television. Information about the degree of grey (or alternatively of the degree of brightness) of each element is sent to the receiving end where it is used to build up a reproduction of the original scene.

Television camera   The first process is the creation of an optical image of the scene to be televised. This is normally formed by the zoom lens of a television camera and the image is focused on the *target* of the *camera tube*. The target is of *photo-emissive* or *photo-conductive* material and generates a pattern of electrical voltages on the rear surface, the voltage at any point being proportional to the brightness of the corresponding point in the optical image on the front surface of the target.

The target is scanned by an electron beam generated in the camera tube, the beam moving over the rear surface of the target in precisely the same way that the eye moves over a printed page in reading. Thus the beam moves from left to right across the image, returns rapidly to the left again and scans a second line immediately under the previous line and continues in this manner until the bottom of the image is reached. The beam now returns to the top of the image and begins the process again.

arrangement of zoom lens and camera tube in a black-and-white television camera

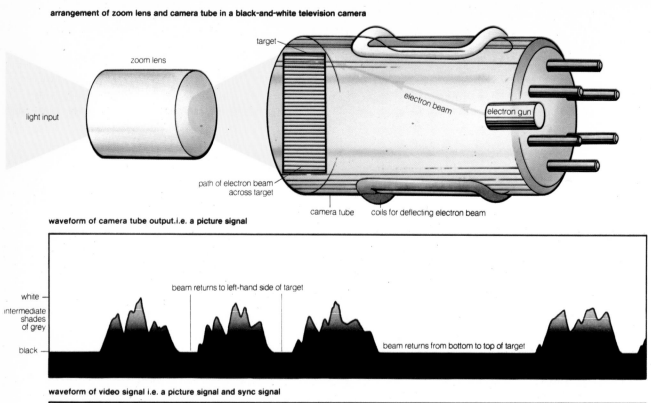

light input

zoom lens

target

electron beam

electron gun

path of electron beam across target

camera tube

coils for deflecting electron beam

waveform of camera tube output.i.e. a picture signal

white

intermediate shades of grey

black

beam returns to left-hand side of target

beam returns from bottom to top of target

waveform of video signal i.e. a picture signal and sync signal

white

black

line sync signals

field sync signal

line sync signal

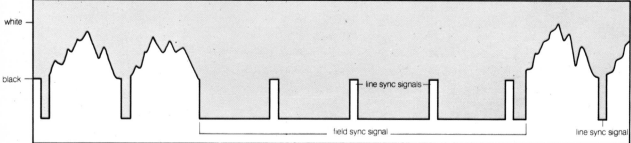

simplified block diagram for a black-and-white television receiver

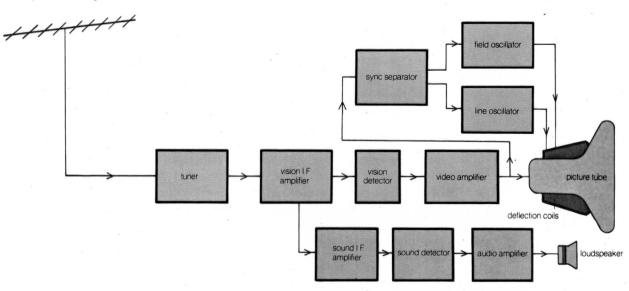

sync separator

field oscillator

line oscillator

tuner

vision I F amplifier

vision detector

video amplifier

picture tube

deflection coils

sound I F amplifier

sound detector

audio amplifier

loudspeaker

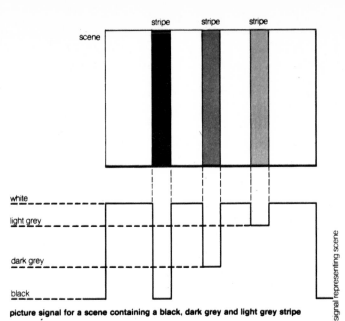

picture signal for a scene containing a black, dark grey and light grey stripe

picture signal representing scene

two types of amplitude modulation used for video signals

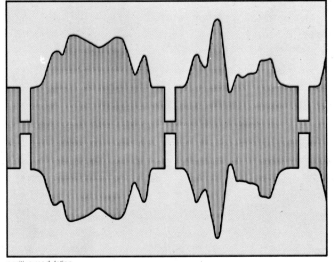

positive modulation

negative modulation

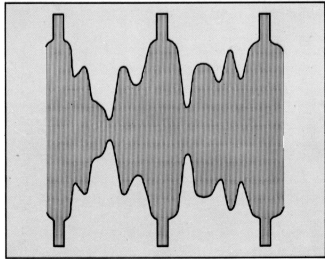

The greater the number of lines used in one complete scan of the target, the higher is the definition of the television system. In Baird's system only 30 lines were used and the definition was poor: in most European countries 625 lines are used and the definition is very good. Each complete scan of the target is accomplished (in Europe) in 1 25th second so that 25 complete pictures are transmitted every second (compared with 24 in movie film). In the USA the picture is composed of 525 lines and 30 pictures are transmitted per second. (In practice, each picture is scanned twice, in alternate stripes like the red and white parts of the American flag, so that a rate of 25 pictures a second involves 50 scans a second, but this is purely a device to reduce flicker.)

As the target is scanned the voltages representing the detail in the optical image are neutralized by the scanning beam and are, in effect, transferred to the output terminal of the camera tube. Thus the tube output consists of a varying voltage which represents the way in which the brightness varies along the scanning lines of the optical image: this varying voltage is known as a *picture signal*. It is interrupted for a brief interval every time the beam returns to the left hand side of the image and, for a longer interval, every time it returns to the top of the image.

To show how the detail in a picture can be represented by such a signal consider a very simple picture consisting of a uniform white background with three separate vertical stripes, one black, the second dark grey and the third light grey. Such a picture has no detail in the vertical direction and the detail along all the scanning (horizontal) lines is the same. Each line begins with a period of white level, then black level when the first stripe is encountered, white level again, then dark grey level at the second stripe, white level again, light grey level at the third stripe and finally white level. Thus the picture signal for this simple scene has a waveform consisting of a series of pulses. The signal is a maximum during white areas, a minimum during black areas with two levels in between for light and dark grey.

In a normal detailed television picture the picture signal is different for each scanning line. Moreover if there is movement in the picture the waveform for any line varies with time.

Picture tube (CRT) To reproduce the picture at the receiver (the television set) a *picture tube* (or cathode ray tube—CRT) is employed which also contains an electron beam focused on a screen, which scans it in exactly the same pattern as the camera tube beam. The picture tube screen has a uniform coating of a material which emits light when struck by the beam, and if the beam density is controlled by the picture signal from the camera, a reproduction of the original scene is built up on the picture tube screen. Complete pictures are received at the rate of 25 per second and hence movement in the original scene is portrayed as effectively as in the movies. It is essential, however, in order

Left: when light from a scene hits the camera target, the target becomes positively charged. An electron beam scans the target in a regular manner and converts this charged pattern into an electrical signal—this is the picture signal. The picture signal varies from its maximum value (white) to its minimum value (black). With line sync and field sync signals added this becomes the video signal containing all the information to control a picture tube (CRT) and reproduce the original scene. For broadcasting, however, the video signal must first be amplitude modulated. The choice of modulation is important when considering colour transmissions. At the receiver, line sync and field signals are separated from the picture signal and control the scanning pattern in the picture tube.

to obtain a satisfactory image at the receiver, that the beam in the picture tube should be exactly in step with that at the camera tube, that is at every moment it should be exactly at the same point of the same line as the camera tube beam. This precision of movement is achieved by the use of *synchronizing signals*. A signal known as the *line synchronizing signal* (abbreviated to *line sync* signal) is sent every time the scanning beam in the camera tube reaches the right hand side of the image and this signal is used at the receiver to deflect the picture tube beam to the left hand side of the screen. Similarly a signal known as the *field synchronizing signal* (abbreviated to *field sync* signal) is sent every time the scanning beam in the camera tube reaches the bottom of the image and this signal is used at the receiver to deflect the picture tube beam to the top of the screen.

To avoid the need to send the sync signals separately to the receiving end, they are combined with the picture signal. By this method the sync signals represent 'blacker than black' and so have no visible effect on the receiver screen. Because the field sync signal occupies the time of several lines, not all the 625 lines appear on the picture tube screen; it also means that the line sync signals must be kept going throughout the duration of the field sync signal so that the receiver line deflection circuits are not interrupted. The combined picture and sync signal is known as a *video signal* and this is the signal which is sent from the camera to the picture tube, either directly by line as in *closed-circuit television* or via radio waves as in the more usual television broadcasting.

Amplitude modulation (AM) is universally used in television broadcasting for transmitting the video signal. In some systems, for example, the British 405 system, *positive modulation* is used: in this the *carrier* amplitude is a maximum for white signals and a minimum for sync signals. In the 625 line system, however, *negative modulation* is used and here the carrier amplitude is a maximum for the sync signals and a minimum for white signals.

Sound accompaniment In many closed-circuit applications of television, the picture provides all the information that is required and there is no need for any sound accompaniment. In television broadcasting, however, sound is essential and is radiated from a separate transmitter which commonly uses the same transmitting aerial as the vision transmitter. Sometimes, as in the British 405 line system, the sound is radiated by amplitude modulation but for the 625 line system, frequency modulation (FM) is used for the sound accompaniment. One advantage of using FM is that the *inter-carrier method* of sound reception can be used in the receiver.

Television channels The sound carrier frequency is placed near that of the vision so that both signals can be amplified simultaneously in the early stages of receivers and the two signals are together regarded as constituting a *television channel*. The *frequency band* occupied by a channel depends on the spacing of the sound and vision carriers and on the frequency band of the vision signal. For the 405 line system channels are spaced at 5 MHz intervals and for the 625 line system at 8 MHz intervals. Sound radio transmissions are usually identified by their *wavelength* (in the medium wave band) or *carrier frequency* (in the very high frequency, VHF, band) but television transmissions are known by their channel numbers.

Black and white receiver Various operations must be carried out on the video signal and in practice may be achieved by one or more valves (vacuum tubes), one or more transistors or by integrated circuits. Early receivers used valves only and later receivers employed a mixture of valves and transistors. Most modern receivers use transistors only or a mixture of transistors and integrated circuits.

Television receivers operate on the *superheterodyne principle*, that is, most of the amplification and *selectivity* of the receiver is provided by an amplifier known as the *inter-*

Above, top: a television camera fitted with an iconoscope tube. The camera lens focuses the image to be transmitted on to a signal plate situated in the glass envelope. The signal plate is scanned by an electron beam generated in the tube leading downwardly and to the right of the main part of the glass envelope. Iconoscope tubes are less sensitive than the newer image orthicon tubes.
Above: a modern colour television camera.

Left: the control room of a modern television studio, showing the monitor screens for the various cameras. During an international telecast of something as complicated as the Olympic games, there may be many more cameras than this in use at any given time, and the producers of the locally televised version must be editors as they work, able to switch from one camera to another at any moment, and correlating the audio signal as well.

mediate *frequency* (IF) amplifier. The carrier frequency of every signal selected by the tuner is changed to the IF value and applied to the IF amplifier. The tuner contains a *frequency changer* stage and a preceding carrier frequency amplifier known as a *radio frequency* (RF) amplifier.

Channel selection in the tuner is commonly controlled by push buttons or multi-position rotary selectors, but continuous tuning is sometimes provided, particularly in portable receivers. The video and sound signals for the selected channel are amplified together in the early stages of the receiver but are divorced later and handled by separate circuits. The video signal is abstracted from the modulated carrier by the *vision detector* and, after further amplification, is applied to the picture tube. The sync signals are removed from the video signal in the *sync-separator* stage and the line sync signals are applied to the line *oscillator* to lock it at the correct frequency. The output of the line oscillator is fed to deflection coils clamped around the neck of the picture tube and these are responsible for horizontal scanning. The frequency of the line oscillator can be adjusted by a control (called line hold or horizontal hold) to bring it into the range in which locking occurs. The field oscillator (responsible for vertical deflection) is similarly locked, the frequency control being labelled field (or frame) hold or vertical hold. The sound signal (assumed amplitude modulated) is abstracted from the modulated carrier by the sound detector and, after amplification, is applied to the loudspeaker.

Colour television Countries which have only recently started their television service can profit from recent advances and can begin with colour: South Africa is an example of such a country. It is more usual, however, for colour television to be introduced into a country which has already had a black and white service for many years. To enable a colour service to coexist with black and white television, a number of technical requirements must be met: first, the colour transmissions must fit into the frequency band of existing black and white channels; second, black and white receivers must give a satisfactory black and white picture when tuned to colour transmissions (known as *compatibility*); and third, colour receivers must give good black and white pictures when tuned to black and white transmissions (known as *reverse compatibility*).

Colour information To satisfy these requirements the colour television system is basically a black and white one to which additional signals have been added to provide information on colour. Indeed, the picture given by a colour television receiver is fundamentally black and white with areas filled in by colour.

To enable a receiver to reproduce the correct colour for each coloured area of the image it must be given two items of information: the basic colour (or *hue*) and its strength (or *saturation*). The hue (whether it is red, yellow, green or whatever) is determined by the position of the colour in the spectrum. Saturation is a measure of the strength or weakness of the colour. If the hue is red, the colour may be crimson, pink or some intermediate shade. In other words it is the extent to which the colour is diluted by white. Crimson is a saturated colour and pink unsaturated.

Information about the hue and saturation of every coloured area of the picture must therefore be sent to the receiving end. This information is not sent directly in the form of measurements of hue and saturation but as follows.

R, G and B signals Green paint can be made by mixing blue and yellow paint, and purple by mixing blue and red. In fact by using only three primary colours it is possible, by varying the proportions of each, to produce practically all known colours. This principle is used in colour printing: it is also used in colour television and the colours chosen for television are red, green and blue (generally abbreviated to R, G and B). It is necessary, therefore, to analyze the image of the original coloured scene and for each coloured area to measure what fraction of its colour is contributed by red, green and blue.

signal representing the black and white content of the picture, as from a single tube black and white television camera and this combined signal, known as the *luminance* signal, is the basic signal transmitted in a colour television system: it is the luminance signal which is accepted and displayed by a black and white receiver tuned to a colour transmission.

Four tube camera To obtain a good luminance signal from a three tube camera, all three scanning beams must, at every instant, be scanning the same element of the same line of the optical image. This is difficult to ensure but the difficulty can be avoided by using a four tube camera.

In this the red, green and blue tubes are used only to give colour information. The fourth tube is fed with the original coloured image and produces a picture signal output as in black and white television: this output is used as the luminance signal. It is, of course, essential for all four beams to be synchronized and for all four optical images to be carefully aligned, but lack of alignment between the luminance and colour tubes does not degrade the quality of the luminance signal.

Transmission of colour information The three outputs of the colour camera tubes must be transmitted to the receiving end because a colour picture tube requires red, green and blue inputs. To transmit information about three varying quantities such as the red, green and blue content of the picture requires three separate signals. These signals need not, however, be the R, G and B picture signals themselves. Any three signals which contain R, G and B will do because from these the R, G and B signals can be obtained by algebraic operations in a circuit known as a *matrix*.

The operation of the matrix is similar to that of the algebra used in solving simultaneous equations. One signal involving R, G and B already exists as the luminance signal, usually represented by Y. Two other signals are therefore required, and the two selected are the (R-Y) and (B-Y) signals, known as *colour difference signals*. A *subcarrier* is used to transmit the two colour difference signals and a number of methods have been devised for modulating the subcarrier by two independent signals and for recovering them at the receiver. To keep the colour transmission within the channel, the colour subcarrier is located within the frequency band of the video signal and its frequency is carefully chosen and maintained to prevent it and the modulating signals from causing interference with the luminance signal.

NTSC system The first colour television system was the NTSC (National Television System Committee) introduced into the USA in 1954 and still used there and in Japan, Canada and Mexico. In this system the colour subcarrier is *amplitude-modulated* by the two colour difference signals by a method known as *quadrature modulation*.

The method involves resolving the carrier wave into two components with a 90° phase difference between them. Each component is then separately amplitude modulated by a colour difference signal. One of the features of the system is that during modulation the subcarrier itself is suppressed. This is permissible because the colour information is contained in the *sidebands*. The suppression of the subcarrier improves the quality of the compatible black and white picture by removing the fine pattern which it would otherwise produce on the receiver screen.

After modulation the two colour difference signals are combined to form the *chrominance signal*. The colour difference signals can be recovered at the receiver in a circuit known as a *quadrature detector*. This requires for its operation a reference signal very accurately locked to the subcarrier frequency. A few cycles of the subcarrier are therefore transmitted immediately after each line sync signal: this is known as the *colour burst*.

During transmission the chrominance signal is superimposed on the luminance signal. The effect of quadrature

This analysis is carried out in the colour television camera. The image is split into its red, green and blue components with *dichroic mirrors*. These are mirrors of special construction which can reflect light belonging to particular regions of the spectrum but permit the remainder to pass through unhindered. The red, green and blue images so obtained are focused on the targets of three identical camera tubes, each containing an electron beam as in the black and white system. The three beams are focused on their respective targets and scan them in exact synchronism. From the tubes three picture signals are obtained, one representing the red content of the picture, another the green content and the third the blue content. By combining these outputs, we get a

Above: a TV program in production. The man controlling the sound boom receives instructions from the control room on headphones.

Opposite page: the colour content of a scene can be recorded and reproduced if information about the red (R), green (G) and blue (B) primary colours is known. In a colour camera this information is obtained from dichroic mirrors (optical filters). For satisfactory black and white reproduction, a luminance signal (Y) is also required. This is obtained from a black and white camera tube. As Y is the sum of R, G and B, only two other independent signals need be transmitted. For convenience, these are the

colour difference signals (R-Y) and (B-Y). For transmission, the Y signal forms the basic signal, for compatibility with black and white transmissions, and the colour difference signals are combined onto a subcarrier frequency. At the receiver (R-Y), (B-Y) and Y are separated and (G-Y) is generated from these. These 3 colour signals difference are applied to the grids of the R, B and G picture guns respectively. Applying Y to all 3 cathodes creates the separate R, G and B scenes. There are two systems for combining the three scene colours—the shadow-mask and phosphor-dot type, and the Trinitron stripe type.

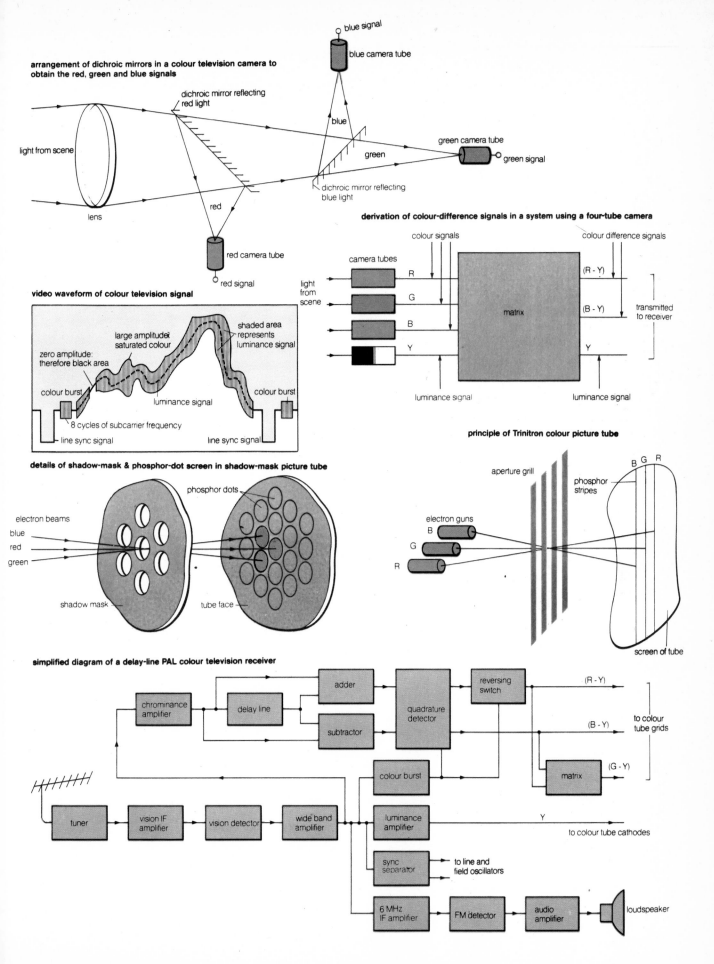

arrangement of dichroic mirrors in a colour television camera to obtain the red, green and blue signals

blue signal

blue camera tube

dichroic mirror reflecting red light

light from scene

blue

green camera tube

green

green signal

dichroic mirror reflecting blue light

lens

red

red camera tube

red signal

derivation of colour-difference signals in a system using a four-tube camera

colour signals

colour difference signals

camera tubes

light from scene

R

G

B

Y

matrix

(R - Y)

(B - Y)

Y

transmitted to receiver

luminance signal

luminance signal

luminance signal

video waveform of colour television signal

large amplitude: saturated colour

shaded area represents luminance signal

zero amplitude: therefore black area

colour burst

colour burst

luminance signal

8 cycles of subcarrier frequency

line sync signal

line sync signal

principle of Trinitron colour picture tube

aperture grill

B G R

phosphor stripes

electron guns

B

G

R

screen of tube

details of shadow-mask & phosphor-dot screen in shadow-mask picture tube

phosphor dots

electron beams

blue

red

green

shadow mask

tube face

simplified diagram of a delay-line PAL colour television receiver

chrominance amplifier

delay line

adder

subtractor

quadrature detector

reversing switch

(R - Y)

(B - Y)

to colour tube grids

colour burst

matrix

(G - Y)

tuner

vision IF amplifier

vision detector

wide band amplifier

luminance amplifier

Y

to colour tube cathodes

sync separator

to line and field oscillators

6 MHz IF amplifier

FM detector

audio amplifier

loudspeaker

modulation and the subsequent combination of the two colour difference signals is to produce a new signal at the subcarrier frequency, the amplitude and phase of which convey the colour information. The hue is represented by the phase of the chrominance signal relative to that of the colour burst, and the saturation is represented by the amplitude of the chrominance signal.

The system works well but has one disadvantage—any variations in the phase of the chrominance signal are interpreted by the receiver as changes in hue. Unwanted phase changes do occur in transmitting equipment (particularly videotape machines) and in the receiver itself. It is therefore essential in an NTSC receiver to have an overriding phase control which can be manually adjusted to give correct hues. This control is normally set to give good rendering of flesh colour (according to the viewer's preference) and all other colours are then automatically correct.

PAL (phase alternating line) system To overcome the effects on reproduced hue of unwanted phase changes in the chrominance signal, the Telefunken Laboratories in Hanover developed a system of automatic compensation which is so effective that PAL receivers do not have and do not need a hue control. This is the colour system used in the United Kingdom, Australia and most of the European countries (except those using SECAM—see below).

The method used is to reverse the polarity of the (R-Y) signal on alternate lines at the transmitting end. The reversal is achieved by an electronic switch and a similar switch (operated at the same instants) is required at the receiver to restore the original polarity.

Suppose there is an unwanted phase change which would cause the hues along a particular line of the displayed picture to be too red. Then, as a result of the polarity reversal, on the next line the reproduced hues would not be red enough. If the picture signals for these two lines are averaged, the resulting picture signal is free of phase error and if the averaged picture signal is displayed the picture so obtained is free of errors in hue. This is the technique used in PAL receivers.

To enable a comparison to be made between the picture signals for successive lines, the receiver incorporates a delay device which introduces a time lag equal to the duration of one line (64 microseconds in the 625 line system). Thus, while the picture signal for a particular line is being transmitted, that for the previous line is being delivered from the delay device. The picture signals for both lines are thus available at the same time and the receiver is able not only to produce the average of the two but to separate the (R-Y) and (B-Y) signals in a most ingenious manner described later. The delay device is usually a glass block and the delay is achieved by propagating ultrasonic waves through it. The chrominance signal (frequency 4.43 MHz) is injected into one end of the block by a piezoelectric transducer and is received from the other end via a similar transducer. The need to introduce a delay device makes a PAL colour receiver more complex and more expensive than an NTSC receiver.

SECAM system The third system of chrominance modulation is that developed in France and used in that country, in the German Democratic Republic, in Hungary, Algeria and the USSR. This is known as SECAM (système en couleurs à mémoire). The (R-Y) and (B-Y) colour difference signals are sent during alternate lines and by *frequency modulation* (FM) of the chrominance subcarrier. The SECAM receiver is simpler than a PAL receiver in that FM detectors are used to recover the colour information but an electronic switch and a line-time delay device are necessary to ensure that (R-Y) and (B-Y) signals are present simultaneously.

The system is immune from the effects of phase distortion but there is a slight loss of definition because half the colour information is not used. Moreover a SECAM picture when reproduced on a black and white receiver is not so satisfactory as from the other two systems: this is because it is not possible with SECAM to suppress the colour subcarrier as in NTSC and PAL.

Shadow-mask picture tube There are several different types of colour picture tube but the most popular is the shadow-mask type. This has three electron guns, one for each of the primary colours, arranged in delta formation, that is, at the corners of an equilateral triangle. The three electron beams are focused on the screen and scan it as in a black and white tube. Very near the screen there is a shadow mask: this is a metal plate containing about half a million small holes arranged in a regular pattern.

Associated with each hole is a group of three *phosphor dots*, known as a *triad*, on the inside of the tube face. One of these dots glows red when struck by the electron beam, another glows green and the third blue. The arrangement is such that the beam from the red gun can strike only the red dots, that from the green gun only the green dots and so on. The intensity of the three beams is controlled by the R, G and B signals generated in the receiver decoder. If there is a saturated blue area in the scene, then the red and green signals go to zero when the corresponding area of the tube face is scanned and only the blue gun operates in this area. If the area is an unsaturated blue then the red and green signals are present also as the area is scanned so as to provide the white light which desaturates the blue: the red and green colours combine with part of the blue colour to produce white which combines with the remainder of the (saturated) blue to produce light (unsaturated) blue.

The dot structure of the tube face coating is too small to be seen at normal viewing distances and the effect is that the tube face is fully filled with picture. If, however, the tube face is examined through a magnifying glass the dot structure can easily be seen.

The shadow-mask tube has three electron beams and can thus produce three images; one red, one green and one blue. Ideally these images should be perfectly superimposed so as to reproduce the colours of the original scene. A very large number of adjustments, known as *convergence* adjustments, are necessary to secure perfect registration of the three *primary* images and special *pattern generators* are used to facilitate these adjustments.

Trinitron picture tube In the Sony Trinitron tube the red, green and blue phosphors are deposited on the inner face of the tube in the form of vertical stripes, several hundred in number. In place of the shadow mask there is a metal grill with vertical slots, one for each group of three phosphor stripes. The electron gun is required to produce three beams in a horizontal line: this greatly simplifies the design which consists of a single gun (compared with three in the shadow-mask tube) with three cathodes arranged in line. This tube can be made very compact and is thus suitable for use in portable colour television receivers. Convergence adjustments are simpler than for a shadow-mask tube. The system is, however, more suitable for small screens than large ones because above a certain size the vertical phosphor stripes become obtrusively visible.

Colour television receiver The PAL colour television receiver (incorporating a delay line) has a tuner, vision intermediate frequency (IF) amplifier, vision detector, sync separator, line and field deflection circuits as in a black and white receiver. Sound is, however, transmitted by frequency modulation of a carrier displaced by 6 MHz from the vision carrier: this makes possible the *inter-carrier* method of sound reception. All the circuits before the vision detector and the amplifier following it have a frequency band wide enough to accept the vision and sound signals. As a result of the detection process the sound signal emerges from the detector as a frequency-modulated carrier of 6 MHz. A sound IF amplifier, tuned to 6 MHz, can thus be used after the wideband amplifier to select the sound signal. An FM detector followed by an audio amplifier are then necessary to provide a signal for the loudspeaker.

From the wideband amplifier the chrominance signal is selected by a filter and is applied to the delay line. The circuit surrounding the delay line is interesting because it provides a most effective method of separating the two colour difference signals while they are still in the form of modulated carriers. The (R-Y) signal is reversed in polarity on alternate lines and so if the direct signal is added to that which has passed through the delay line, the two signals cancel, leaving only the (B-Y) signal. The (B-Y) signal is not subjected to polarity reversal and thus the direct (B-Y) signal is in phase with that which has traversed the delay device. Cancellation of the (B-Y) signals can thus be achieved by subtracting the direct and delayed signals, leaving only the (R-Y) signal.

Associated with the quadrature detector is the (R-Y) polarity-reversing switch which eliminates any phase errors. The switch is operated from the colour burst as indicated. The only effect of this phase-error cancelling circuit is a small, usually imperceptible, reduction in colour saturation in reproduced pictures.

It is possible to use a matrix which accepts the (R-Y), (B-Y) and Y signals and derives from them the corresponding R, G and B signals which can be applied directly to the input of the colour picture tube. It is more usual, however, to design the matrix to produce the third colour difference signal (G-Y). The picture tube *grids* are now fed with the (R-Y), G-Y) and (B-Y) signals and the *cathodes* are fed with the Y signal. The Y signal is thus common to each grid and cathode and so cancels: in effect, therefore, the grids are fed with the R, G and B signals as required. This elaborate method of driving the tube is adopted in the interests of reverse compatibility. If the colour receiver is tuned to a black and white transmission, a circuit detects the absence of the subcarrier and 'kills' the colour difference circuits so that no signals are fed to the picture tube grids. The Y (luminance) signal is still applied to the tube cathodes as in a black and white receiver: thus the colour receiver reproduces a black and white signal in black and white.

THERMOS FLASK

The Thermos flask, also known as the Dewar flask or vacuum flask, was invented by the Scottish chemist and physicist Sir James Dewar in the 1890s for the purpose of storing liquid gases at very low temperatures. 'Thermos flask' is a proprietary name applied to a form of vacuum flask protected by a casing. The basic function of the vacuum flask is to thermally insulate the contents and prevent heat flowing either in or out.

The flask is a glass vessel with double walls, the space between which is evacuated, and it is primarily this feature which hinders the transfer of heat to or from the container. The vacuum is created by pumping the air in the wall cavity out through a glass tube, which is an integral part of the outer wall during manufacture. This tube is then sealed by being melted when the desired degree of vacuum has been achieved.

There are only three ways heat can be transferred, namely *convection*, *conduction*, and *radiation*, and a vacuum effectively stops the first two of these as it is a non-conductor. Radiation is reduced to a minimum by silvering the glass, generally on the two internal faces, so that radiant heat waves are reflected. The chief path by which heat can be communicated either to or from the interior of the inner vessel is at the vessel's neck, which is the only junction of the walls, and it is consequently made as small as possible.

Uses There is a wide variety of uses to which the vacuum flask is put, its chief scientific uses being in the field of low temperature studies where it is used to store liquid gases at very low temperatures or to reduce apparatus to very low temperatures by immersing it in liquid nitrogen contained in a vacuum flask.

Probably the best known use is the flask in which hot (or cold) drinks are stored, and if used properly a glass flask will keep a fluid near its original temperature for a very long period. For best results, the flask should always be kept upright so that the fluid does not come into contact with the stopper. If this happens, the heat transfer to the stopper is greatly accelerated as the air cushion between it and the liquid is eliminated.

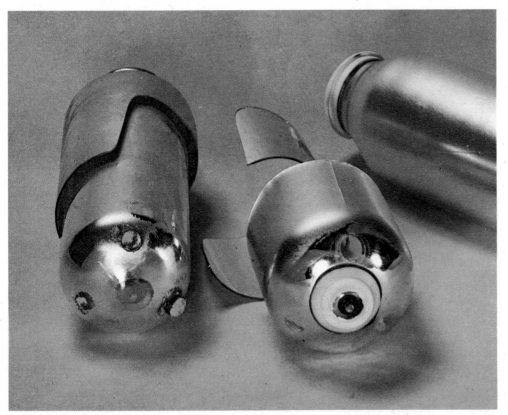

Left: a dismantled vacuum bottle, showing the construction, and a complete one on the right.

THERMOSTAT

A device that controls the temperature of a circulating fluid, or of an enclosed space, is a thermostat. Such a device has a sensing element, which responds to the temperature, and a control element, which regulates a heating or cooling process in such a way as to correct any departures from the desired temperature. Automatic temperature regulation is essential in domestic refrigerators, central heating and air conditioning systems, in cars' cooling systems, and in countless industrial processes.

The essential features of any thermostat are well illustrated by one of the earliest known examples, designed about 1660 by Cornelius Drebbel, a Dutchman living in London. The device regulated a furnace which heated an incubator. Hot furnace gases rose around a water jacket surrounding the incubating space. The sensing element, placed in the incubating space, was a vessel filled with alcohol, which expands on heating. One end of this vessel communicated with a U-shaped tube containing mercury. If the temperature of the water jacket rose, the expanding alcohol pushed down on the mercury column in one branch of the tube. The mercury rose in the other branch and pushed up a rod, which in

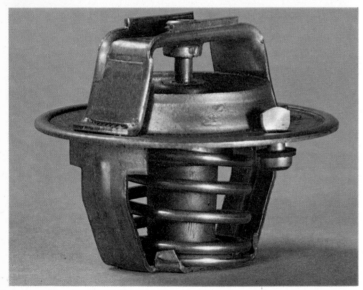

Top: a central heating thermostat with a bimetallic strip. Springy metal makes the contacts move abruptly, preventing sparks.
Above: a car's thermostat prevents water flow until the engine warms up, melting the wax. Then water pressure keeps it open.

turn raised one end of a lever. The other end of the lever descended and lowered a damper over the flue through which the furnace's gases vented. This restricted the draught to the furnace and slowed its rate of burning. When the temperature fell as a result, the alcohol contracted, and the damper was raised, increasing the strength of the draught and the rate of burning.

The device would undoubtedly have 'hunted'; that is, the rate of supply of heat would have alternated between being too great and too small. The water jacket, however, would have been slow to change in temperature and would have smoothed out the fluctuations. The device included a screw regulator with which the average temperature actually maintained could be set.

Such devices were used only in isolated instances until the late nineteenth century, when they were applied to the furnaces and radiators of heating systems in large apartment houses. The control industry received great impetus from the development of central heating and refrigeration in the ordinary home after World War II.

Bimetallic strips One class of thermostat sensing elements depends on the expansion of materials with rise in temperature, as did Drebbel's alcohol-filled vessel. One such device, called the *bimetallic strip*, is made of two strips of metal welded together. The two metals have different *coefficients of thermal expansion*—that is, they expand by different amounts when heated. Brass, for example, expands by one fifty-thousandth of its length for each degree Celsius that it is heated through. Copper expands by about 90% of this. Hence the two components of a bimetallic strip expand by different amounts when heated and, small through the difference is, the strip is forced to bend towards the side that lengthens less. This bending can be used directly to operate a valve or damper of some kind, or it can be used in a make-and-break electrical circuit. When the strip is touching an electrical contact, the circuit is closed and current flows; when it is not touching the contact, the circuit is broken.

A bimetallic strip is used in one common form of room thermostat for central heating systems. The strip is in the form of a U, one end of which is fixed. The other end makes an electrical circuit when the temperature has fallen to some preset level. The position of the contact that the strip touches controls the oil or gas burner and the circulating pump. When the water reaches its highest safe temperature, the burner is switched off and the pump is started, if it is not already active. Normally the water in the boiler is at a lower temperature than this, and both the burner and the pump are regulated by the room thermostat.

All forms of thermostat depend on the principle of feedback; that is, they feed information from a process's output (in this case, a temperature) back to the inputs that control it—the mechanical controls of the heating or cooling process.

TICKET MACHINES

Hand operated ticket punches, or 'nippers', which validated or cancelled tickets by punching a hole of a certain shape, were in use in the 19th century. Ticket issuing machines came into use in the early 20th century, for convenience and economy. The range of tickets could be extended by providing means for printing variations of value or time. Ticket machines are cashier operated or coin (or banknote) operated; the issuing mechanism is basically the same, but the cashier's lever or keyboard is replaced by machinery for inspecting coins. As labour has become more expensive and marketing requirements more clear, coin operated machines have become more specialized, and are used in car parks, public transport facilities, and other places.

Issuing mechanism Tickets themselves fall into three categories: preprinted, partially preprinted and printed at

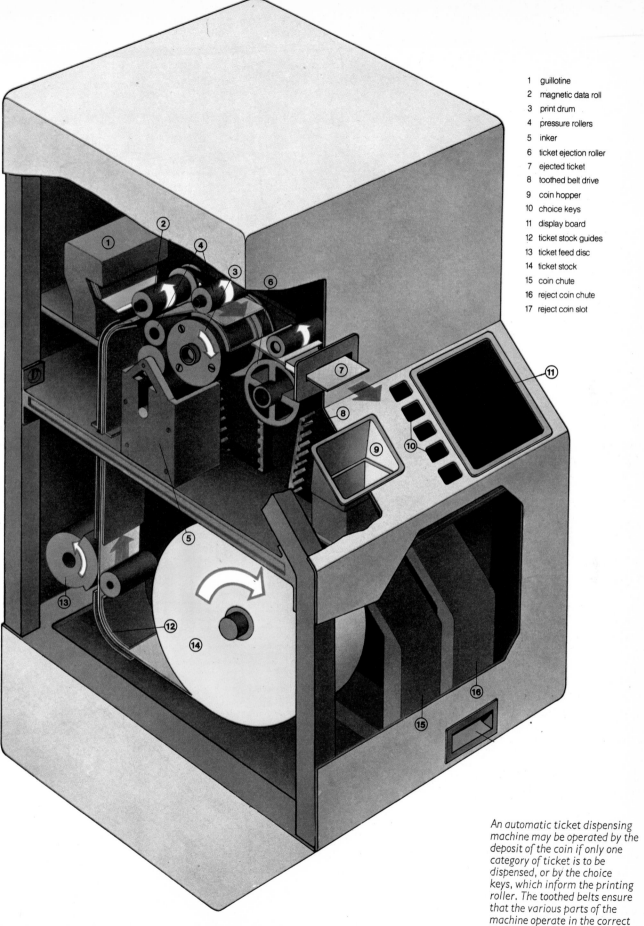

1 guillotine
2 magnetic data roll
3 print drum
4 pressure rollers
5 inker
6 ticket ejection roller
7 ejected ticket
8 toothed belt drive
9 coin hopper
10 choice keys
11 display board
12 ticket stock guides
13 ticket feed disc
14 ticket stock
15 coin chute
16 reject coin chute
17 reject coin slot

An automatic ticket dispensing machine may be operated by the deposit of the coin if only one category of ticket is to be dispensed, or by the choice keys, which inform the printing roller. The toothed belts ensure that the various parts of the machine operate in the correct timing sequence.

Right: a ticket machine in a London Transport tube station. The blank ticket stock, coated on the back with an oxide which operates the gates, is printed by the machine with the fare value, station of origin, the date and a serial number. Two tickets are printed ahead of time so the ink dries before the customer touches it. The red-tipped devices at the top of each machine are magnets which operate the machine when tripped by coins; the machine on the left has an extra one which was for the old sixpenny pieces, no longer used.

the time of issue. The dispensing machinery increases in complexity accordingly.

The fully preprinted ticket invariably has *feed holes* punched in it, usually at the division between each ticket. These holes engage with pins projecting from the periphery of a *feed wheel* which is turned through an arc corresponding to one ticket length for each operation. A *feed gear* having the same number of teeth as there are feed pins is mounted on the same shaft as the feed wheel. This gear is prevented from turning backwards by a detent pawl abutting the drive gear, or a separate stepped detent disc. Engaging the feed gear is a *rack* member (a straight length of metal with teeth on one edge). By withdrawing the rack a predetermined distance, set by manually adjustable stops, the forward movement of the feed wheel is controlled .Upon release of the feed rack, a spring brings it back to its original position and the selected number of tickets is moved forward. At the conclusion of the operation a guillotine knife snaps forward to sever the issued tickets from the reel of stock. This type of machine is installed in virtually every cinema in the UK and many other countries.

Printing tickets Partially preprinted tickets are issued by a similar mechanism, but the tickets are moved one at a time, so that the printing device can be aligned with the printed matter already on the ticket. The rack takes the form of a *ratchet pawl*, which in this case is essentially a rack having only one tooth. The time during which the ratchet pawl is being moved back is enough time for the *variable printing* to take place. The action is repeated for any number of tickets.

The printing operation usually takes the form of engraved drums mounted over the blank area of the ticket, and adjusted by knobs or gears, manually or electro-mechanically, to turn them to the desired value. A pressure pad or platen presses the ticket against the drum with an inked ribbon interposed. When the printing pressure is removed, the issuing cycle takes place.

Tickets fully printed at the time of issue are produced by a variety of means, the most common of which are variations in reduced size of a conventional printing press. A typical design takes a strip of plain paper from a roll, cuts it off, feeds it between inked engraved type rotating at the same speed as the paper is moving, and delivers it.

The ticket stock is fed between two engaging rollers, the larger of which is driven and revolves once for each ticket. Part of its outer surface is cut away, leaving a raised rim of length equal to the length of a ticket, so that one ticket is moved forward and stops, although the roller may keep turning. The length of the gap on the roller allows time for the machine to cut the ticket and for the printing drum to grip the ticket between the engraved type faces and a pressure roller. The type faces have been inked by a revolving inked roller just before contact with the ticket surface.

The actual printing of the ticket is an intermittent action because the forward movement of the ticket depends on the pressure between the type face and the pressure roller. The distance between the bottom of one type face and the top of another may be one inch (25.4 mm) measured on the drum, but only 0.156 inch (4 mm) on the ticket, because the ticket is not moving when there is no printing contact. Often the ticket is still inaccessible to the customer or clerk when the printing is finished, and another pair of rollers is necessary to deliver it.

Other methods of printing are used, depending on the variety or quantity of printing required. *Fly printing* is a method where the print drum is engraved with rows of numbers and letters around its circumference, amounting to perhaps ten or twenty columns. The drum runs continuously at high speed and the printing is achieved by tiny electromagnetic hammers hitting one selected character in each column as it 'flies' past. *Stylus printing* is another method, in which a group of electromagnetically operated needles are operated in various sequences to form characters, with an inked ribbon interposed. Electronic operation is necessary in these methods because of the high speed and the variety of selection.

Coin or banknote operated machines require extra equipment to examine the money, by means of magnets, weighing and measuring coins, and so forth. If the machine gives change, the coins are stored in a hopper or coin storage tube and released as necessary. The coins taken in payment are stored within the machine in a security vault, which automatically locks when it is removed from the machine.

TYPEWRITER, manual and electric

The first known patent for a typewriter was issued in England by Queen Anne to Henry Mill in 1714, but it is not known how the machine worked or if it was ever built. Many designs were produced during the first half of the 19th century, but the first practical typewriter was built by Christopher Sholes and Carlos Glidden in the USA in 1867.

Development of the Sholes and Glidden machines was taken up by Remington and Sons, a firm of gunsmiths, and the first typewriters made by them went on sale in early 1874. By the end of the 19th century many companies were engaged in the design and production of typewriters, and following the introduction of electric machines, which in turn led to the development of a wide range of typewriter-based equipment, typewriter manufacture has become a major industry.

Print quality Among the factors affecting the production of a good printed image are the need for a thin, even film of ink covering the typeface, a firm even impression of the typeface against the paper, and a resilient base against which the paper is pressed.

The quality of inking of the typeface has improved as new and better inks have been developed, along with improved ribbon materials to hold the ink. For example, the finer the mesh of a silk or nylon ribbon, the more even the inking of the typeface. One of the most important advances in this area has been the introduction of polyethylene ribbons coated on

Below: a relatively simple dispenser of tickets for a car park. In this type of machine the ticket can be either wholly pre-printed or printed only with the time or date issued.

one side, rather like carbon paper, with a layer of ink. This ink contains a far higher density of carbon than could be used on a fabric ribbon.

The base against which the paper is pressed by the type in a printing press is called the *platen*, and this is also the name given to the rubber roller that performs a similar function in a typewriter. The platen provides a base that is resilient and yet sufficiently flexible to give a little under pressure, and this aids a complete transfer of ink from the ribbon to to the paper. The design of the platen, and of the moving *carriage* that carries it, must ensure that the paper is kept in perfect alignment as it moves both horizontally and vertically.

Typebar operation The character images are cast on small pieces of metal called *typeslugs*, which are fixed to the ends of the *typebars*. The typebars are mounted on a slotted semicircular casting called a *segment*, and the complete set of typebars is called the *typebasket*.

The basic principle is that the downward motion of the keylever on the keyboard is converted by a series of linkages into a downward and forward motion of the lower end of the typebar. The typebar is pivoted a little way above its lower end, and so this motion moves the other end of the typebar (carrying the typeslug) upwards and rearwards to strike the ribbon against the paper and print the character.

The typeslugs usually have two characters cast on them, one above the other. In the case of alphabetical characters,

the upper character on the slug is the capital (*upper case*) letter and the lower one is the small (*lower case*) letter. The design of the typeslug and the curvature of the paper around the platen ensure that only one of these characters can print at a time. The whole typebasket is mounted on vertical guides, and when it is at the top of the guides the lower case characters are printed. Operating the *shift* key moves the basket to the bottom of the guides so that the upper case characters are printed. A *shift lock* enables the typist to lock the machine into its upper case position when required.

As the typebar moves up under the action of the keylever other linkages bring the *escapement* and ribbon mechanisms into operation. The *ribbon feed* advances the ribbon a small amount to bring a fresh area of ribbon into use, and the *ribbon lift* mechanism lifts the ribbon up into place in front of the paper. When the machine is being used for typing stencils, the ribbon control lever is set in the 'stencil' position and this prevents the ribbon lift operating so that the typeslug hits the stencil directly.

The escapement mechanism controls the horizontal motion of the carriage, which is pulled to the left by a coiled spring that is rewound each time the carriage is moved back to the right to begin a new line. The escapement mechanism is a form of ratchet which allows the carriage to move one letter space to the left each time a character is printed or for each operation of the *spacebar*, which is used to provide the spaces between the words. The *backspace* mechanism pulls the carriage one space to the right when the backspace button is depressed, for example when making corrections.

The left and right hand margins are controlled by movable stops mounted on a rack on the carriage, which engage with a fixed stop on the frame of the machine. A warning bell rings a few spaces before the carriage reaches the right hand margin, and a lockout device is usually provided which locks the character keys and the spacebar when the margin is reached.

Carriage return Carriage return is operated by means of a lever extending forward from the left hand end of the carriage. The initial movement of this lever drives a link which has a tooth-shaped end that engages with a ratchet mounted on the left hand end of the platen, and this rotates the platen to give the desired vertical spacing of the paper. When the platen has been rotated (*indexed*) the correct amount, the drive link contacts a stop which prevents it moving any further, and so the pull on the carriage return lever is transferred to the carriage itself, and the carriage moves to the right. Operating the carriage return lever when the carriage is at the left margin results in indexing of the platen (also called *linespacing*) with no movement of the carriage.

Tabulation The *tabulation* (tab) mechanism allows the carriage to be moved rapidly leftwards to preset positions aong the writing line. A rack carrying a set of movable stops, one for each letter position along the line, is mounted on the carriage. To set a tab stop, the carriage is moved to the position required and the tab set button is pressed, which moves the tap stop for that position into its 'set' position.

When a tab operation is performed, the escapement ratchet is released and the carriage moves to the left under the action of the escapement spring. When the carriage reaches the point at which the tab stop was set, the stop contacts the tab mechanism fixed to the frame of the machine. This stops the carriage and brings the escapement ratchet into operation.

Electric typewriters The first commercially successful electric typewriter was developed by International Business Machines (IBM) in 1935, but it took many years and much technological development before the electric machine began to compete effectively with the manual machines. Few of the other typewriter manufacturers produced electric machines before about 1950.

The design of electric typewriters has undergone radical change in the last thirty years. Although the early machines

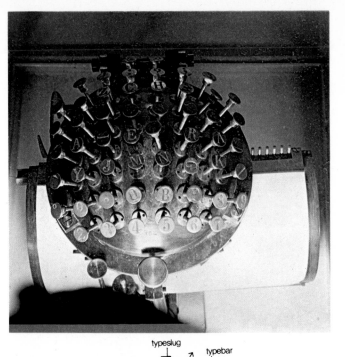

Opposite page:
An Italian machine, the 'Cembalo Scrivano', built in 1857. This machine prints capital letters and a few symbols.
Left: this typewriter, built in Denmark in 1872, had the type and the keylevers mounted on a hemispherical frame above the paper.
Below: the mechanism of the typewriter transforms pressure on the keybutton into a sharp impulse that flings the typebar forward. The keylever is also connected to linkages that move the carriage on one space and reel the ribbon from spool to spool. The shift button lowers the typebasket so that the upper section of each typeslug strikes the paper, thus printing a capital letter.

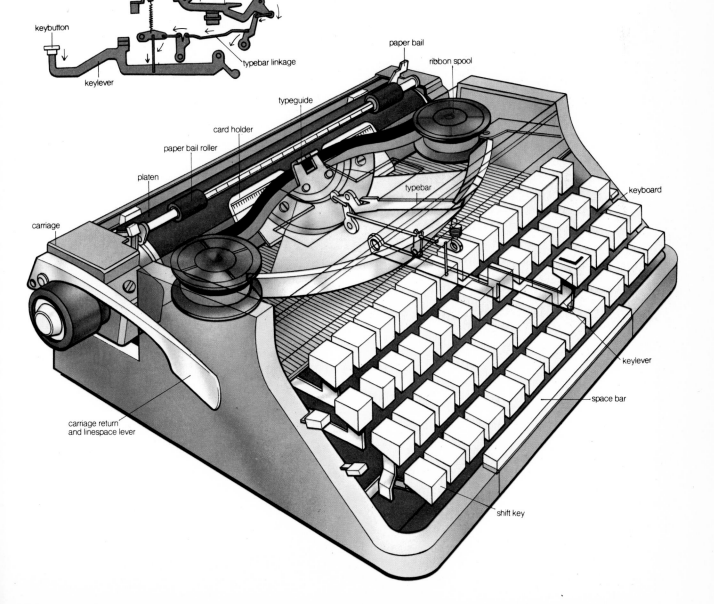

were successful in using an electric motor to power the type-bar motion, they failed to control this movement adequately. This resulted in excessive vibration of the typebar as the character was printed, causing a certain amount of blurring of the image. Much of the early development work was aimed at eliminating this vibration by reducing the number of linkages in the mechanisms, making them more rigid and better controlled. One design, by Underwood, divided the linkage into two parts, each acting independently. One controlled the upward movement of the typebar, the other the return.

Drive systems There are many different arrangements for using electric motors to drive the typebars and functional controls of electric typewriters. Most, however, follow the general pattern of using an AC electric motor mounted at the rear of the machine and driving, via a belt drive, a main power shaft or roller running across below the type-basket.

Single element Although the concept of single element typing has been around since the end of the 18th century, the first successful single element machine, the IBM 'Selectric', was not introduced until the beginning of the 1960s.

The main difference between most single element typewriters and conventional typebar machines is that the carriage assembly remains stationary, while the typing element moves along the typing line with each successive key stroke.

Instead of the type characters being carried on a set of typebars they are all carried on the single typing element. In the case of the IBM machines the element is often referred to as a 'golfball' because of its rounded shape and its size; the more cylindrical elements used on Adler machines are known as 'cores'. The single element machine has the advantage that its typestyle can be changed quickly and simply by removing the element and replacing it with one of a different typestyle, and this can be done by the operator in a matter of seconds.

Another advantage of single element machines is that, unlike the typebar machines, they are not prone to damage or misalignment of the characters due to typebar clash. With all the characters located on one element only one character can be selected at one time.

Character selection   The typing element carries the 44 lower case characters on one half and the 44 upper case characters on the other half. Shifting from lower case to upper case means turning the element through 180°. The characters on the nickel-plated plastic element are arranged in four horizontal bands (*tilt positions*), making a total of 22 vertical columns with four characters in each.

When a character is to be printed, the element must be turned to the *rotate position* of that character and tilted to the appropriate tilt position, and then driven against the ribbon to print the image on the paper.

With the upper case characters on one side of the element or *typehead* and the lower case characters on the other, a total of 11 rotate positions are required. The rotate motion is transmitted from the selection mechanism to the type-head by a thin steel tape, and the tilt motion is transmitted by a tape (IBM) or by a shaft running across the machine in line with the platen (Adler). Once the typehead has been moved to the correct position it is *detented* (locked) in place to ensure accurate alignment during printing.

The selection mechanism is driven by a set of cams, and the amount of motion transmitted to the typehead is controlled by a set of levers and latches. When a key is pressed, it operates a mechanism which sets up the correct combination of latches or levers to give the required amount of rotate and tilt motion for that character.

Print cycle   The typehead is carried along the typing line by the *carrier* assembly, which travels along a shaft (the *print shaft*) in front of the platen. This shaft has a groove machined along its length, and it passes through a hollow sleeve inside the carrier. This sleeve has a set of cams mounted on it, and a metal peg or key which fits into the groove on the shaft so

that when the shaft rotates during a print operation the sleeve and its cams turn with it.

The ribbon, together with the ribbon lift and ribbon feed mechanisms, is mounted on the carrier so that it moves along with the typehead.

The cams on the sleeve provide the motion to drive the typehead to the platen and print the character, to operate the detents which hold the head in its selected position, and to drive the ribbon lift and feed. The print shaft is driven by the motor via a clutch assembly which is actuated when a key-lever is depressed, and it makes one revolution for each print operation.

Escapement   After the character has been printed, the carrier moves along to the next printing position. This move-ment is powered by a spring which is rewound when the carrier is moved to the left to start a new line. On most machines the spring turns a drum which winds on a nylon cord attached to the carrier. The amount the carrier moves is controlled by a pawl assembly mounted on it which engages with a toothed rack running along the machine below the platen. Other machines use a *leadscrew*, turned by a spring, to drive the carriage along; in this case the amount the lead-screw rotates when a character is printed determines how far the carrier moves.

On these two pages are pictures of some unusual typewriters. The machine at the top of the opposite page must be the smallest ever made; it was shown at an exhibition in Brussels in 1959. It is 6½ inches square, two inches high and weighs two pounds. Below it is the world's largest typewriter, made by apprentices in a German typewriter factory in the 1950s. It took them 10,000 working hours to build it, and it weighs 450 pounds. On this page, a shop window displays an ordinary type-writer, and below it, a Japanese language machine which has several hundred typeslugs stored in the tray at the front. To print a character, the carriage is positioned above the required typeslug, which is then picked up and driven against the paper.

Above, top: three late 19th-century attempts to produce an efficient single-element typewriter, by Lambert, Columbia and Blick. The Blick, on the right, is a true single element machine, having a keyboard and a cylindrical typehead.
Above: the typehead carrier and ribbon cassette of an IBM Model 82.

Applications Apart from its original use as an office typewriter, the single element machine now forms the basis for a wide range of other equipment. Its advantages of having no moving carriage and a higher operating speed than a typebar machine have made it suitable for use in data processing terminals, and as the input-output device of many minicomputers. Because it has no moving carriage, the single element machine takes up less space than a typebar machine, and for on-line terminal use they can be left running unattended by using continuous stationery.

The use of the single element typewriter as an input or output device means that much simpler circuitry is required to connect it to the processor. An ordinary typebar machine, with 44 typebars, needs a separate solenoid for each typebar, plus one to operate the shift mechanism. A single element machine such as an IBM 72, on the other hand, requires only six solenoids to operate the character selection latches, plus one for the shift.

In recent years an increasing number of word processing systems have been introduced, and because of its advantages in terms of automatic operation the single element typewriter is particularly suited to this application. Original typing of drafts, reports, or normal correspondence is recorded via the typewriter keyboard and suitable circuitry on to some form of storage media such as magnetic tape, magnetic cards or paper tape, or simply entered into an electronic memory. The typist can originate this material at draft speed, since any mistakes can be instantly corrected without re-typing an entire word, line or page. This also means that any other amendments can be made easily and quickly. Once the document has been edited, the final version can be typed out automatically by the machine at speeds from often in excess of 150 words per minute.

A more advanced form of single element machine has been developed to produce lithographic master plates for the printing industry. This machine has proportionally spaced escapement to enable the characters to take an amount of

Left: the IBM Selectric Composer is an advanced kind of single element machine which produces high-quality lithographic master plates for printing work.
Below: the typehead is tilted and rotated to bring the required character into position by two sets of latches, the tilt and rotate latches. A specific combination of latches is selected for each character, and these are pulled downwards by the latch bail. The amount of motion transmitted through the bellcranks depends on which latches have been selected.

space appropriate to their size, and the *indexing* or *leading* (inter-line spacing) can be adjusted to suit the typestyle being used. This machine can also produce *justified* copy, that is, copy with a straight right hand margin.

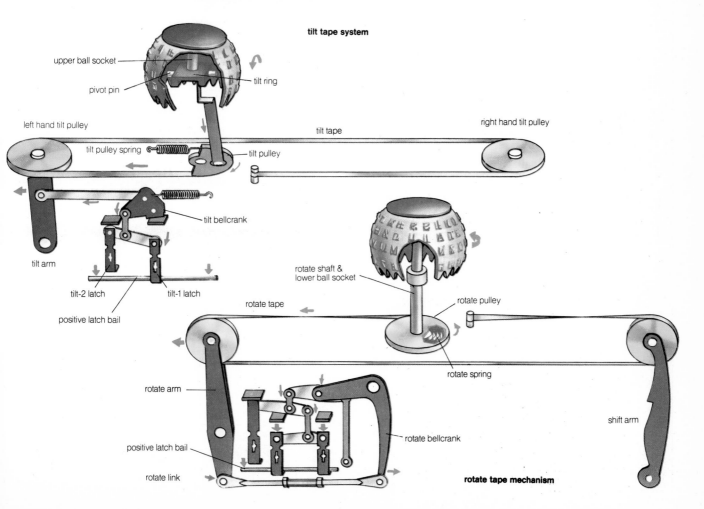

tilt tape system

upper ball socket
tilt ring
pivot pin
left hand tilt pulley
tilt tape
right hand tilt pulley
tilt pulley spring
tilt pulley
tilt bellcrank
tilt arm
tilt-2 latch
tilt-1 latch
positive latch bail

rotate shaft & lower ball socket
rotate tape
rotate pulley
rotate spring
rotate arm
shift arm
rotate bellcrank
positive latch bail
rotate link

rotate tape mechanism

VENDING MACHINES

All vending machines consist of two basic parts: a coinage system that identifies and totals the inserted coins, and a system for selecting and dispensing the product required.

Coinage systems The simplest coinage systems are those that require only one coin to operate the machine. These are commonly found in many of the smaller beverage machines and wall-mounted column-type snacks machines. Larger vending machines generally have sophisticated totalizing-type coinage systems which are capable of accepting 3, 4 or even 5 different denominations of coins and adding these up to achieve the required prices. Frequently, totalizing channels are also multi-price to enable the various commodities offered to be priced differently.

All coinage systems have a means for checking the validity of the inserted coins before crediting them. The most common form of checking is mechanically whereby the diameter, thickness and weight of the coins are measured. Accepted coins then roll down an incline past the face of a magnet. The magnet causes eddy currents to be generated in the coin as it rolls by, which will create drag to slow the coin down. The magnitude of the eddy currents and the consequent drag will depend on the metallic properties of the coin. Coins leaving the end of the incline must follow a certain trajectory to be accepted, for which they must be travelling at the right speed. Coins of the wrong metallic composition will be travelling at the wrong speed and will therefore be rejected. Some devices also brush the coin to see if it has a hole in it. If it has a hole it is a washer and is rejected.

Although the mechanical means of checking coins is by far the most common, there is an increasing trend towards the use of electronic methods. One electronic system used in Britain utilizes the same principle of rolling coins down an incline past a magnet. The speed of the coin is measured by the time it takes to pass photoelectric transistors. Photoelectric transistors are also used to measure the diameter.

Hot drinks Hot drinks machines form one of the largest sectors of the vending market. Instant ingredients are usually used to give coffee, tea, chocolate and soup drinks all of which can be dispensed by the same machine. These instant ingredients are often the same as those found in the home.

Nearly all cycles start with the dispensing of a cup into the

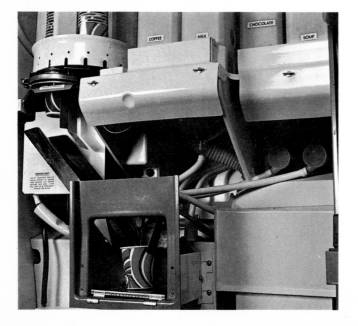

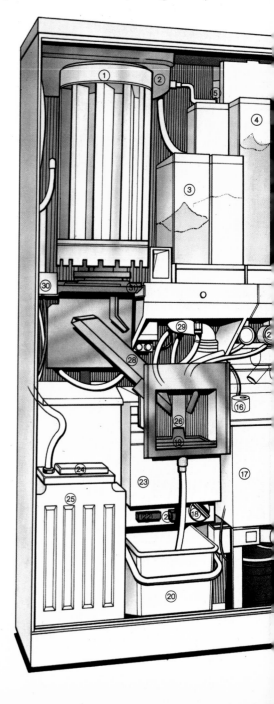

cup station ready for the acceptance of a drink. The cups are stored in stacks in the dispenser mechanism, with the bottom cup being separated from the stack each time a drink is required. The cups are made of either plastic or paper.

Cold water is taken from a normal mains water supply and is heated in a boiler inside the machine to a temperature of approximately 180°F (82°C). This will give a final drink temperature of about 165°F (74°C), suitable for immediate consumption. The mixing of drinks takes place in funnel shaped mixing bowls. Hot water is injected at the top of the bowl in such a way as to cause it to swirl around the bowl to the outlet at the bottom. Measured amounts of the ingredients are dropped into the swirl of water where they are dissolved to make the required drink before passing from the mixing bowl into the cup. Chocoate and soup drinks usually pass from the mixing bowl into a whipper chamber to ensure a better mix before passing into the cup.

For those who prefer more conventional methods, machines using tea leaves or coffee grounds are available. The leaves or grounds are held for a short time in hot water to allow infusion to take place before dispensing.

Cold drinks The coinage, selection, and cup dispensing systems of cold drinks machines are identical to those in hot drinks machines. The ingredients are stored in the machine in the form of concentrated syrups which, on dispensing, are mixed with still or carbonated cold water.

Mains water is chilled by a refrigeration unit to a temperature of approximately 38° to 42°F (3° to 6°C). The cold water can then be mixed with syrup in the correct proportions to give a still drink. For carbonated drinks, the cold water must be passed first into a *carbonator* to allow carbon dioxide gas to be added.

In the carbonator chilled water is injected under high pressure into a chamber filled with carbon dioxide (itself under pressure). The stream of water is broken up into fine droplets and absorbs the carbon dioxide to give carbonated water. A typical carbonation level for a fairly fizzy drink is 3.5 volumes of carbon dioxide for every volume of drink (at normal temperature and pressure). Other types of drink in the same machine usually require a lower carbonation level (orange would be at about 2.5 volumes) and this is usually achieved by diluting the highly carbonated water with still water.

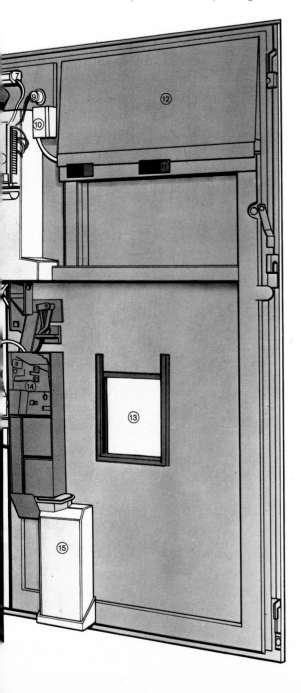

Far left, top: a machine for vending perfume, Paris, 1893. The coin completes the mechanical linkage.
Far left, below: a machine for vending drinks, showing the cup dispenser and drinks containers.
Left: a drawing of a modern vending machine, for instant drinks, powdered soups and so forth. Some machines can make fresh coffee, by punching holes in the coffee tin and using it as the top of a percolator.

1 cup dispenser unit
2 water reservoir tank
3 ingredient container (small)
4 ingredient container (tall)
5 water filter
6 beverage selector mechanism
7 total vends counter
8 whipped chocolate pre-set counter
9 display sign light starters
10 cold drinks counter
11 display sign light chokes
12 display sign lights hinged mounting
13 clear view vend door
14 coin selector mechanism
15 coin box
16 cooler unit water filler plug
17 cooler unit
18 hot water tank thermostat
19 grille
20 waste bucket
21 waste bucket overflow cut-out switch
22 immersion heater
23 hot water tank
24 fuse box
25 electronic control box or standard relay box
26 cup station
27 whipper units
28 cup chute
29 main mixing bowl
30 mains power supply isolation switch
31 main and whipped ingredients
    bowls flush switches

WATCH

The springwound watch has a history of some 500 years. Over this period it has become a nearly perfect mechanical device, and its accuracy has only been superseded by the quartz clock, and finally the atomic clock. No other mechanical device must work for 24 hours a day, 365 days a year for years on end; furthermore, if its error rate is 20 seconds a day, this is an error of only 0.023%. Such accuracies are normally found only in scientific instruments, which do not suffer the rough treatment received by the wrist watch.

Components The mechanical watch is made up of several main sections. The *mainspring* is a coiled steel spring contained in the *spring barrel*, and is the source of motive power for the watch. The power is transmitted by a *train* (series) of gear wheels to the *escapement*, the device which checks the forward motion of the gear wheels and uses this energy to give the impulses which drive the *balance wheel*, which is the controller or governor of the watch. It turns in alternate directions at a fixed rate, and controls the timekeeping of the watch in the same way as the pendulum does the clock.

Development The coiled spring was first used as a source of power in portable clocks in the late 15th century, probably in Nuremberg. By the early 16th century the size of these timepieces had been reduced, and the introduction of the *hog's bristle* made the first truly portable watches possible. These early watches used the *verge* escapement and a simple balance wheel in place of the *foliot* balance used in clocks of the period. The hog's bristle, which was a crude form of balance spring, consists of one or two lengths of stiff bristle arranged to act as buffers for the balance at the limit of its 'arc of vibration' (the extent of the to-and-fro rotation of the balance).

These early watches, often enclosed in fine pierced or gem-set cases, were really little more than expensive toys. The timekeeping errors must have been in the region of 15 minutes per day; in fact only an hour hand was used in this era. The minute hand did not appear on watches until the late 17th century.

The period from 1675–1800 saw a revolution in the design of watches as well as clocks; from the primitive verge watch to fine chronometers which had an accuracy of better than 2 seconds per day (strictly the term 'chronometer' applied at this time only to the very accurate timepieces used for navigation at sea, but pocket versions were produced). The first step in the improvement of the verge watch was the introduction of the balance spring (sometimes called the 'hairspring'); this was first devised by the English scientist Robert Hooke in 1658, and was developed later in the century, by Huygens and the Abbé Hautevill, into the form of a fine spiral steel spring of five to ten coils. The centres of the spring is secured to the balance wheel, and the outside end is pinned to the balance cock (a removable bridge which holds the upper bearing of the balance wheel). The balance spring is alternately wound and unwound as the balance rotates, reversing its direction usually five times per second. The time taken by the balance to make each turn or 'vibration' is controlled principally by the strength of the balance spring and the weight of the balance wheel.

It was soon realized that the improved balance could be a very accurate time controller if it were allowed to turn freely, with as little interference as possible from the escapement. The old verge escapement was discarded in favour of the cylinder escapement, which was perfected by George Graham in about 1725 (earlier experiments were carried out by Tompion and Booth in the late 17th century). This escapement allowed the balance to turn much more freely than the verge, although a considerable amount of friction was still involved; a well-made cylinder watch could be regulated to within 2 or 3 minutes per day.

An important break-through came with the invention of the lever escapement by Thomas Mudge in 1759; this allows

the balance to turn without any interference except for a short period during each swing when it receives its impulse. The lever escapement, with a few minor improvements, is the one used in all high grade 'jewelled lever' watches today.

Temperature compensation A great problem which had to be overcome by the 18th century watchmakers was temperature error. The trouble with the steel balance spring is that at low temperatures it becomes stronger and at higher temperatures weaker, this causes a change in the 'rate' of the watch of about 8 seconds a day per °C. A solution to this problem was urgently needed in order to produce accurate timepieces for marine navigation; each year thousands of tons of shipping were lost because no accurate means were available to determine longitude.

John Harrison (1693–1776) produced a series of 'marine timekeepers' (now at the Greenwich Maritime Museum), which would stand the rigours of a long sea voyage and still maintain a high accuracy. His solution to the temperature problem was the *compensation curb*. This was essentially a *bimetallic strip* of steel and brass fused together; the brass expands at a faster rate than the steel and causes the strip to bend at different rates according to the temperature. This device was used to shorten and lengthen the effective length of the balance spring, thus compensating for changes in temperature. In modern watches the balance spring is made of an alloy, which is not affected by temperature changes.

Other developments Jewels in watches have been used since the 18th century; these are bearings, made originally from natural rubies, but synthetic rubies have been used for the past 70 or 80 years. Before methods of drilling rubies

Far left: an early example of a self-winding pocket watch. The swinging weight is on the left; the ratchet (on the right).
Near left: a springwound clockwork watch, the most nearly perfect of mechanical devices. As an example of the kind of precise work done by watchmakers, a tiny screw used in watches has $317\frac{1}{2}$ threads per inch, and a major diameter of 0.0118 inch (0.3 mm).
Below: a temperature-compensated balance spring.
Bottom: exploded view of a fully jewelled watch movement with lever escapement.

were discovered, brass bearings had to be used; these were subject to a high rate of wear. It soon became general practice to use jewel bearings for the balance wheel, escapement and train wheels. The modern fully jewelled watch has upwards of 17 jewels, which give protection against wear at all important points of friction.

Since World War I mass production techniques, developed largely in Switzerland, have revolutionized the watch market; it now became possible to produce accurate, reliable, small sized watches comparatively cheaply, in order to cater for the new demand for wrist watches. The fashion for wearing watches on the wrist made the production of 'automatic' watches a practical proposition. These are wound by wrist movement; an eccentric rotor pivoted behind the watch movement is free to swing to and fro with each motion of the wrist, and this energy is used to wind the watch through reduction gears.

The first patent for an automatic wrist watch was taken out by John Harwood in 1923, although a similar principle had been used to produce self-winding pocket watches in the late 18th century by Perrelet, Breguet and others. The idea was taken up in the 1930s by several Swiss companies, and since World War II many highly successful models have been produced. Apart from the obvious advantage of not having to wind the watch manually, the accuracy of the watch is improved by the mainspring being kept in a fully wound condition most of the time, resulting in a much more even transmission of power to the escapement.

Electronics Electronic watches of various types are now being produced in very large numbers, at prices comparable with high grade mechanical watches. The first watches to use electric power instead of a coiled spring were produced during the 1950s; they were made possible by the introduction of small batteries (similar to the type used in hearing aids), lasting upwards of a year in operation.

The elementary electric watch uses a balance wheel and balance spring as a motor; instead of the balance being driven by the mainspring through the escapement, as in a conventional watch, the motive power comes from an electrical coil mounted on the balance wheel. For a short period during each 'vibration' of the balance this coil becomes energized, and repels the balance from a small permanent magnet

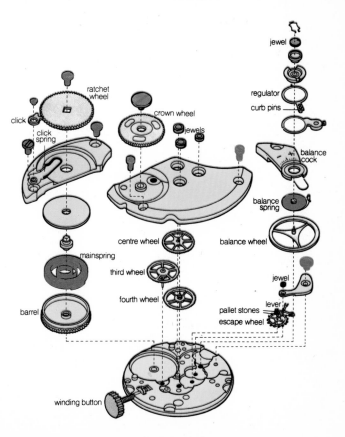

Right: the Bulova 'Accutron' was the first tuning fork watch. Back (left) and front (right) views are shown. The copper cylinders at the top are the coils at the ends of the tines of the fork, which cause the fork to vibrate. The silver disc is the battery. It relies for its accuracy on the precise resonant frequency of a tuning fork. The fork is pulsed by electromagnets situated at the heads of the fork. One of the electromagnet coils is split in two—one part acting as a sensing coil. The head movement induces a voltage in this which activates a transistor through a resistor-capacitor circuit, thus pulsing the tuning fork at the correct moment.

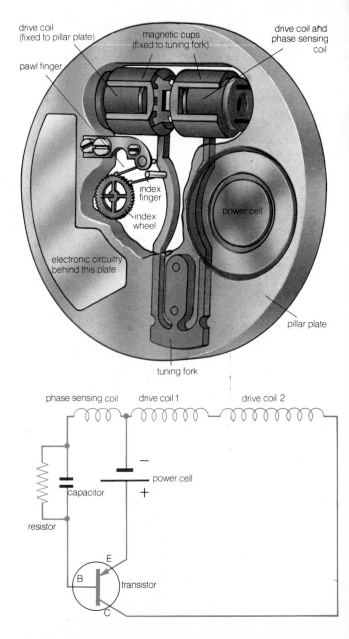

mounted nearby. The current is switched on and off by a light spring contactor, which engages with a pin mounted near the axis of the balance wheel; the hands of the watch are driven through an adapted lever escapement.

Later versions of this design have dispensed with the contactor, replacing it by an electronic switching system. The coil on the balance passes over a permanent magnet, inducing a small current in the coil which is used to trigger a transistor. The circuit containing the transistor operates a switch, sending pulses of current to propel the balance wheel. The system has proved very successful, and several manufacturers are producing watches of this type.

Tuning fork watches The best accuracy that can be attained by a conventional wrist watch, using a balance wheel, is in the order of 3 or 4 seconds per day (the highest Swiss chronometer rating is an average daily variation of 4 seconds). In an effort to find an alternative to the balance wheel, watch designers began experiments with tuning forks in the 1950s, the first successful tuning fork watch coming on to the market in the early 1960s.

The advantage of the tuning fork is that it vibrates at a very precise frequency: the problem in using it as a time standard is to keep it vibrating, and to use its very small amplitude to drive the hands of a watch. To maintain the vibration of the tuning fork an electromagnetic impulsing system is used, controlled by a transistor switch similar to the type used in the electronic balance wheel watches described above. At the end of each tine of the fork is mounted a cone shaped magnet; these are free to vibrate in and out of coils mounted in a stationary position. A small section of one of the coils is used as a *phase sensing coil*; during each cycle of the fork, the magnet comes close to the phase sensing coil, inducing a small voltage in it, which is used to trigger the transistor switching circuit. This sends a pulse of current to the main *drive coils*, which repel the magnets on the tuning fork, pushing the tines inwards, and thus giving the impulse to

Left: quartz watches are particularly suited to digital displays, since the combination gives a device without moving parts. The four light-emitting diodes, here appearing to read 8888 because they are switched off, are two-thirds of the way up. Below: the simplest form of electric watch resembles a clockwork one, but the balance wheel is driven by an electromagnet.

maintain the vibrations of the fork. This cycle takes place in the original type of tuning fork watch 360 times per second; higher frequency tuning forks have been used in more recent watches.

The hands of the watch are driven by reduction gears from an *index wheel*; this is a small wheel with 300 teeth cut into its outer rim and is mounted close to the tuning fork. A straight spring tipped with a tiny jewel is attached to one tine of the fork; this engages with the teeth of the index wheel, and advances the wheel one tooth for each complete oscillation of the tuning fork. The index wheel therefore makes a complete revolution in a little under one second. The tuning fork watch has an accuracy of within two seconds per day; because of its small size and reliability it was used extensively in satellites and spacecraft.

Quartz watches Developments in the field of microelectronics have made possible the production of *quartz watches*; these work in essentially the same way as the earlier quartz clocks. Two main types have been evolved: the first uses the quartz oscillator to provide an alternating current which drives a small motor; this drives the hands of the watch through reduction gears. The second type has no moving parts; the time is shown by a digital display, which is usually of either the liquid crystal display or the LED (light emitting diode) type. The liquid crystal display uses reflected light, and therefore it cannot be seen in the dark, but it has the advantage of a very low power consumption. The LED display produces its own light, but it has a comparatively high power consumption. Watches using this display are provided with a button on the case which illuminates the figure for a short period, leaving the display screen blank at other times.

Quartz watches using *solar energy* have recently been developed in the USA. Two small solar cells are mounted on the upper face of the watch, which collect light to charge the power cells. These will keep the watch running for at least six months in total darkness, once fully charged. As LED display shows the hours, minutes and seconds, and also the day and month, the calendar is programmed to allow for the length of each month, and leap years until the year 2100. Because the watch does not use ordinary power cells, which have to be changed periodically, the movement can be sealed in a waterproof case, which is filled with a shock absorbing gel. The time setting controls are operated by magnets mounted outside the movement capsule which activates switches embedded inside the capsule.

The accuracy of high grade quartz watches is generally better than one minute per year; as larger numbers of these watches are being produced, the prices are becoming comparable with high quality mechanical and tuning fork types. Clearly in the years to come the electronic watch will dominate the market, new mass production techniques in electronics making these accurate and reliable watches available at prices which are competitive with all but the cheapest mechanical watches.

Above: a self-indicating balance for weighing gold coins, built in 1800. The instrument, about twelve inches tall, has three scales and three pointers, indicating the value in gold; the troy weight, in which one pound equals twelve ounces or 5760 grains; and avoirdupois weight, in which one pound equals 7000 grains. The coin being weighed is compared by means of gravity to the weight of the pendulum.

WEIGHING MACHINES

The principle of the balance, a freely suspended beam which is balanced by known and unknown weights, has been known from primitive times up to the present day, and is still used both for simple weighing machines and in highly accurate scientific instruments.

In primitive wooden balances, the pivots were simply cords passed through vertical holes in the beam. These were inaccurate, firstly because of the difficulty of locating the holes accurately and secondly because the cords could wander in the holes. The Egyptians improved on this by bringing the cords out of the ends of the beam. The lever arm lengths could then be equalized by shaving down the ends of the beam, and the cords were held firmly against the beam ends by the loads. The Egyptian cord pivot was still in use in many Eastern markets until well into this century.

The first attempts to replace cord pivots with metal pivots were probably made by the Greeks and Romans; curiously, they repeated the error of the primitive cord pivot by using a horizontal hole with a ring through it. This form of pivot remained in use in Europe until the Renaissance, when the lessons of the Egyptian beam were relearned, and the wing was passed through a hole in a trumpet-shaped end.

Modern history The evolution of the modern knife-edge pivot, which has the prime requisites of low friction and high dimensional accuracy, took place during the 16th and 17th centuries.

At the end of the seventeenth century, Roberval's concept of *static enigma* represented a significant forward step in weighing by allowing the goods and weight plates to be positioned above the beam, and by maintaining the plates level so that the weights and goods could be placed anywhere on the plates without affecting accuracy. (This is the principle of the familiar but old-fashioned greengrocer's balance). An earlier important idea from Leonardo de Vinci was to use the principle of the pendulum, not only to weigh but also to indicate the weight of the load. Thus the possibility existed of a self-indicating scale which did not require an operator to manipulate loose weights or sliding *poises*. These outstanding innovations seem not to have been combined until the 19th century, when the introduction of the penny post in Britain provided the incentive.

At the beginning of the 20th century scalemakers saw they could combine the pendulum, the Roberval mechanism and a more robust design invented by Joseph Beranger, a 19th century French scalemaker. These combinations remain in use today in the common type of retail scales with an indicator needle moving around an arc-shaped dial.

Other types Around 1743 John Wyatt added a new type of weighing machine to his list of inventions. Until this time all goods to be weighed were suspended from the weighing machine. In Wyatt's cart weighbridge, the weigh plate was supported by a system of levers. All mechanical platform scales and weighbridges are still based on this compound lever system. Wyatt's weighbridge used proportional weights to balance the goods being weighed, and despite his claim that the weighbridge 'would weigh a load of coal or a pound of butter with equal facility, and with nearly equal accuracy', the method he used to do so was not accurate.

The performance of weighbridges was greatly improved in the early 19th century when the proportional weights were replaced by a *steelyard*, having two or more moving poises to achieve balance. By the end of the century, attempts were being made to produce a self-indicating system. One of the most successful types was the *hydrostatic* indicator in which a counterweight was suspended in a tank of water. As load was applied to the scale, a lever connected to the counterweight rotated, cuasing it to be raised out of the water until loss of buoyancy increased its weight sufficiently to balance the load. A chain connected to the lever was passed round a drum and so caused a pointer to move. In winter it was not uncommon

Left: an old Avery platform machine with double pendulum resistants. A downward pull on the yoke at the bottom is transformed through steel ribbons and the cams at the top into an upward movement of the pendulums and a rotation of the indicator needle.
Below on the left, a common type of balance with Roberval linkage, as seen on market stalls. The pans always remain horizontal and accurate wherever the weight is placed. On the right, a pendulum weighing machine of the type often found in food stores. The pan linkage has the same advantages, but the weight can be read off directly from the scale (up to the maximum reading) without the need to put standard weights on another pan.

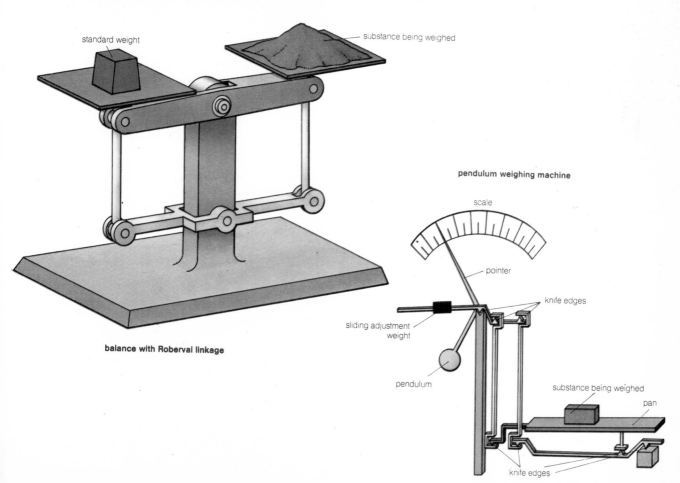

standard weight

substance being weighed

balance with Roberval linkage

pendulum weighing machine

scale

pointer

knife edges

sliding adjustment
weight

pendulum

substance being weighed

pan

knife edges

Right: an electronic machine which uses a gyroscope. A downward force on the gyro is turned by precession into a twisting movement whose speed is proportional to the force, allowing it to be measured.

to have to thaw the ice in the water tank with a fire before weighing could take place.

At the beginning of the 20th century the pendulum resistance mechanism was successfully applied to industrial weight indicators, using a rack and pinion to give 360° pointer movement, and having a number of proportional weights, which effectively break the full weighing range into a number of chart movements. Weighbridges with this type of weight indicator are capable of an accuracy of 1 part in 6000, at least twice as accurate as the dispensing beam scale used by pharmacists. The proportional weights in many modern indicators are automatically added and removed, giving a fully self-indicating machine.

The weighing machines described so far all operate on the principle of balancing an unknown against a known weight. Another form of weighing, the *spring balance*, uses the linear relationship between a spring's deflection and the load supplied. The spring deflection is either measured directly or magnified by a rack and pinion and measured by means of a pointer on a circular dial. References to the spring balance are found in 17th century literature, and it therefore precedes all other forms of self-indicating weighing machines by at least two centuries. The accuracy obtained with spring balances was inferior to that obtained from weighing devices which compared known and unknown weights, but recent improvements in materials science have reduced the effects of temperature and *hysteresis* (the lagging of the effect behind the cause) sufficiently to allow scale makers to make spring mechanisms that compare with such self-indicating systems as the pendulum.

Electronic weighing The rapid development of electronics technology over the last two decades has brought considerable changes to weighing machines. The availability of the transistor and, more recently, the integrated circuit has allowed the improvement of many weighing systems and the invention of new ones. The integrated circuit (IC) allows complex circuit functions to be carried out at high speed yet with the consumption of minute amounts of electrical power. In an electronic weigher, signals are generated that can be used, not only to indicate the weight value, but to calculate the selling price of the goods and, in some instances, produce a printed ticket with this information.

The retail food industry benefited from the use of electronics in weighing machines developed specifically for use in weighing and pricing pre-packed goods for supermarkets. One of the first such machines made use of a spring combined with a static enigma-type mechanism. Instead of a pointer moving over a graduated weight scale, the spring deflection caused a coded disc to rotate, allowing a light source to activate photoelectric detectors according to the pattern on the disc. Each value of weight is uniquely coded. The signals from the encoder exist in the form of a voltage being present or absent on a terminal, and typically there would be ten such output terminals, giving the possibility of 1024 weight values.

With suitable circuits, these signals can operate digital indicators displaying values in Arabic numerals. Additionally, the weight signals can be multiplied by a factor corresponding to the unit weight price being charged for the goods and the result fed to a display or ticket printer. One method of multiplying is by successive addition. If a number, x, of blocks of pulses is fed into an accumulating counter and the number of pulses in each block is y, then when x blocks have been fed in, the accumulator contains the product x times y. If x represents the weight and y the price per pound, the result is the cost.

Another form of electrical weight transducer makes use of the strain which is produced when load is applied to a material. The strain, which is proportional to the applied load, is measured by electrical resistance strain gauges. Suitably connecting a number of gauges to an electrical supply produces an output voltage level proportional to load. Such a device is known as a *load cell* and a number of cells would be used to support a weighbridge platform and provide signals representing load.

The low signal level from the load cell is amplified before

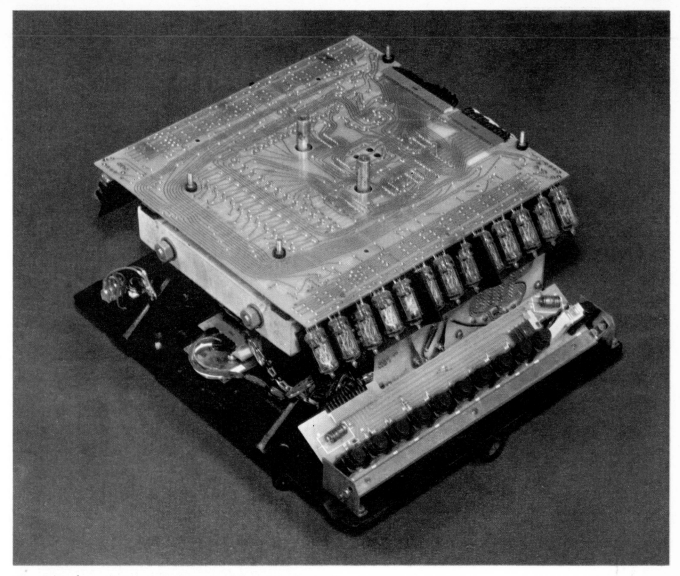

Above: an electronic scale employing strain gauge load transducer. The platform is removed to show the printed circuit board; price setting touch keys and numerical indicating tubes can be seen.

conversion for a digital readout device. Several methods of converting the analog signal to a digital form are available. One method is to amplify the weight voltage and use this amplified voltage to create a current flow in a stable resistor, and then balance this current with an adjustable current that can be varied by known amounts. The balancing current is controlled by electronic switches, which are turned on or off by the action of a null sensing amplifier which determines whether more or less balancing current is required. When balance has been achieved, some of the switches controlling the balancing current will be turned on, while others will be off. As the current controlled by each consecutive switch is twice that controlled by its predecessor, the patterns of on and off switches represent the weight signal in digital form. This method of digitizing allows any changes in weight to be followed as they occur.

Having obtained the digital weight signal, various arithmetic functions can be applied if required, such as a system to indicate the position of the centre of gravity of a container.

Another method of digitizing is known as the *dual-ramp*, in which the value of the weight is repetitiously sampled for short time periods. There are many different designs, but essentially a voltage is allowed to increase at a rate proportional to the weight for a fixed time period. The time taken to reduce this voltage level at a known fixed rate to zero voltage is a measure of the unknown weight. By using digital pulses to provide the timing periods, the output appears in digital form.

ZIPPER

The fashion for buttoned boots in the 19th century prompted an American engineer and part time inventor, Whitcomb Judson, to invent an alternative type of fastener based on the hook and eye principle, which could be fastened by hand or with a moveable guide. It was patented in 1893 and over the next 12 years Judson took out several patents for improved designs.

It was a brilliant idea, but in practice was less dazzling, as the fastener had an annoying tendency to spring open. Furthermore they had to be hand manufactured because the development of suitable machines proved insuperable.

In 1905 Judson invented another type which had the individual fastener elements attached to a tape instead of each other as in a chain. But in spite of this improvement the design was still not satisfactory and Gideon Sundback, a Swedish electrical engineer who had been employed by Judson's company, set about improving the design. In 1913 he invented the hookless fastener in which the individual

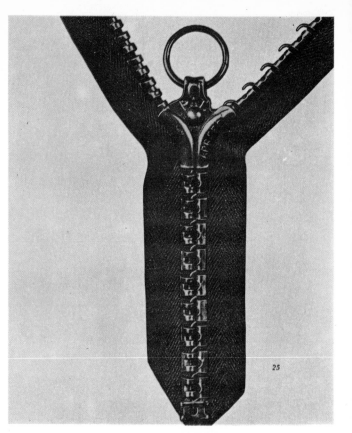

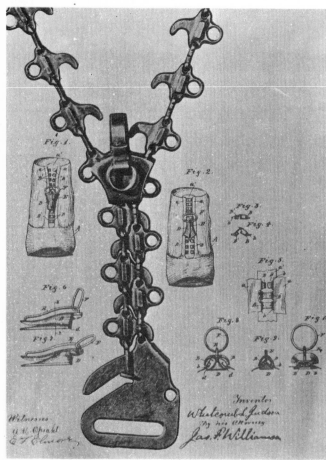

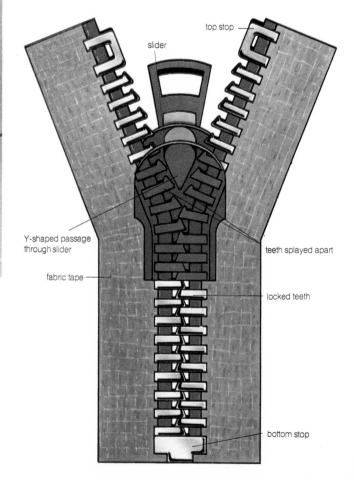

Above, left to right: the original 'zip', introduced in the 1890s: a series of hooks and 'eyes'; a 1905 design called the 'C-curity'; a forerunner of the modern zipper, made in 1913. Right: a modern zipper. The teeth are clamped to a tape and shaped so they will not pull apart when parallel. The slider curves the line, creating extra space between the teeth so that they can join or separate.

top stop

slider

Y-shaped passage through slider

teeth splayed apart

fabric tape

locked teeth

bottom stop

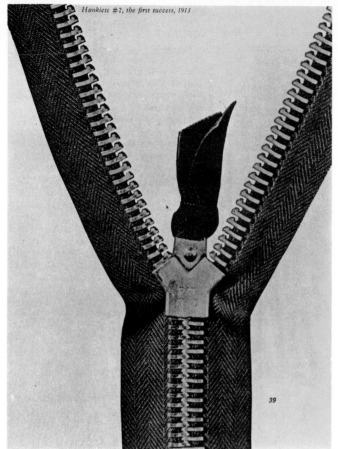

Hookless #2, the first success, 1913

ZOOM LENS

One of the drawbacks of an ordinary camera lens is that a subject at a particular distance gives an image of a fixed size on the photographic film. The scale of the image can be varied by moving the camera nearer to or further from the subject, but this may not always be possible. A movie camera or television camera moving closer to a subject would also have to be continuously refocused to keep the image sharp.

In the *zoom* lens, a development of the telephoto lens, the effective focal length can be varied by moving some of the individual lenses (components) which comprise it. Since the magnification of the image depends on the focal length, this arrangement gives a continuously variable image scale, and the effect of 'zooming in' to a subject can be achieved without moving the camera. The effect is often dramatically useful in films and television, although it can become merely distracting if used too frequently.

The simplest design of zoom lens consists of a diverging lens between two converging lenses. Moving the diverging lens by a small amount then causes a relatively large change in the focal length of the combination. The image would not stay in focus, however, if there were only one moving lens, and so the front lens must also be moved (by a different amount). Most zoom lenses incorporate many more component lenses than just three as described here, and the whole combination is more robust if the other moving lens is not the front lens, but another of the lenses within the assembly. In early designs, the amount of movement required to keep the image in focus differed for the two moving lenses, and a cam mechanism was employed to link the two movements. This *mechanical compensation* is not ideal, because any wear on the cam results in a loss of image quality.

Modern zoom lenses have an arrangement of lenses designed so that the two moving lenses require exactly the same amount of movement, and they can be rigidly linked together and moved as one unit. (This arrangement is known as *optical compensation*.)

Zoom lenses are also available for still cameras, where they are especially useful for the longer focal lengths. A still

elements or teeth were identical and interchangeable. He also invented the machines to stamp out these parts and attach them to the tape.

They were first used on garments in 1918 when a clothing manufacturer with a contract to supply flying suits ordered several thousand. The usefulness of fasteners began to catch on and in 1923 B F Goodrich put fasteners on their galoshes. The name itself did not appear until 1926 when an enthusiastic marketing man demonstrating the fastener's advantages declared, '*zip* it's open, *zip* it's closed!'

Modern designs Modern zips consist of a series of teeth clamped along the edge of a strong textile tape, which are interlocked with another series of teeth on a tape opposite, by means of a slider. Stops are fitted at the top and bottom of the tapes to prevent the slider from slipping off.

Each tooth has a small protrusion on its upper face and a hollow on its lower face. Teeth on opposite tapes are staggered so that the protrusion of one fits into the hollow of the opposite tooth on the adjoining tape, but once it had meshed cannot slip out because the protrusion is too large to slip through the gap between the teeth. The teeth are interlocked by the slider, which consists of two channels that diverge at the top and converge at the bottom. It works by splaying the teeth out as it runs, allowing the head of a tooth to pass through the gap between a pair of teeth opposite, and vice versa. They are interlocked by the narrow part of the slide, drawing them together in precise contact. The working components of the zip may be either of metal or nylon, and some zips are designed so that the slide may be disengaged from one side at the bottom to allow the article to be opened out flat, as with anoraks [parkas], jackets and so on.

Some types of zip do not have individual teeth, but rather a fine plastic spiral of loops that interlock.

In the clothing industry, the latest development is automatic zip machines which assemble a zip from a continuous roll and sew it into the garment, simultaneously.

Above: a 16mm camera for documentary or television film use, fitted with a zoom lens. This design is particularly compact, yet it has a ratio of 10:1 at the extremes of its settings.

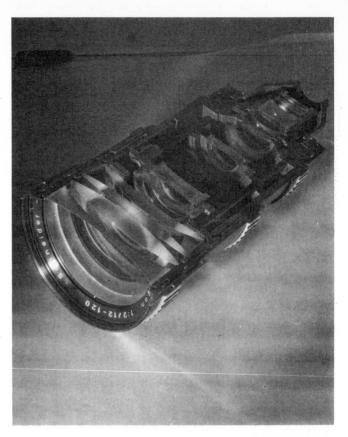

Right: cutaway of a zoom lens with focal length variable from 12 mm to 120 mm, as used on 16 mm movie cameras.

camera does not actually require a continuous variation in focal length, but a zoom lens can be used instead of 2 or 3 different telephoto lenses. For shorter focal lengths, still camera zoom lenses are comparatively bulky, and they usually operate only over a rather limited focal length range 2:1 or 3:1, as compared with 20:1 for a professional movie camera), so they have relatively little advantage over a fixed focal length lens.

Any type of camera using a zoom lens must be equipped with a reflex viewfinder, where the image is seen through the main lens as it will appear on the film, so that the zoom lens can be adjusted to give the required image size, unless the viewfinder is compensated in some way.

Zoom eyepieces constructed on the same principle can be used in microscopes and telescopes to alter the magnification of the final image, while zoom projector lenses allow the operator to fill the screen with the projected picture without having to move the projector.

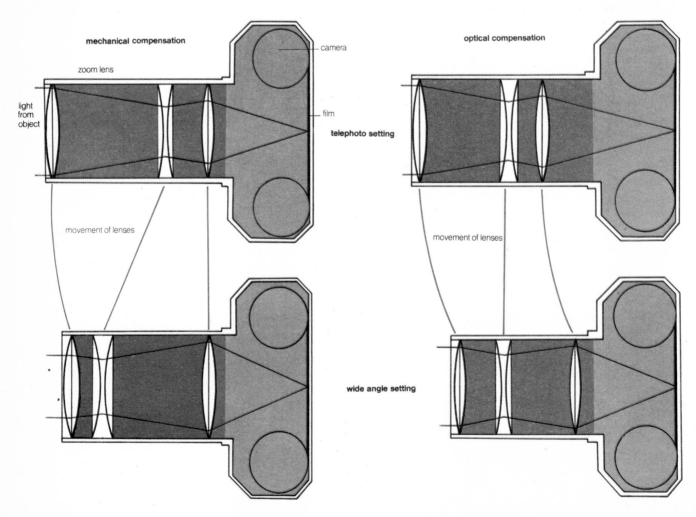

mechanical compensation

camera

zoom lens

light from object

film

movement of lenses

optical compensation

telephoto setting

movement of lenses

wide angle setting

INDEX

Bold type indicates picture